Paper *and* Composites *from* Agro-Based Resources

Edited by

Roger M. Rowell
Raymond A. Young
Judith K. Rowell

Aquiring Editor	Joel Stein
Project Editor:	Albert W. Starkweather, Jr.
Marketing Manager:	Greg Daurelle
Direct Marketing Manager:	Arline Massey
Cover design:	Denise Craig
PrePress:	Carlos Esser
Manufacturing Assistant:	Sheri Schwartz

About the Cover: Scanning electron micrographs at various magnifications depict kenaf cell walls.

Library of Congress Cataloging-in-Publication Data

Paper and composites from agro-based resources / edited by
Roger M. Rowell, Raymond A. Young, and Judith K. Rowell
p. cm.
Includes bibliographical references and index.
ISBN 1-56670-235-6 (alk. paper)
1. Paper. 2. Composite materials. 3. Plant fibers. 4. Biomass chemicals. I. Rowell, Roger M. II. Young, Raymond Allen, 1945- . III. Rowell, Judith K.
TS1109.P1713 1996
620.1′18—dc20 96-8877
CIP

Direct all inquiries to CRC Press, Inc., 2000 Corporate Blvd., N.W., Boca Raton, Florida 33431.

Lewis Publishers is an imprint of CRC Press

International Standard Book Number 1-56670-235-6
Library of Congress Card Number 96-8877
Printed in the United States of America 1 2 3 4 5 6 7 8 9 0
Printed on acid-free paper

Editors

Roger M. Rowell, Judith K. Rowell, and Raymond A Young

Roger M. Rowell, United States Department of Agriculture, Forest Service, Forest Products Laboratory, One Gifford Pinchot Drive, Madison, WI 53705.

Dr. Rowell is a project leader at the Forest Products Laboratory of a research work unit on Modification of Lignocellulosics for Advanced Materials and New Uses, a Professor of Carbohydrate Chemistry in the Department of Forestry and a Professor of Biobased Composites in the Department of Biological Systems Engineering at the University of Wisconsin–Madison, and Research Professor, Department of Technical Chemistry, Chalmers University of Technology, Göteborg, Sweden.

His research involves the enhancement of agro-based materials properties through the chemical modification of cell walls, polymers, and recycling. He currently serves as team leader for an International Team on Property Enhancement of Wood Composites Through Chemical Modification, as a team leader for a National Team on Property Enhanced and Recycled Bio-based Composites. His research includes cooperative studies in the U.S., Brazil, New Zealand, Australia, Japan, China, India, Malaysia, Indonesia, Great Britain, Denmark, France, Germany, Poland, Switzerland, Finland, and Sweden.

For part of 1985, he was a guest professor in the Forest Chemistry Department, Beijing Forestry University, Beijing, China, and, for part of 1985 and 1986, was a National Science Foundation Exchange Professor at the Wood Research Institute, Kyoto University, Uji, Japan. For part of 1986, he was a guest professor in the Department of Wood Science, University college of North Wales, Bangor, United Kingdom. For part of 1988, he was a guest research fellow at the Forest Research Institute, Rotorua, New Zealand. For part of 1991, he was a guest scholar at Kyoto University, Kyoto, Japan.

In addition to authoring more than 300 publications, editing seven books, and receiving 22 patents, Dr. Rowell has presented numerous papers at national and international scientific meetings, organized national and international symposia, and has been active in consulting and technology transfer of his research world wide.

Dr. Rowell is a fellow in the International Academy of Wood Science, a member of the Sigma Xi Honorary Research Society, Materials Research Society, International Society for Controlled Release, Society for Wood Science and Technology, New Uses Council, Association for the Advancement of Industrial Crops, vice-chair of IUFRO Division 5.05, Advanced Industrial Materials Program (AIM), and the American Chemical Society, where he served as chairman of the Cellulose, Paper and Textile Division in 1980, and became a Fellow of the Division in 1991.

Dr. Rowell received his B.S. degree in mathematics and chemistry in 1961 from Southwestern College in Kansas. He received his M.S. and Ph.D. in biochemistry from Purdue University in 1963 and 1965, respectively. He was a Post Doctorate Fellow at Purdue University in 1965–66 and was a Honorary Research Fellow in the Chemistry Department at Birmingham University, Birmingham, United Kingdom in 1967.

Raymond A. Young received his B.S. and M.S. degrees from the State University of New York College of Environmental Science and Forestry at Syracuse and his Ph.D. from the University of Washington. He received two Fulbright Scholarships, one for graduate studies at the Royal Institute of Technology in Stockholm, Sweden, (1973) and another as a Senior Research Scholar at the Aristotelian University in Thessaloniki, Greece (1989). Dr. Young spent two years at the Textile Research Institute, Princeton, NJ, as a postdoctoral fellow and staff scientist.

He also was employed by Kimberly-Clark Corp. in Niagara Falls, NY, as a process supervisor in Pulp and Paper Production. Since 1975, Dr. Young has been a professor in the Department of Forestry, University of Wisconsin–Madison. He teaches courses in pulp and paper and cellulose and polymer chemistry, consults with industry and international aid organizations and conducts a research program in pulp and paper, cellulose and lignin chemistry, and plasma technology.

Professor Young has published more than 100 original scientific papers, 10 book chapters, two encyclopedic contributions, five books, and holds several patents. He is currently working on another book entitled *Environmentally Friendly Technologies for the Pulp and Paper Industry.* He lives with his wife, Kathryn, on the shores of Lake Mendota and dreams of life on a yacht on the Florida Intercoastal Waterway.

Judith K. Rowell is a biologist and freelance editor for both technical and non technical articles and books. She was coeditor on the book *Materials Interactions Relevant to Recycling of Wood-Based Materials* for the Materials Research Society. She is a volunteer for the USDA, Forest Service, Forest Products Laboratory in Madison, Wisconsin, and is active in studies on recycling biobased resources into composites. She is chairman of the Board for the United States section of Alcell, a Swedish based company dealing with advanced biobased composite materials.

She is a member of an International Team on Property Enhancement of Wood Composites Through Chemical Modification and has worked on cooperative studies in the U.S., Brazil, New Zealand, Australia, Japan, China, India, Malaysia, Indonesia, Great Britain, Denmark, France, Germany, Poland, Switzerland, Finland, and Sweden.

She received her B.S. in biology from Southwestern College in Winfield, Kansas, and did her graduate work in biology at Purdue University in West Lafayette, Indiana.

Preface

Agro-based fibers have been utilized for thousands of years for paper and composites. The use of straw for production of paper and reinforcement of mud huts was established well before the advent of wood-based papers and modern fiber-reinforced composites. In more recent times the use of agro-based fibers in these types of products has increased dramatically and a considerable amount of literature has been published on these applications. However, the information on the use of agro-based fibers for paper and composites is widely distributed and has never been brought together and comprehensively treated as we have done in this book.

We realized the need for such a book from a variety of experiences, from research projects with our students on utilization of agro-based fibers, through our work with international philanthropic organizations and from our interactions with industrial and governmental programs in developing countries. Many countries are facing increasing pressure to find alternate sources of agro-based fibers for renewable products and often the choice of additional harvesting of forests is no longer a viable alternative. The use of a much wider spectrum of sustainable agro-based fibers is, therefore, the desired recourse and we have attempted to address the major issues for use of these fibers in this book.

The first section of the book is devoted to the growth and inventory of agro-based fibers. The current and future availability of agro-based plants is discussed in the first chapter which demonstrates the tremendous volume of agro-based fibers available for use in biobased products. Changes in fiber properties during growth and potential improvement of fiber crops with genetics and biotechnology are also treated in this section.

To properly utilize agro-based fibers in biobased products it is necessary that there be a good understanding of the properties of the fibers. This is given thorough treatment in the second section of the book, both for physical and chemical properties in two separate chapters. Standard procedures for chemical analyses and an extensive compilation of fiber properties is given in tabulated form in Chapter 5.

A major use of agro-based plants is for the production of pulp and paper. Chapter 6 provides an extensive discussion of the various methods utilized for pulp and paper production from agro-based plants. The agro-based materials can be readily pulped to give a wide variety of paper grades. Specific pulping conditions and properties of paper from a wide variety of agro-based plants are described in this chapter.

The use of biobased composites has rapidly expanded in recent years and there is tremendous potential for future growth in this area. The uses range from automotive interior components to geotextiles. A broad range of agro-based fibers is utilized as the main structural components or as fillers/reinforcing agents in these composite materials. The methods and approaches utilized for production of the many different types of composites based on agro-fibers is thoroughly described in the last section of the book. The properties of both high and low fiber content thermoplastic and thermosetting based composites are described in several chapters and methods for chemical modification of agro-based fibers for property improvement are described in another chapter.

The book gives a unique compilation and treatment of past and current information on utilization of agro-based fibers. Certainly it conveys the broad range of uses and the great potential for expanded applications of this fibrous resource. The book will serve as a handy reference source for both students and professionals interested in the utilization of agro-based fibers for paper and composites.

List of Contributors

Dilpreet Singh Bajwa, Department of Forestry, University of Illinois, W-503 Turner Hall, Urbana, IL 61801

S.S. Bisen, Tropical Forest Research Institute, Mandla Road, Jabalpur 482 021, India

Daniel F. Caulfield, United States Department of Agriculture, Forest Service, Forest Products Laboratory, One Gifford Pinchot Dr., Madison, WI 53705

Poo Chow, Department of Forestry, University of Illinois, W-503 Turner Hall, Urbana, IL 61801

Charles G. Cook, United States Department of Agriculture, Agricultural Research Service, 2413 E. Hwy. 83, Weslaco, TX 78596

Brent English, United States Department of Agriculture, Forest Service, Forest Products Laboratory, One Gifford Pinchot Dr., Madison, WI 53705

Sharon T. Friedman, USDA, Forest Service, Research Coordination, Resources Program and Assessment Staff, 14th and Independence, SW, Washington, D.C. 20090-6090

Thomas E. Hamilton, United States Department of Agriculture, Forest Service, Forest Products Laboratory, One Gifford Pinchot Dr., Madison, WI, 53705

James S. Han, United States Department of Agriculture, Forest Service, Forest Products Laboratory, One Gifford Pinchot Dr., Madison, WI 53705

Rodney E. Jacobson, United States Department of Agriculture, Forest Service, Forest Products Laboratory, One Gifford Pinchot Dr., Madison, WI 53705

Andrzej M. Krzysik, United States Department of Agriculture, Forest Service, Forest Products Laboratory, One Gifford Pinchot Dr., Madison, WI 53705

Timothy La Farge, United States Department of Agriculture, Forest Service, Southern Region, 1720 Peachtree Road, N.W., Atlanta, GA 30367

Theodore L. Laufenberg, United States Department of Agriculture, Forest Service, Forest Products Laboratory, One Gifford Pinchot Dr., Madison, WI 53705

Roger Meimban, Department of Forestry, University of Illinois, W-503 Turner Hall, Urbana, IL 61801

Timothy G. Rials, United States Department of Agriculture, Forest Service, Southern Research Station, Alexandria Forestry Center, 2500 Shreveport Hwy., Pineville, LA 71360

Jeffrey S. Rowell, United States Department of Agriculture, Forest Service, Forest Products Laboratory, One Gifford Pinchot Dr., Madison, WI 53705

Roger M. Rowell, United States Department of Agriculture, Forest Service, Forest Products Laboratory, One Gifford Pinchot Dr., Madison, WI 53705

Anand R. Sanadi, Department of Forestry, University of Wisconsin, 1630 Linden Dr., Madison, WI 53706

Frank Werber, United States Department of Agriculture, Agricultural Research Service, National Program Staff, Room 221, Building 005, Barc-West, Beltsville, MD 20705

George A. White, United States Department of Agriculture, Agricultural Research Service, National Clonal Geoplasm Repository, University of California–Davis, Straloch Road, Davis, CA 95616

Michael P. Wolcott, Department of Civil and Environmental Engineering Wood Materials, Engineering Laboratory, Washington State University, Pullman, WA 99164–2910

Raymond A. Young, Department of Forestry, University of Wisconsin, 1630 Linden Dr., Madison, WI 53706

John A. Youngquist, United States Department of Agriculture, Forest Service, Forest Products Laboratory, One Gifford Pinchot Dr., Madison, WI 53705

Contents

Paper *and* Composites *from* Agro-Based Resources

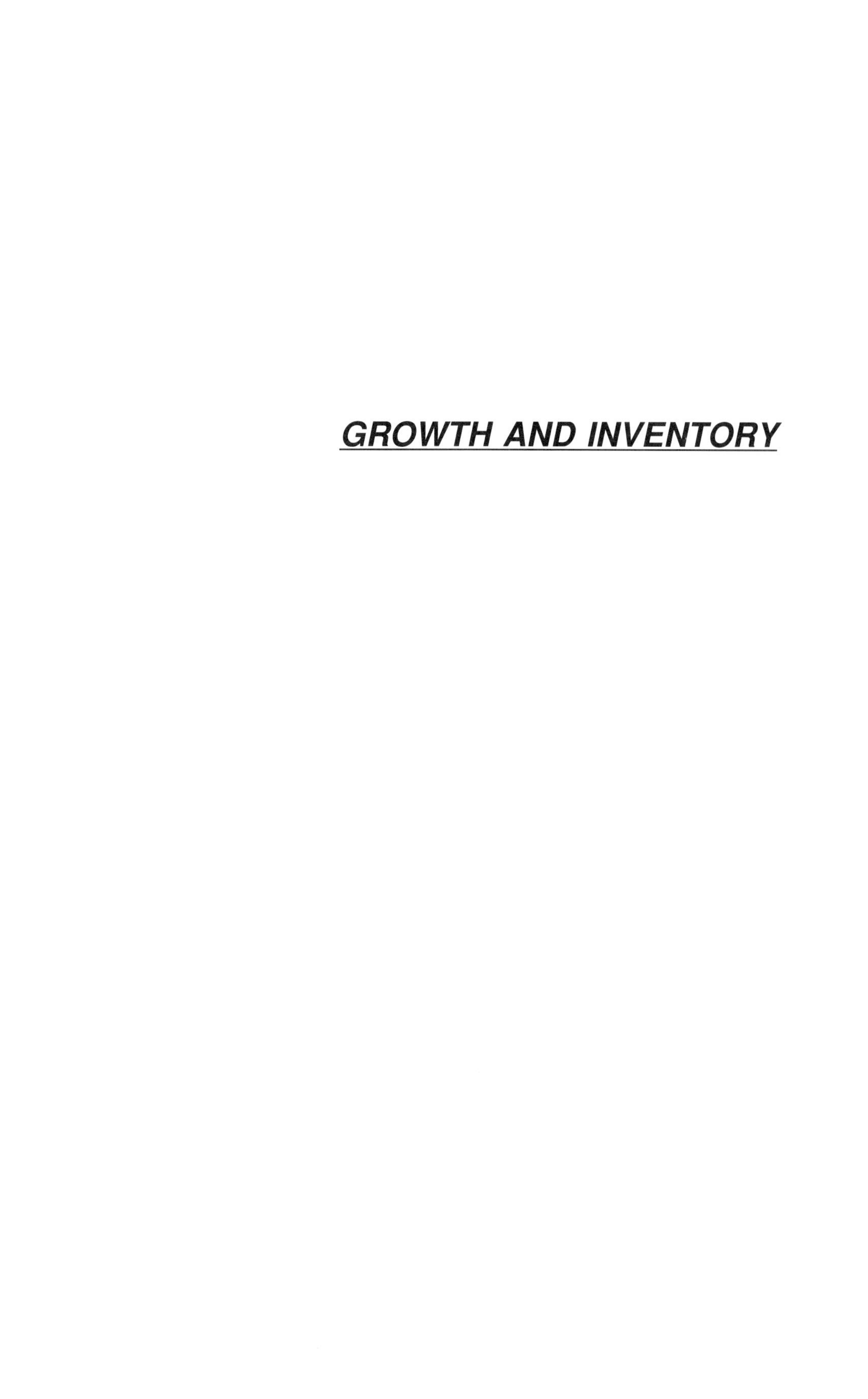

GROWTH AND INVENTORY

INTRODUCTION

Sustainable Fiber Supply

Frank Werber and Thomas E. Hamilton

CONTENTS

1. INTRODUCTION

Sustainable development is an important concept underlying many of today's renewable resource policies. While definitions of sustainable development vary, the essence of the concept is that we must sustain ecosystems and we must sustain people. This means that we will manage and use our ecosystems in ways that do not put them at risk, and we will not abandon people's needs for fiber. To sustain all resources as a goal, we need a course of action to get us there. That course is ecosystem management, which is an approach to the management of natural resources that strives to maintain or restore the sustainability of ecosystems and to provide present and future generations a continuous flow of multiple benefits in a manner harmonious with ecosystem sustainability (Unger, 1994).

Agro-based resources such as wood have been used since the beginning of man's existence for shelter, fuel, decoration, and protection. Wood has been a very important and convenient source for construction materials, pulps, fuel, and composites because it is low in cost, easily stored in the form of a tree, available for harvest at any time of the year, easy to store after cutting, compact, widely distributed, and renewable. We have a wealth of knowledge and experience invested in wood technology. Wood, while very important, however, is only one type of agromass in the vast world of photosynthetic resources. A universal plan for renewable and sustainable resource management that seeks to develop strategies for the production of paper and composites using sound ecosystem management principles should consider and evaluate all types of agro-based resources.

It is important to start this book with a clear understanding of what is included in the "agro-based resource." Agro-based resource is used to mean the plant material produced as a result of photosynthesis. This resource is sometimes referred to as photomass, phytomass, agromass, solar mass, or photosynthetic mass. Biomass is another term that is often used but this is misleading as it includes resources such as bone, protein, lipids, and other biological components. Another general term used is lignocellulosics, which is accurate; however, it is a term meaning lignin- and cellulose-containing, and some of the information covered in this book will not include lignin. Cotton is almost pure cellulose, as is a chemical and bleached pulp, and both are included in this book. The book covers plant based resources including wood, agricultural crops and residues, grasses, recycled agro-based resources, and components from these sources.

2. SUSTAINABLE FIBER SUPPLY

In developing a strategy for assuring a sustainable future fiber supply, there are significant changes that must be taken into account. In the past, we have depended on our forests for most of our fiber for paper and composites. This will likely continue in the United States, but "wood-poor" countries such as India are already utilizing large quantities of agro-based fibers *in lieu* of wood fibers. There are also serious concerns about the health and use of forests. Topics such as multiple use, ecosystem management, conservation, preservation, jobs, and endangered species (such as the spotted owl) have sensitized the public about the "proper" use of forest lands.

Chapter 1 gives an inventory of the various types of agro-fibers that are grown in the United States. While accurate inventories have been developed for several types, such as hardwoods, softwoods, kenaf, cotton, and others, an estimate of the volume of many potential fiber sources does not exist. There is a need to develop such data in two forms: what is now grown and what can be grown in each geographic area of the United States and globally.

As responsible custodians of agricultural land, farmers must rotate crops intelligently. For the long term, this means growing a mix of crops and livestock to preserve the productivity of the land. Also, depending on the local soil and climatic conditions, farmers should have alternatives to the major staple crops — soybeans, corn, wheat, rye, cotton, etc. In parallel to this driving force, for more than 50 years the U. S. Department of Agriculture, by Act of Congress, has had as an important objective the development of new products and new crops, to extend markets for the American farmer. An important objective, then, has been diversifying farm production to an increasing percentage on agricultural-based products for industrial as opposed to food uses. In turn, compared to wood fiber, agro-fibers are at an advantage in certain processing characteristics. For instance, fibers of sufficient length can be moved continuously as a coherent web from blending through "needling" to a compression-molding press, compared to wood fiber which cannot be efficiently entangled. There are also many opportunities to combine long agro-based fiber with wood fiber to produce a wide variety of composite products.

Considering all agro-based fiber, the entire spectra of fiber properties such as fiber length, fiber diameter, chemical composition, density, and morphology can be used to make decisions for making paper and composites. All of the agro-based resources should be marketed to take advantage of the unique properties, not just because we have a desire to promote one fiber over another. Unless one particular fiber has some advantage in the market, it will be replaced with whatever resource has the market advantage. That market advantage can be based on many things such as availability, price, or performance, but there must be some reason to use one specific fiber over any other natural, renewable resource. Desire does not drive markets! Producers and manufacturers of agro-fibers must explore common interests and, where possible, prepare an enterprise-driven long range strategic plan for development and promotion of an agro-fiber industry. This is very important for the success of any single agro-based fiber.

The use of non-traditional fibers for the existing paper and composite industries will require new techniques for growing, harvesting, collecting, separating, sorting, storing, shipping, and handling fibers as well as designing and developing new manufacturing systems. Elements of economy of scale, pollution control, energy requirements, water use, processing, disposal, life cycle assessment, and recyclability are critical issues that must be considered for successful new ventures.

Recycling will also increase the fiber supply by reusing agro-based products. The pulp/paper and composite industries can work very well together in a combined strategy for recycling. Some grades of paper will be recycled directly into new paper products while some waste paper, including all inks, adhesives, and inorganics can be used directly to make composites. An integrated paper recycling program will allow recycled fiber to flow into the highest possible market with the minimum of reprocessing.

Creating opportunities for using a wide variety of agro-based fibers will require a carefully designed blend of research and development. It will also depend on partnerships that integrate traditional fiber supplier and users with those new to this industry. Finally, it will depend on meeting the objectives of multiple use of our natural resources, and improving both the environment and the economy.

There are many exciting opportunities for obtaining fiber from agro-based resources for paper and composites that will require a critical examination of existing paradigms and careful development of new technologies that ultimately may lead to new and exciting uses.

REFERENCES

Unger, D. G., The United States Department of Agriculture Forest Service Perspective on Ecosystem Management, *Symposium on Ecosystem Management*, Burlington, VT, 1994.

CHAPTER 1

Inventory of Agro-Mass

George A. White and Charles G. Cook

CONTENTS

1. INTRODUCTION

The concurrent increases in population, with much of it in wood-poor developing countries, and the rising demand for a wide range of fiber-derived products portend a bright future indeed for agro-fiber sources (Atchison, 1987b). Significant yield increases in presently grown fiber crops should be possible through improved management and exploitation of genetic diversity and new biological techniques (see Chapter 3). However, consideration should also be given to the use of all types of plant species, especially high-yielding legumes, grasses and grass hybrids. For example, White et al. (1974) demonstrated tremendous biomass yield differences among only a few sorghums. Several fast-growing subtropical–tropical grasses (Schank et al., 1993) and tall-growing exotic corn could also be utilized. The continued improvement of collection and storage methods for fibrous crop residues should enhance their commercial usage.

One needs to consider the ramifications of ever-increasing use of annually renewable and herbaceous perennial sources of agro-based fibers. Good management of tillable crop land calls for crop rotation, return of a portion of residues to the soil for organic matter enrichment and minimizing erosion, reduction of reliance on chemical pesticides, and use of proper fertilization practices. In addition, in some areas there will be competition for crop residues for animal feed, bedding, fuel, and land reclamation.

According to MacLeod (1988), approximately 93% of the 145 million metric tons (mt) of world pulp production comes from wood, and only 2% of the 10 million tons of pulp from agro-based sources is produced in the United States. Atchison (1994) has updated data on agro-based pulp capacity in the world, which indicates that straw comprises 46% of all agro-based papermaking pulp capacity, bagasse 14%, bamboo 6%, and the remaining 34% from miscellaneous fibers such as sisal, jute, hemp, kenaf, cotton fibers–linters, etc. He further states that the 1993 agro-based pulp capacity was almost 21 million mt or 10.6% of worldwide papermaking pulp capacity.

A rapid increase in the use of agro-based fibers for pulp products is occurring, especially in wood-deficient countries such as China and India. Fourteen countries now pulp only agro-based fibers, and more than 20 produce 50% or more of their pulping capacity from agro-based fibers. This upward trend is expected to continue through this century and beyond. By 1998, as much as 11% of the world's pulping capacity may be garnered by agro-based fibers compared to 6.9% in 1975. Atchison (1992a) states the startling statistic that only 10% of the world's supply of straw could produce 3.0 million mt of pulp per year.

Our objective in this chapter is to provide inventory estimates of agro-based fiber sources from fiber crops and residues of field crops. There is a vast reservoir of renewable raw materials for pulping that will be utilized increasingly in the years ahead. We also briefly discuss harvest and storage methods which are critical to the commercial use of agro-based fiber for paper and composites.

2. AGRO-BASED FIBER CROPS

2.1 Crop Species

Crops grown primarily for fiber or having significant fiber components are listed in Table 1.1. Some of them are dual or multipurpose crops such as flax (fiber, linen, linseed oil), sugarcane (sugar, fuel, fiber), and bamboo (fiber, furniture, poles, construction, blinds, etc.). The crops listed are mostly ancient ones with highly variable production levels. The degree of crop improvement among them also varies greatly. Many of the problems associated with harvesting, transporting, and storing the raw materials have been solved especially in developed countries. However, high labor requirements, retting, remoteness from processors, unreliable quantities and variable quality and trash content present major barriers to commercial expansion. This group of fiber-producing plants encompasses many species not covered in Table 1.1. For these, reliable estimates of availability are difficult to obtain and some production estimates lump various species. Jeyasingam (1988) mentions special problems and considerations for various plants including coconut palm trunks, palmyra palm leaves, date palm stalks, banana stems, pineapple leaves, esparto grass, sabai grass, illuk grass, papyrus, and reeds.

Table 1.1 Agro-based Fiber Crops

Crop	Scientific name	Fiber component	Leading country producer
Abaca	*Musa textilis* Nee	leaf	Philippines, Ecuador
Bamboo	*Bambusa vulgaris* L. other species	stem (culms)	China, India
Coir	*Cocos nucifera* L.	fruit	Sri Lanka
Flax	*Linum usitatissimum* L.	stem	Czechoslovakia, Russian Federation
Hemp	*Cannabis sativa* L.	stem	India, China
Jute	*Corchorus capsularis* L.	stem	India, Bangladesh, China
Kapok	*Ceiba pentandra* (L.) Gaertner	fruit	Indonesia, Thailand
Kenaf	*Hibiscus cannabinus* L. (& roselle, *H. sadariffa* L.)	stem	China, India, Thailand
Ramie	*Boehmeria nivia* (L.) Gaudich	stem	China
Reeds	*Arundo donax* L.	leaf	Brazil
	Phragmites communis Trin.	stem	China, Hungary
Sisal	*Agave sisalana* Perrine	leaf	Brazil

Separation of bast from cortical fibers remains a problem in processing stem fibers from dicotyledonous plants (also see Chapter 6). However, in the United States, the development of efficient separation machines has largely solved this problem. In fact, the mechanical separation of bast from the inner core fibers of kenaf has opened new market outlets for the inner woody core (White et al., 1994). These include oil absorbents, mats for seeding or bank erosion control, chicken litter, potting media, etc. Successful mechanical separation allows the longer bast fibers to be available for various high quality pulps or for recombination in controlled proportions with the inner core fibers.

2.2 Inventory Estimates

Estimates of potentially available feedstocks of agro-based fiber crops are given in Table 1.2. The estimates are based on data from FAO statistics (FAO, 1994) and from Atchison (1994). It is difficult to discern exact production because of varying statistical combinations of stem-fiber species and related species of the same genus. There were no good estimates of the available tonnage of bamboo and reeds other than estimates by Atchison (1994). Atchison's estimate of just over 76 million mt

Table 1.2 Estimated Annual Worldwide Availability of Agro-based Fiber Crops (mt: 000s)

		FAO statistics[2]	
Fiber crop	**Atchison[1]**	**1993**	**Average, 1989-93**
Abaca (manila hemp)	80	107	104
Bamboo	30,000	—	—
Coir	—	116	111
Flax	2,000	1,290[3]	690
Kapok	—	101	105
Reeds	30,000	—	—
Sisal (& allied species)	500	357	431
Stem fibers[4]	13,700	15,591	15,596
Ramie	—	335	461
Jute & jute-like	—	14,730	14,565
Hemp		526	570
Miscellaneous fibers	—	408	406
Totals	76,280	17,970	17,443

[1] After Atchison, J. E., 1994.
[2] Statistics Division, FAO, Rome (1993, 1994).
[3] Includes non-FAO1993 estimate of seed flax straw from Canadian and U.S. production.
[4] Includes ramie, jute, kenaf, hemp, etc. Estimates are stalk weights based on FAO statistics for bast fiber (= 23% of dry stalks weight per J.E. Atchison, personal communication).

and ours of just under 78 million mt (including 60,000 for bamboo and reeds) are nearly identical.

2.3 Harvesting and Storage

These operations tend to be labor intensive, especially for leaf and fruit fibers which may be hand-stripped and for reeds that are largely hand-harvested. Even much of the bamboo is selectively hand-cut. Only bagasse is readily available as a by-product of the cane sugar process. Decorticating equipment is available for many of the stem-fiber crops such as jute, ramie, kenaf, roselle, sunn hemp, and hemp (see Chapter 6). Various harvesting schemes have been tried experimentally in the U.S. for kenaf and sunn hemp. Decortication of air-dry whole stalks followed by baling of bast fibers has worked well. A major problem is getting the stalk moisture content to a storable level. Since the separation of the bast from the inner woody core has opened up several market outlets for the core, efficient collection and storage procedures for the core need to be developed. In countries with high labor costs, mechanization is essential to success in using agro-based fibers. For example, strip cutting by machines of commercial bamboo plantings is preferred for regrowth of plants and low labor inputs. High density balers are desirable to facilitate storage of bulky materials. In the case of bagasse, today's economics favor bulk storage and handling. The book *Secondary Fibers and Non-wood Pulping* (Hamilton and Leopold, 1987) contains a wealth of information about collection, storage, and initial cleaning/pulping of agro-based fibers. Included are papers on agro-based fibers (Atchison, 1987a), bagasse (Atchison, 1987c), bamboo (Bhargava, 1987), kenaf

(Touzinsky, 1987), cereal straw (Misra, 1987), reeds (Wiedermann, 1987), and other fibers (McGovern et al., 1987). This subject is also covered in Chapter 6.

2.4 Kenaf – A Case Study

Kenaf, a native, annual plant of East-Central Africa, is cultivated from seed for the bast and core fibers produced in the stem (Dempsey, 1975; White et al., 1970). Bast fibers (ca. 2.5 mm in length) are produced in the outer bark, and the shorter core fibers (ca. 0.6 mm in length) are located in the pith or interior portion of the stem.

Kenaf gained attention in the early 1940s as an alternative fiber source when jute supplies were curtailed during World War II. A cooperative kenaf research program was established in Cuba between the United States Department of Agriculture (USDA) and the Cooperative Fiber Commission (CFC) to research kenaf as a cordage crop. Subsequent research in the 1950s focused on the potential use of kenaf as a fiber source for paper and pulp products. Recent research has identified several new products which can be made from kenaf fibers, including oil-spill adsorbents, potting soil amendments, livestock bedding, packing material, particle and insulation boards, filters, dry-formed containers, turf and wildlife seed mats, and poultry litter (USDA, 1993). Kenaf leaves, which contain 20–30% crude protein, also may be harvested as a livestock feed source.

Kenaf is adapted to a fairly wide range of climates and soils. However, it is extremely sensitive to frost, and planting should occur at dates which avoid freezing temperatures while maximizing the growing season. In the United States, kenaf is most adapted to the growing conditions of the southern tier of states from Florida to Texas and California. The best yields generally are obtained on fertile, well drained soils. In the United States, kenaf is grown as a row crop using standard farm equipment. Rows generally are spaced 76–102 cm apart with seeding rates used to provide 170,000–370,000 plants per hectare. Dense plant populations generally produce greater total and bast fiber yields, reduce weed competition, and improve harvesting efficiency.

The kenaf cultivars presently grown in the United States are photoperiod-sensitive types i.e., vegetative growth increases until daylength becomes less than 12 h and 30 min. Floral initiation then occurs and vegetative growth rate declines. The most widely grown cultivars in the United States include Cubano and Cuba 108 from the USDA and CFC program in Cuba; Everglades 41 and 71 from the USDA program in Belle Glade, Florida; and Tainung 1 and 2 from the Fiber Crops Experiment Station in Taiwan. The greatest yield-limiting biotic factor in the United States is the southern root-knot nematode, *Meloidogyne incognita* (Kofoid and White) Chitwood. Although no high levels of resistance have been discovered, kenaf germplasm which has good tolerance to the root-knot nematode/soil fungi complex has been identified (Cook and Mullin, 1994). Crop rotations which include non-host species such as sunn hemp, combined with cultivars which possess good levels of tolerance, should significantly reduce yield losses associated with nematodes.

Kenaf is harvested in the United States with modified sugarcane equipment (Figure 1.1) and self-propelled forage chopping equipment. Prior to the fiber separation processing, stalks harvested with sugarcane equipment typically are stored as

Figure 1.1 Harvesting kenaf with a modified sugarcane harvester in south Texas.

whole stalks, whereas kenaf harvested with forage chopping equipment generally is stored in compressed modules. After fiber separation, the processed bast fibers are baled for storage and shipping. The core material is processed through a hammermill and bagged.

In the United States, the total number of hectares of kenaf harvested in 1993 was estimated at 640. The total area harvested and average yield were 285 ha and 7.8 mts/ha in Mississippi, 200 ha and 13.4 metric tons/ha in Texas, 110 ha and 11.2 mts/ha in Louisiana, and 45 ha and 10.6 mts/ha in California. Data compiled by the junior author from personal contacts show that kenaf has been established in the United States on a very small commercial scale. Atchison (1987b, 1991, 1992a, 1992b) frequently addresses its potential as an annual pulp source. As he mentions (1992b), kenaf production could be rapidly increased as markets expand. Ahmed (1993) discusses various aspects of kenaf production, uses and products, and economics of production and processing. Sabharwal et al. (1994) indicate that a new bio-pulping method involving fungal treatment in the refiner mechanical pulping phase could enhance the paper making potential of kenaf.

3. FIBER FROM FIELD CROP RESIDUES

While wood dominates as a pulping raw material, agro-based fiber sources will become increasingly important. There exists a large, annual worldwide volume of crop residues that are suitable for various pulp-related products. While blends with wood are still common, more 100% pulp furnishes from agro-based fibers can be anticipated. Excellent quality printing and writing papers are produced from 100% bagasse pulps in Colombia, Argentina, and Indonesia. Annually renewable resources are especially attractive these days as more concerns about environment, endangered animal and plant species, destruction of ecosystems, etc. are expressed and legislated. Bamboo and crop residues such as bagasse and cereal straws are the major sources of agro-based fibers for pulping because of broad availability and good pulp quality. The largest, mainly untapped, reserves of fibrous raw materials are the straw of cereal crops and stalks of corn and sorghum. However, estimates of the availability of these and other fibrous residues are probably slightly high because of practical aspects of collecting and transporting to pulp mills, isolation of raw materials, other uses such as bedding, feed, plowdown, etc. Economics and mechanization of collecting, storing, and processing also enter prominently into the picture.

Much of the technology for use of agro-based fibers was developed in the United States. This technology, developed primarily for sugarcane bagasse, can be applied to other agro-based fibers. Plans are being developed to establish fiberboard processing plants based on straw in the United States and Canada for up to 175,000 mt/year capacity.

3.1 Crop Species

The crops listed in Table 1.3 are of primary interest because of significant production of straw, stalks, and other fibrous components. Several grasses, such as ryegrass (*Lolium* spp.), fescues (*Festuca* spp.), orchard grass (*Dactylis* spp.), blue grass (*Poa* spp.), and others produce substantial quantities of straw from grass seed production in Oregon (see discussion). These are not listed, but rather considered as a whole in inventory data.

Table 1.3 Field Crops as Sources of Agro-based Fiber from Residues

Crop	Scientific name	Fiber component	Leading country grain/lint producers
Barley	*Hordeum vulgare* L.	straw	Former USSR, Germany, Canada, France
Corn (maize)	*Zea mays* L.	stalks	United States, China
Cotton	*Gossypium hirsutum* L.	lint, linters, stalk	China, United States
Oats	*Avena sativa* L.	straw	Former USSR, United States, Canada
Rice	*Oryza sativa* L.	straw	China, India, Indonesia
Rye	*Secale cereale* L.	straw	Former USSR, Poland, Germany
Sorghum	*Sorghum bicolor* (L.) Moench	stalks	Former USSR, United States, Nigeria
Wheat	*Triticum aestivum* L.	straw	China, former USSR, India

3.2 Inventory Estimates

Atchison (1991, 1992a, 1994) has published estimates of worldwide production of various crop residues suitable for pulping. He used hectarage and yield estimates for straw and stalks to estimate worldwide availability (Atchison, 1987a). We have compiled estimates based on FAO and USDA Agricultural Statistics and compared them to Atchison's estimates.

In most cases, only grain production data are available. For residue estimates, we used, after reviewing research papers (Deloughery and Crookston, 1979; Prihar and Stewart, 1990, 1991; and Steiner et al., 1994) and consulting with crop specialists, the ratios or harvest index (HI) of grain to total aboveground biomass as follows:

barley, rice, wheat	0.43	**corn (maize)**	0.42
oats and rye	0.40	**sorghum**	0.40

The HI values can vary with environment, cultivar, plant densities, and cutting height.

According to a crop specialist, tropical maize could have a HI as low as 0.33 compared to a high of 0.45–0.48 for temperate cornbelt types. Computations of HI from yield data reported by Steiner et al. (1994) give a similar HI range for sorghum. Exotic corn and sorghum have very high biomass yield potential; but, generally seed production would be much lower than for shorter hybrids. To estimate sugarcane bagasse availability, we used 14% of cane as harvested for processing.

Reasonable estimates of cotton stalk residues are difficult to calculate because of wide variations among cultivars and species (short staple, long staple, tropical types) and one must account for lint, linters, stalks and mote. Mote is a fuzzy, lightweight material composed of short fibers, leaf fragments, etc., from the ginning operation of which about 50% is fibrous. Crop and equipment specialists provided advice as how to estimate the potential supply of the fibrous components of the cotton plant. The conversion factors with lint production as the base factor used are: lint × 2 = stalk yield; 13% of lint yield = linter yield and 5% of lint minus 50% = mote yield. These conversion factors for grain and cotton crops are arbitrary and are largely based on United States estimates modified slightly for worldwide production. Availability estimates are presented in Table 1.4.

3.3 Harvesting and Storage

Factors that consistently hamper increased use of agro-based fibers in pulps are economics and problems and costs associated with the collection and storage of the raw materials. For straws and stalks of corn, cotton and sorghum, baling appears the most practical. Baling equipment is available for conventional-sized bales of 45 kg or less up to equipment that produces bales of about 0.45– 0.9 mt (Figure 1.2). Also, there is a large volume round baler that wraps each bale in plastic for field or outside storage. Unwrapped large round bales are frequently stored outside for several months, but there can be some loss from water damage and weathering except in relatively arid climates. Many general farm balers do not compact these bulky raw materials sufficiently for economic storage. This is especially true for stalks. Large round bales seem better for stalks such as sorghum and coarser grasses. Storage

Table 1.4 Estimated Worldwide Tonnage of Fibrous Raw Materials from Field Crops (000mt)[1]

		Atchison[2]		Agricultural statistics, USDA[3]	
Crop	Plant component	United States	Worldwide	United States	Worldwide
Cereals	straw				
barley		7,000	195,000	12,000	218,500
oats		5,000	55,000	6,000	50,800
rice		3,000	360,000	7,500	465,200
rye		400	40,000	400	41,900
wheat		76,000	600,000	78,900	739,700
Total straw		91,400	1,250,000	104,800	1,516,100
Corn (maize)	stalks	150,000	750,000	300,800[4]	727,300
Cotton	lint	3,500	18,300	3,500	18,000
	linters	500	2,700	500	2,300
	stalks	4,600	68,000	7,100	35,900
	mote	—	—	200	900
Flax (oilseed)	straw	500	2,000	700[5]	—
Grass (seed)	straw	1,100	3,000	900	—
Sorghum	stalks	28,000	252,000	33,700	104,700[6]
Sugarcane	bagasse	4,400	102,200	3,000	100,200
Totals		284,000	2,448,200	455,200	2,505,400

[1] Numbers rounded to nearest hundred.
[2] Atchison, 1994.
[3] Computed from production data for 1992/93 per Agricultural Statistics 1993 (USDA).
[4] Based on grain production only.
[5] Estimate for 1993 Canadian and United States production (personal communication).
[6] Worldwide production not available from Agricultural Statistics, 1993; FAO estimate for grain production used.

of air dry bales on farms, at a central storage site, or at the pulping mill poses no particular problem except for the space requirements for bulky materials. Costs of baling and transportation to storage sites are critical economic considerations. Transportation costs can become equal to or surpass the value of these low density commodities.

In the seed flax producing areas of Minnesota, North Dakota, and Canada, a pulp company sends portable decorticating machines to central straw storage sites (Figure 1.3) and directly to farms of the larger producers.

After decortication, the bast fibers are densely baled for storage and processing. The resultant shives (inner stalk material) are used for bedding or returned to the soil. Because the seed oil types of flax are short-growing, the yields of straw and hence of decorticated fiber are low. This same type of decorticating equipment has been successfully used, as observed by the senior author, in pilot tests for kenaf and sunn hemp.

The quality of crop residues for pulp products is greatly affected by the timeliness of harvesting and baling, weather-induced deterioration, and the presence of weeds and extraneous leaves, soil particles, etc. in the baled materials. Silica in straws has long been recognized as a problem to the recovery of pulping chemicals. A further discussion of the effect of these contaminants on processing to pulp and paper is given in Chapter 6.

Figure 1.2 A high volume baler in operation that produces rectangular bales of up to one short ton each. This equipment is suitable for hay, straw, and stalks of corn, cotton, sorghum. Photo by Hesston Corporation. (With permission.) Names are necessary to report factually on available data; however, the USDA neither guarantees nor warrants the standard of the product, and the use of the name by USDA implies no approval of the product to the exclusion of others that may also be suitable.

4. DISCUSSION

Many plant species have been tested as possible agro-based fiber sources, but relatively few will ever be used commercially because of unfavorable economics, although products derived from them may have desirable characteristics. For commercialization, there must be a reliable and adequate supply of the raw material that stores well with minimal deterioration. Because of bulkiness, production in the proximity of the processing mill is essential to keep transportation costs low. The fiber must be of high and uniform quality. Since processing some of the leaf and fruit fibers is so labor intensive, there will likely be little expansion of them and perhaps a decline in their usage. The demand for leaf and fruit fibers in the short term will persist because of the unique characteristics of derived products. The use of straw and sugarcane bagasse for pulp products will likely expand rapidly because of overall abundance and known fiber qualities. The expansion of the use of bagasse will be tempered by fuel costs for processing cane into sugar (Atchison, 1987c). If fossil fuel costs should rise sharply, the price of the bagasse will rise accordingly,

Figure 1.3 Grab stacking a load of flax straw during a fall purchasing season. Photo by Kimberly-Clark, Inc. (With permission.) Names are necessary to report factually on available data; however, the USDA neither guarantees nor warrants the standard of the product, and the use of the name by USDA implies no approval of the product to the exclusion of others that may also be suitable.

and processors may return quickly to using bagasse as their main fuel source. Atchison (1991) believes that there will only be moderate expansion in the use of bamboo for pulp because of the wide diversity of bamboo-derived products. One could also predict the lessened use of wild reeds and grasses because of rising labor costs, the probable decline of harvesting areas, and the growing worldwide concerns about the environment. Hence, expanded use of straws and the production of annually renewable fiber crops seem to offer the greatest opportunity for expanding the use of agro-based fibers in the long term. Our estimates and those of Atchison (1994) show the possible availability of almost 2.5 billion metric tons of crop residues on an annual basis. The use of HI ratios may result in high estimates because stubble height after harvest is not considered. Our estimates in Table 1.4 do not include the production of corn for silage or forage sorghum which could be diverted into pulping raw materials instead of animal feed.

The discrepancy between our estimate and that of Atchison (1994) in Table 1.4 for worldwide production of sorghum stalks is difficult to explain. We used FAO 1992 data because Agricultural Statistics did not include worldwide figures. The FAO 1992 grain production estimate of 69,797,400 mt converts to 104,696,100 mt of stalks through use of an HI factor of 0.40. This estimate is substantially lower than 252,000,000 mt per Atchison who estimated stalk availability by multiplying the estimated production area times an estimated stalk yield per hectare. It should be noted that our estimate (33,800,000 mt) of stalks from U.S. grain sorghum exceeds that of Atchison (28,000,000 mt); thus the main difference relates to the foreign production.

An attractive potential crop residue for pulping in the United States is the straw from grass seed production in the Willamette Valley of Oregon. Producers in Oregon account for 60–70% of the grass seed production in the United States. Approximately 0.91 million mt of straw are generated annually from about 162,000 ha (Donald Churchill, personal communication). In the past, the fields were burned to remove the straw and destroy pests, but state laws have been enacted to phase out most of the burning. Many grass seed producers have storage facilities, and suitable baling equipment is readily available. Some straw is exported. The raw material is generally comparable to wheat straw for pulp products, and should be readily available at modest cost. Atchison (1992a) estimates the availability of straw from grass seed production at 1.0 million mt for the United States and 3.0 mt worldwide.

Breeding emphasis, that perhaps could be accelerated through the use of new biological techniques, on annual leguminous species such as *Crotalaria juncea* (sunn hemp), *Sesbania* spp., and others could, in the long term, pay big dividends. Sunn hemp for example is an excellent fiber source, is highly resistant to root knot nematodes (a production nemesis to kenaf), is fast drying, and is a nitrogen-fixing plant (White and Haun, 1965). Yet a paltry amount of breeding to improve yield and other agronomic traits has occurred. Jeyasingam (1988) mentions that a mill near Calcutta, India utilizes sunn hemp for producing cigarette tissues. A limited breeding effort is underway in the United States in conjunction with studies on sunn hemp as a rotational crop with kenaf.

There are a number of plant species not included herein that could be considered as raw materials for pulping and composite materials. Youngquist et al. (1993), for example, state that technically, sunflower hulls and stalks can be used to make composition panels. They indicate, however, that bagasse, cereal straw, and kenaf appear to have the most promise in panel development.

While adequate technology exists to convert most agro-based fibers into quality paper and board products, economics will dictate how quickly and how much the use of agro-based fibers for pulp products will expand during the remainder of the 20th century and into the 21th century. The cost of production, collection, storage, and processing, coupled with the value of the saleable products must result in a reasonable investment return, and be competitive with other fiber sources (wood and recycled papers). The mechanization of handling sugarcane bagasse, annual fiber crops, and straw/stalk residues is a reality in developed countries and some developing countries (e.g., Colombia, Peru, Indonesia). Machines are slowly replacing hand labor with these crops in other developing countries.

5. SUMMARY

While wood continues to dominate as the major fiber raw material, agro-based fiber usage will expand well into the 21st century. How fast this expansion will occur will depend on the growth of the paper industry globally, international trade regulations, and the growth in the use of recycled fibers. Cereal straws, sugarcane bagasse, and annual fiber crops such as kenaf will be the feedstocks that account for much of the future expansion. Mechanization of all operations from production to storage of raw materials is possible in developed countries and in practice in some developing countries and will gradually reduce the labor intensity of handling and processing fiber crops and crop residues in other developing countries. High density balers are available that improve the economics of transportation and storage of the bulky fibrous raw materials.

Accurate estimates of availability of agro-based fiber crops and crop residues are difficult to calculate because of so many variables that affect total quantities. The authors relied on FAO Statistics and estimates by Atchison (1994). Combining these estimates showed the possible availability of more than 76 million metric tons of agro-based fiber crops. For estimating the availability of fibrous crop residues, we used harvest indices for the cereal straws and for corn and sorghum stalks. Percentages of lint, as recommended by cotton specialists, provided estimates of linters, stalks, and mote. These residue estimates were more than 2.5 billion metric tons of potentially available agro-based raw materials. Atchison (1992a) had previously estimated availability at about 2.3 billion metric tons.

Opportunities exist in the United States for expansion of kenaf production and perhaps other new fiber crops, for increased use of cereal straws, and the utilization of grass seed straw especially from Oregon.

ACKNOWLEDGEMENTS

The authors express appreciation to the following persons who provided oral and/or printed information for use in this paper: Joseph E. Atchison, Donald Churchill, FAO staff (Statistics Div.), Bob Granaas, Arnel Hallauer, Bruce Maunder, Bill Mayfield, Bill Meredith, Terry Rothenbuehler, and Bobby Stewart.

REFERENCES

Ahmed, I., *Kenaf: Background and current commercialization*, Institute for Local Self-Reliance, Washington, D.C., 10, 1993.

Atchison, J. E., Data on non-wood fibers, *Secondary Fibers and Non-Wood Pulping, Vol. 3: Pulp and Paper Manufacture*, Hamilton, F., Leopold, B., (eds.), *TAPPI*, Atlanta, GA, 1987a, 4.

Atchison, J. E., The future of non-wood fibers in pulp and papermaking, *Secondary Fibers and Non-Wood Pulping, Vol. 3, Pulp and Paper Manufacture*, Hamilton, F., Leopold, B., (eds.), TAPPI Press, Atlanta, GA, 1987b, 17.

Atchison, J. E., Bagasse, *Secondary Fibers and Non-Wood Pulping, Vol. 3, Pulp and Paper Manufacture*, Hamilton, F., Leopold, B, (eds.), TAPPI Press, Atlanta, GA, 1987c, 22–70.

Atchison, J. E., Update on progress in non-wood plant fiber pulp and papermaking and prospects for the 1990's, China Paper, '91 Conference, Beijing, China, 1991.

Atchison, J. E., Non-wood pulping progress, *Asia Pacific Pulp & Paper,* 19, May, 1992a.

Atchison, J. E., U.S. non-wood fiber potential rises as wood costs escalate, *Pulp & Paper*, 139, September, 1992b.

Atchison, J. E., Present status and future prospects for use of non-wood plant fibers for paper grade pulps, Presentation at American Forest & Paper Association (AF&PA) 1994 Pulp and Fiber Fall Seminar, Tucson, AZ, 1994.

Bhargava, R. L., Bamboo, *Secondary Fibers and Non-Wood Pulping, Vol. 3 Pulp and Paper Manufacture*, Hamilton, F., Leopold, B. (eds.), TAPPI Press, Atlanta, GA, 1987, 71.

Cook, C. G. and Mullin, B. A., Growth response of kenaf cultivars in root-knot nematode/soil borne fungi infested soil, *Crop Sci.*, 34, 1455, 1994.

Dempsey, J. M., *Fiber Crops*, University Presses of Florida, Gainesville, FL, 1975.

DeLoughery, R. L. and Crookston, R. K., Harvest index of corn affected by population density, maturity rating, and environment, *Agron. J.*, 71, 577, 1979.

FAO, Jute, kenaf and allied fibers, *Quarterly Statistics Food and Agriculture Organization (FAO) of the United Nations*, Rome, Italy, 1993.

FAO, FAO Statistical reports on production data for non-wood fiber crops, FAO, Rome, Italy, 1994.

Hamilton, F. and Leopold, B. (eds.), Part I, Non-Wood Pulping, *Secondary Fibers and Non-Wood Pulping, Vol. 3, Pulp and Paper Manufacture*, TAPPI Press, Atlanta, GA, 1987.

Jeyasingam, J. T., A summary of special problems and considerations related to non wood fibre pulping worldwide, *1988 Pulping Conference Proceedings*, TAPPI Press, Atlanta, GA, 571, 1988.

MacLeod, M., Non-wood fiber: number 2, and trying harder: An interview with Dr. Joseph E. Atchison, *TAPPI J.*, Atlanta, GA, 50, 1988.

McGovern, J. N., Coffelt, D. E., Hurter, A. M., Ahuja, N. K., and Wiedermann A., Other fibers, *Secondary Fibers and Non-Wood Pulping, Vol. 3, Pulp and Paper Manufacture*, Hamilton, F., and Leopold, B., eds. TAPPI Press, Atlanta, GA, 1987, 110.

Misra, D. K., Cereal straw, *Secondary Fibers and Non-Wood Pulping, Vol. 3, Pulp and Paper Manufacture*, Hamilton, F., and Leopold B., (eds.) TAPPI Press,Atlanta, GA, 1987, 82.

Prihar, S. S. and Stewart, B.A., Using upper-bound slope through origin to estimate genetic harvest index, *Agron. J.*, 82, 1160, 1990.

Prihar, S. S. and Stewart, B.A., Sorghum harvest index in relation to plant size, environment, and cultivar, *Agron. J.*, 83, 603, 1991.

Sabharwal, H. S., Blanchette, R. A., and Young, R. A., New bio-pulping method could enhance potential of paper making from kenaf, a non-wood plant, *Ag. Industrial Materials & Products*, March, 1994.

Schank, S. C., Chynoweth, D. P., Turick, C. E. and Mendoza, P. E., Napiergrass genotypes and plant parts for biomass energy, *Biomass and Bioenergy*, 4(1), 107, 1993.

Steiner, J. L., Schomberg, H. H., and Morrison, J. E., Residue decomposition and redistribution, *Crop Residue Management to Reduce Erosion and Improve Soil Quality: Southern Plains*, Stewart, B. A. and Moldenhauer, W. C., (eds.) ARS Conservation Report Series, 1994.

Touzinsky, G. F., Kenaf, *Secondary Fibers and Non-Wood Pulping, Vol. 3, Pulp and Paper Manufacture*, Hamilton, F., Leopold, B., (eds.), TAPPI Press, Atlanta, GA, 1987, 106.

USDA, New industrial uses, new markets for U.S. crops: status of technology and commercial adoption, *Cooperative State Research Service, Office of Agricultural Materials*, Washington, D.C., 46, 1993.

White, G. A., Clark, T. F., Craigmiles, J. P., Mitchell, R. L., Robinson, R. G., Whiteley, E. L., and Lessman, K. J., Agronomic and chemical evaluation of selected sorghums as sources of pulp, *Econ. Bot.*, 28, 136, 1974.

White, G. A., Cumins, D. C., Whiteley, E. L., Fike, W. T., Greig, J. K., Martin, J. A., Killinger, G. B., Higgins, J. J., and Clark, T. F., Cultural and harvesting methods of kenaf–an annual crop source of pulp in the Southeast, *USDA Prod. Res. Rep., 113*, 1970.

White, G. A., Gardner, J. C., and Cook, C. G., Biodiversity for industrial crop development in the United States, *Industrial Crops and Products*, 2, 259, 1994.

White, G. A. and Haun, J. R., Growing *Crotalaria juncea*, a multi-purpose legume, for paper pulp, *Econ. Bot.*, 19(2), 175, 1965.

Wiedermann, A., Reeds,*Secondary Fibers and Non-Wood Pulping, Vol. 3. Pulp and Paper Manufacture*, Hamilton, F., Leopold, B., (eds.), TAPPI, Atlanta, GA, 1987, 94.

Youngquist, J. A., English, B. E., Spelter, H., and Chow, P., Agricultural fibers in composition panels, *Particleboard Symposium (No. 27)*, 133, 1993.

CHAPTER 2

Changes in Fiber Properties During the Growing Season

Roger M. Rowell, James S. Han, and S. S. Bisen

CONTENTS

1. INTRODUCTION

It is well known that different parts of a plant have different chemical and physical properties. That is, the chemical composition and fiber properties of plant tissue taken from the roots, stem, trunk, and leaves are different. What is not so well known is that the chemical composition and fiber properties of plant tissue are also different at different stages of the growing season.

Plants have, in general, five stages in their life cycle: germination, growth, flowering, seed formation, and death. Annual plants go through these stages in one growing season. Biennials have a two year cycle where the second year's plants grow from the root system of the first year's plants. Perennial plants have the same cycle as annual plants except growth, flowering, and seed formation occurs many times before the plant dies.

Various industries harvest plants for products at different times during the plant life cycle. For example, the food industry harvests young sprouts such as beans just after germination. Crops such as lettuce and asparagus are harvested during the early growing part of their cycle. The cut flower industry harvests the plant flowers at the bud stage or shortly thereafter. Seeds used for food or oil production are harvested after the flowering stage but before the seeds are allowed to drop from the seed pod. Many crops, however, are allowed to compete their life cycle before harvesting. For example, annual grain crops are allowed to field ripen and dry before harvesting.

In general, annual plants used for fiber for paper and composites are harvested at the end of the growing season, allowed to dry in the field, and then processed into fiber. Fiber from trees for paper and composites is derived from logs of various ages, and fiberized by one of several methods.

There may be an advantage in harvesting fiber for paper and composites at some time earlier than from a mature plant. For example, fiber from an immature plant may be low in lignin which could be used for paper since there would be little chemical pulping required to remove the lignin. While the yield may be lower, there may be an advantage in chemical and energy use to harvest early. In the case of annual plants, it may be possible to harvest two crops in one season to give the same yield of fiber but with much less lignin. Fiber from juvenile plants such as jute and kenaf are reported to be "silklike," i.e., fine texture, very flexible, and thin. Again, the yield may be lower, based on the traditional end of the season yield, but a fiber that could be used for textiles may command a higher price resulting from such early harvesting.

This chapter is a literature review, and a report on early results of growing kenaf, of the changes in chemistry, and fiber properties as a function of the growing season.

2. EARLY RESEARCH

The earliest reported literature on changes in chemical composition as a function of the growing season was done with wheat in the 1930s (Phillips et al., 1931). They found that the cellulose content was highest in the early part of the growing season and that lignin and ash content varied with the amount of fertilizer used.

In a study using several varieties of flax, Overbeke and Mazingue (1949) found that both cellulose and lignin content increased with plant age but pectins, hemicellulose, and ash content followed no systematic progression with age.

A great deal of research was done on cotton during the 1950s. Usmanov et al. (1957, 1958) found that the degree of polymerization (DP) of cellulose increases up to the 15th day after bloom, then slows and stops on approximately the 40th day. The DP 3 days prior to the opening of the boll, and 0, 1, 5, 7, 10, 12, 15, and 17 days after the opening of the boll was 2920, 3758, 4286, 4564, 4406, 4282, 4082, 3622, and 3326 respectively. Strength and DP of the cellulose were at a maximum at 5 to 6 days after the boll opened and it was recommended that the cotton should be picked 12 to 15 days after the opening of the boll. Using the electron microscope, Usmanov and Nikonovich (1960) went on to show that the most rapid accumulation of crystalline microfibrils took place between 17 and 20 days of growth and reached a maximum at 40 days. After that, the main growth was across fibers which led to the formation of the primary wall.

Ono and Sato (1957) studied the relationship between cellulose, nitrogen, phosphorus, ash, wax, and pectin as they related to maturity. They reported that immature lint contained a large amount of non-cellulosic substances and hence lost a large amount of weight during purification. Immature fiber was harder to bleach and dye as compared to mature fiber. The reducing sugar content was higher in immature lint and decreased as the plant aged.

3. FIBERS

3.1 Jute

The University of Manchester, the Shirley Institute, and the British Textile Technology Group in the United Kingdom have spent years working on jute. While some of the research has been published, the results relating to the changes in properties of jute fiber as a function of the growing season were done for the International Jute Organization in Bangladesh and never published (Ozsanlav, 1992). The research records are stored in Bangladesh and attempts to gain access have failed. Personal communications concerning these results indicate that juvenile jute fiber looks and feels like silk but this has never been documented in print.

Chatterjee (1959) working at the Technological Jute Research Laboratories in Calcutta, India first reported the changes in chemical composition at different stages of jute plant growth. Table 2.1 shows a summary of his results. These results show that there is little difference in cellulose, holocellulose, and lignin content but that xylan, ash, and iron content decrease as the plant matures. The aggregate fiber length increases as the growing season progresses. Without defining what is meant by "best," Chatterjee reports that the best fiber is obtained at the bud stage.

Later, Mukherjee et al. (1986) working at the Indian Jute Industries' Research Association in Calcutta, studied characteristics of jute fiber at different stages of growth. They found that at the early stages of growth, there was an incomplete formation of the middle lamella in the cell wall and the parallel bundles of fibrils were oriented at an angle with respect to the fiber axis that gradually decreased with growth. After about 35 days of growth, the fibrils run parallel to the fiber axis. In that mature plant, a few helically oriented fibrils in the Z-direction were observed just below the primary cell wall layer.

3.2 Bamboo

The earliest work on chemical changes in bamboo during its growing season was done by Migita in 1947. He reported that the α-cellulose content of Madake was almost constant from a week to three years but that lignin increased steadily from 8.4% to 24.0% during the same period. Taniguchi (1956) showed that there

Table 2.1 Changes in Chemical Composition of Jute at Different Stages of Plant Growth

	Stage of plant growth %				
Component	**Pre-Bud**	**Bud**	**Flower**	**Small Pod**	**Large Pod**
α-Cellulose	58.3	57.6	59.4	58.7	59.1
Holocellulose	86.8	87.8	87.3	87.1	86.8
Xylan	15.5	14.8	14.4	13.7	13.9
Lignin	12.7	12.1	12.4	12.0	12.0
Ash	0.57	0.53	0.47	0.67	0.47
Iron	0.020	0.018	0.009	0.011	0.008
Reed Length	(6.6 ft)	(9.1 ft)	(9.3 ft)	(9.6 ft)	(10.7 ft)

Reported on 100 g of dry material.

were ultrastructural changes in bamboo during its growth, particularly as it related to its fine structure. Normura and Yamada (1974b) studied the changes in crystallinity using X-ray analysis in bamboo internodes during the growing stages and found that the amorphous pattern observed in juvenile nodes gradually changed to orientation of the fiber in a mature node. Crystallinity increased slightly during the formation of the first 10 internodes but then decreased over the next 30 nodes and finally leveling off after 35 nodes. They also showed that the internode length increased up to 6 weeks of growth and then decreased back to a value reported at 5 weeks in a mature plant.

Normura and Yamada (1974a) followed the variation of amino acids in bamboo using X-ray analysis and reported that tyrosine was at the highest concentration (about 30% based on dry weight of the plant) in the fourth week of growth or about the 17th internode. Higuchi and Shimada (1969) had found earlier that tyrosine was present in the highest amount of any stage of growth as compared to all other amino acids present. Higuchi et al. (1953, 1966, 1967) studied the formation of lignin in bamboo and reported that lignification increased with plant age.

Fujii et al. (1993) has done the most complete chemical analysis of bamboo as a function of its growth. Their results are shown in Table 2.2.

Table 2.2 Changes in Chemical Composition of Bamboo at Different Stages of Plant Growth

	Stage of plant growth (by node position) %					
Component	**1–11**	**12–16**	**17–20**	**21–24**	**25–27**	**28–40**
Moisture Dry Weight	79.4	80.0	84.6	88.5	90.3	90.3
Matter Cold Water	40.3	25.6	15.6	9.2	6.4	2.9
Extractives	15.3	16.7	25.2	38.9	46.9	48.7
Lignin	8.8	8.2	5.4	2.6	0.6	0.3
Holocellulose	47.3	48.8	37.6	32.6	23.0	13.9
α-Cellulose	80.6	81.1	71.0	65.2	53.1	39.8
Pentosan	24.5	34.3	28.0	21.9	11.4	5.1
Protein	5.0	5.8	7.6	11.7	19.6	30.6
Ash	2.0	2.1	3.1	5.0	7.5	10.8
Crystallinity Index	44.1	43.2	40.7	32.2	23.2	8.5

Cold water, alcohol/benzene, and 1% NaOH extractives, ash content and protein decreased as the plant aged while cellulose, holocellulose, pentosan, lignin, and crystallinity increased with plant age.

3.3 Wood

The first wood substance laid down by a tree is known as juvenile wood. It differs both in physical and chemical properties as compared to mature wood. Usually, the percent of juvenile wood is low in a large mature tree but can represent a very large percentage of small young trees. For example, a 10-year-old Sitka spruce tree may contain up to 90% juvenile wood whereas a 100-year-old Sitka spruce tree may contain only 1% juvenile wood at its center.

Juvenile wood usually grows in wider annual rings, has a higher ratio of early-wood (wood substance laid down in the spring and early summer) as compared to latewood (wood substance laid down in the late summer and early fall), lower basic density, and higher moisture content as compared to mature wood (Panshin and de Zeeuw, 1980). Mature wood shrinks very little in the longitudinal (growing) direction whereas juvenile wood can shrink a great deal in the growing direction.

At the ultrastructural level, juvenile wood fibers have a larger microfibril angle in the S_2 layer as compared to mature wood (20), which is why juvenile wood shrinks more in the longitudinal direction (0.6%) than mature wood (less than 0.1%) (Rowell, 1984).

The fiber length of juvenile wood is much shorter (3.0 mm) than mature wood (4.2 mm), while the lumen size is larger (42.3 μm) in juvenile wood compared to mature wood (32.8 μm). Cell wall thickness is about 3.9 μm for juvenile wood compared to about 8 μm for mature wood. Cell diameters are about the same for both (juvenile, 50 μm, mature, 49 μm). The breaking strength of mature wood is about 30% higher and the compression strength parallel to the grain is about 20% higher than juvenile wood.

In general, juvenile wood from both hardwoods and softwoods has less cellulose (lower pulp yields) and more lignin and hemicellulose content than mature woods. (Taniguchi, 1956; Panshin and de Zeeuw, 1980). There is little difference between the types and amounts of hemicelluloses in both hardwoods and softwoods in juvenile and mature wood.

The extractives can be very different in juvenile wood as compared to mature wood. Extractives from juvenile wood are often more toxic and in higher concentration. This may account for the decreased digestibility of new growth in birch trees (Palo et al., 1985). The concentration of phenolic acids in the extractives is higher in juvenile wood just after leafing has started compared to mature wood.

3.4 Grass

Theander (1991) recently completed a study of the changes in chemical composition of several Swedish grasses as a function of growing time. He found that cellulose and lignin increased as the plant matured while protein decreased in reed canary grass (*Phalaris arundinacea*). Hemicellulose content and pectin remained constant during the entire plant life.

Otoguro et al. (1991) reported on the amino acid content as a function of plant growth and also found that cellulose and hemicellulose content decreased with plant age.

3.5 Sisal

Chand and Hashmi (1993) found that the density of plant fibers decreased from 1.18 to 1.16 g/cm^3 as the plant aged from 2 to 5 years, but then increased to 1.27 when the plant reached age 9. Hemicellulose content was found to be the highest when the plant was 5 years old. Tensile strength increased from 285–393 MPa with an increase in plant age from 2 to 9 years. This change in strength was explained in terms of plant fiber structure.

3.6 Kenaf

Clark and Wolff (1969) carried out the first studies on the changes in chemical composition of kenaf as a function of the growing season. They also studied the chemical differences along the stem and between leaves and stem. The values shown in Figure 2.1 were taken from the bottom (all but the top 0.66 meter of the plant) and show that the pentosans, lignin, and α-cellulose content is increasing with age while the protein and hot water extractives content is decreasing with age. Data taken from the top part of the plant shows similar trends but the top part has less cellulose, pentosans, and lignin but higher hot water extractives and protein than the bottom part of the plant (Table 2.3).

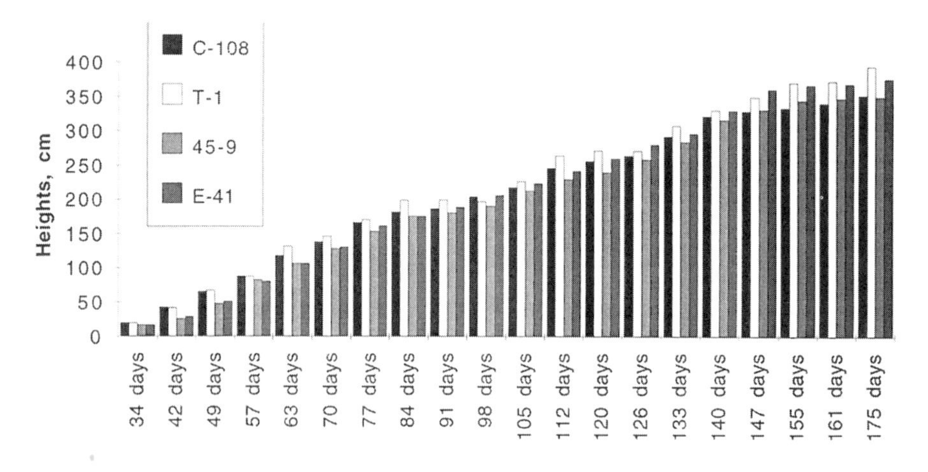

Figure 2.1 Heights of kenaf in 1994.

Clark et al. (1967) also studied the changes in fiber properties during the growing season. Table 2.4 shows that the bast single fibers are longer than core fibers and

Table 2.3 Changes in Chemical Composition of Kenaf at Different Stages of Plant Growth

Component	90 dap[1]	120 dap	138 dap	147 dap	158 dap	244 dap
Hot Water Extractives	37.4	39.0	35.2	31.6	30.6	12.8
Lignin	4.5	3.9	6.4	7.2	7.4	11.4
α-Cellulose	10.6	14.5	18.5	18.1	20.6	29.8
Pentosan	5.0	12.5	16.1	17.0	16.7	20.1
Protein	25.0	17.9	16.1	13.3	14.9	11.1

(Data from the top 0.66 m of the plant, % by weight).
[1] dap = days after planting.

both decrease in length with age. Core single fibers are twice as wide and have twice the cell wall thickness as bast single fibers and both dimensions decrease with age. Finally the lumen width is wider in pith fibers as compared to bast single fibers and both decrease with age.

Table 2.4 Changes in Fiber Properties of Kenaf at Different Stages of Plant Growth

	Stage of plant growth			
Component	90 dap[1]	120 dap	150 dap	180 dap
Bast Fiber				
Length (mm)	3.34	2.28	2.16	2.42
Width (microns)	18.3	14.5	13.6	15.1
Lumen Width (microns)	11.1	5.4	6.8	7.7
Cell Wall				
Thickness (microns)	3.6	4.6	3.4	3.7
Core Fiber				
Length (mm)	0.55	0.54	0.45	0.36
Width (microns)	36.9	31.2	32	31.6
Lumen Width (microns)	22.7	14.8	18.6	18.7
Cell Wall				
Thickness (microns)	7.1	8.2	6.7	6.4

[1] dap = days after planting.

In a recent study, Han et al. (1994) reported changes in kenaf as a function of the growing season. Their data do not necessarily agree with that of Clark and Wolff. The most critical difference between Han et al. and Clark and Wolff was the difference in fiber lengths. The average length of a bast and core (stick) fiber increased as the plant aged in contrast to that of Clark and Wolff.

Han et al. (1994) studied changes in chemical composition during the growing season for four varieties of kenaf: C–108, Tainung 1, and Everglade 45–9, and Everglade 41. Samples were collected weekly starting about 50 days after the planting (dap) to the end of growing of about 170 dap. Chemical analyses procedures used for this study can be found in Chapter 5.

X-ray diffraction of kenaf samples were used for crystallinity values (Table 2.5). Crystallinity values decreased as the plant matured. Ash contents of fibers and cores were determined before and after extraction (Table 2.6). Ash content decreased as the plant reached maturity.

Table 2. 5 Crystallinity Values

	C–108		T–1		E–41		45–9	
Growth	Fiber	Core	Fiber	Core	Fiber	Core	Fiber	Core
56 dap	84.25	78.38	80.77	87.72	78.4	81.08	80.53	82.35
84 dap	78.87	72.13	76.43	73.85	81.48	72.22	77.62	76.92
112 dap	78.91	78.87	73.94	70.63	80.43	71.79	78.32	75.41
140 dap	80.71	72.31	80.29	79.39	78.17	66.93	78.10	68.18
168 dap	73.57	66.93	77.77	65.19	70.90	64.00	74.64	67.19
196 dap	72.34	68.18	72.86	73.11	70.63	69.53	68.38	70.83

Table 2.6 Ash Content (T–1), % Dry Basis

	Fiber		Core	
Growth	Unextracted	Extracted	Unextracted	Extracted
49 dap	13.08	9.54	14.52	7.77
98 dap	7.52	6.83	5.10	4.21
147 dap	6.24	6.00	4.55	4.44
175 dap	3.76	3.84	2.84	2.39

Table 2.7 Protein Contents (T–1), % Dry Basis

	Fiber		Core	
Growth	Unextracted	Extracted	Unextracted	Extracted
49 dap	10.60	7.65	12.75	8.00
98 dap	4.25	3.25	4.85	4.65
147 dap	2.05	3.35	3.40	3.20
175 dap	1.43	1.23	4.15	3.28

Protein content of fibers and cores were determined before and after the extraction (Table 2.7). Protein content decreased as the plant reached maturity.

Extractives, lignin, and sugar contents were also determined (Table 2.8). The values are averages of four different cultivars. Klason lignin analysis was done after the extraction. The Klason lignin values increased from ca. 4% at the beginning to 10% at the end of growing season (Bagby et al. [1971] reported about 10% using Florida kenaf). This value is significantly lower than that of softwood (26–32%) and hardwood (20–28%) (Sjöström, 1981). The actual value of Klason lignin could be lower than it appears to be due to the presence of protein in the kenaf. Kjeldahl determination of protein was performed (Han et al., 1995) on several batches of combined Klason lignin samples and the amount of protein in the Klason lignin was measured. The protein content of kenaf is between 4 to 14% of the Klason lignin depending upon the age of the plant. Only 38% of the protein was found in the Klason lignin and the rest was found in the hydrolysate (unpublished Forest Products Laboratory [FPL] Data). In general, protein content decreased with plant age.

Solvent extractive content varied as a function of growth. In general, they were high at the beginning, decreased during the first part of the growing time, and then increased again. L-Arabinose, L-rhamnose, L-galactose, and D-mannose content

Table 2.8 Chemical Composition of Kenaf Fiber (% Oven-dry Basis)

Growth dap	Extractives %	Klason Lignin %	Polysaccharides Content (% anhydro sugars on ovendry basis)					
			Arabinan	Rhamnan	Galactan	Glucan	Xylan	Mannan
35	14.87	4.32	3.95	2.72	0.78	28.86	6.54	1.76
42	8.80	6.00	3.18	1.82	0.62	33.20	7.31	1.63
57	5.13	8.32	2.21	1.46	0.55	35.45	8.08	1.59
63	4.34	7.74	2.43	1.48	0.62	37.08	8.61	1.53
70	4.63	8.70	2.02	1.25	0.46	40.53	9.37	1.47
77	4.99	9.23	2.05	1.36	0.39	40.52	9.16	1.34
84	5.07	8.33	2.27	1.63	0.49	39.88	9.39	1.53
91	5.68	9.38	1.91	1.35	0.42	42.82	9.98	1.31
98	2.42	8.81	2.13	1.43	0.48	41.60	9.69	1.35
133	8.03	8.94	1.67	1.15	0.48	41.98	9.72	1.31
155	7.83	9.99	1.27	0.87	0.38	46.39	11.20	1.19
161	11.51	10.22	2.54	1.52	0.56	39.22	9.75	1.33
168	12.31	9.74	2.18	1.37	0.47	41.41	10.36	1.39
175	8.23	9.69	1.40	0.87	0.36	49.33	12.29	1.02

Table 2.9 Fiber Length (mm)

	C–108		Tainung–1		45–9	
Growth	Bast	Core	Bast	Core	Bast	Core
50 dap	2.3	0.7	2.2	0.7	2.2	0.7
60 dap	2.7	0.7	2.8	0.7	2.7	0.7
77 dap	2.9	—	3.0	—	3.0	—
84 dap	3.4	0.8	3.1	0.8	3.7	0.8

decreased as a function of growth while D-glucose and D-xylose content increased over this same period of time.

The fiber length increased as a function of growth (Table 2.9). The core fiber lengths were ca. 0.8 mm at the end of 84 growing days with an average diameter of ca. 0.5 mm.

Weight ratios between fiber and core (core/fiber) increased as the growing days advanced (Table 2.10). The maximum of 1.8 was reached at 175 dap in T–1. Holocellulose content was measured after the extraction (Table 2.11). The juvenile samples had low holocellulose values and gradually increased as the plant aged.

Plant height increased with plant age at an even rate (Figure 2.1). This is a function of the growing conditions and would change with different moisture and sun conditions. The diameter of the stalk also increased gradually with age until reaching 160 dap (Figure 2.2). At the end of 160 dap, the rate of growth became more significant. However, this dramatic increase in volume is indicative of an increase in core and not bast fiber. Growth of the plants versus temperature and rainfall is shown in Figure 2.3. A maximum weekly growth of 30 cm was achieved during high temperature and a good rainfall.

Scanning electron microscopy (SEM) studies were done on plants harvested at 63 dap, 71 dap, 84 dap, and 112 dap. The bast portion was separated from the core

Table 2.10 Weight Ratio of Fiber vs. Core

Growth	C–108	T–1	45–9	E–41	C/F Ave
53 dap	0.88	0.90	0.87	0.74	0.85
57 dap	0.93	1.17	1.02	1.03	1.03
63 dap	1.02	1.10	1.11	1.05	1.07
70 dap	1.29	1.10	1.31	1.14	1.21
77 dap	1.19	1.03	1.15	1.31	1.17
84 dap	1.10	1.33	1.12	1.25	1.20
91 dap	1.42	1.23	1.22	1.41	1.32
98 dap	1.14	1.33	1.03	1.22	1.18
105 dap	1.23	1.94	1.09	1.22	1.37
112 dap	1.16	1.36	1.03	1.80	1.34
120 dap	1.23	1.30	1.01	1.29	1.21
126 dap	1.21	1.62	1.19	1.41	1.36
133 dap	1.28	1.70	1.25	1.30	1.38
140 dap	1.29	1.82	1.41	1.78	1.57
147 dap	1.31	1.31	1.45	1.24	1.32
155 dap	1.40	1.71	1.39	1.98	1.62
161 dap	1.57	2.12	1.69	1.64	1.76
175 dap	1.31	1.81	1.33	1.83	1.57

Table 2.11 Holocellulose Content

Growth, dap	Holocellulose, %	Growth, dap	Holocellulose, %
35	58.97	42	64.47
49	68.24	57	69.02
63	73.91	70	75.69
77	76.52	84	74.15
91	76.13	98	74.87
105	73.65	126	75.40
133	75.27	140	78.50
147	76.18	161	73.88
168	76.27	175	78.60

manually and cross sections were prepared of the constituent tissues. The prepared specimens were dehydrated in a series of ethanol concentration, critical point dried in CO_2, and gold palladium coated for observation. SEM observations indicate that the bast fiber bundles are thin walled at the 63 dap and are in the process of thickening. The middle lamella is not well formed as suggested by the weak bonding seen in Figures 2.4-1 and 2.4-4. The fibers' tendency to gelatinize may be due to the wind effect which often results during the bending of plants. At this stage, the parenchyma bands separating the bast fiber bundles are well formed and occupy a considerable area of tissue system.

At 71 dap, the plants became comparatively stronger as bast fiber bundles occupy more area, the fiber wall thickened, and lignification of middle lamella (Figure 2.4-5) becomes apparent. Parenchyma cells tend to crush due to the development of fiber bundles, thus allowing more area to be occupied by fiber bundles (Figure 2.4-3).

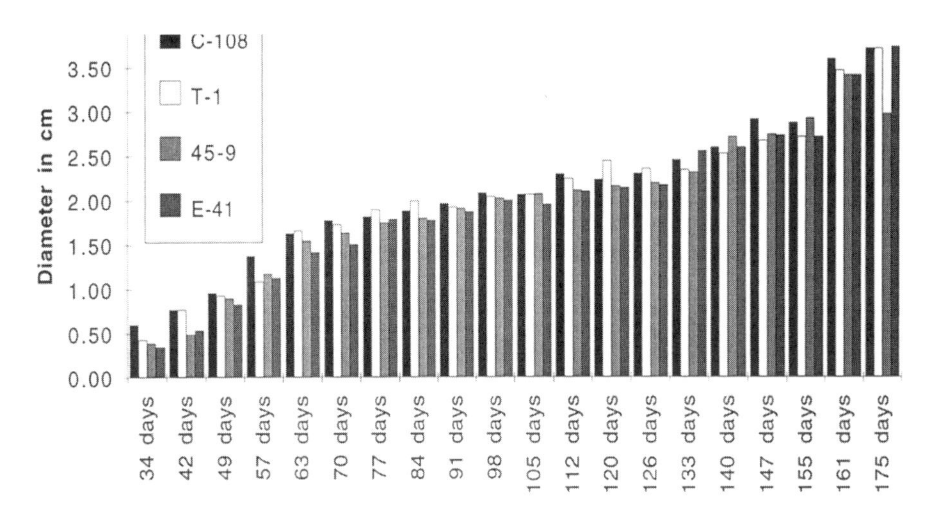

Figure 2.2 Diameter at the bottom of kenaf plants in 1994.

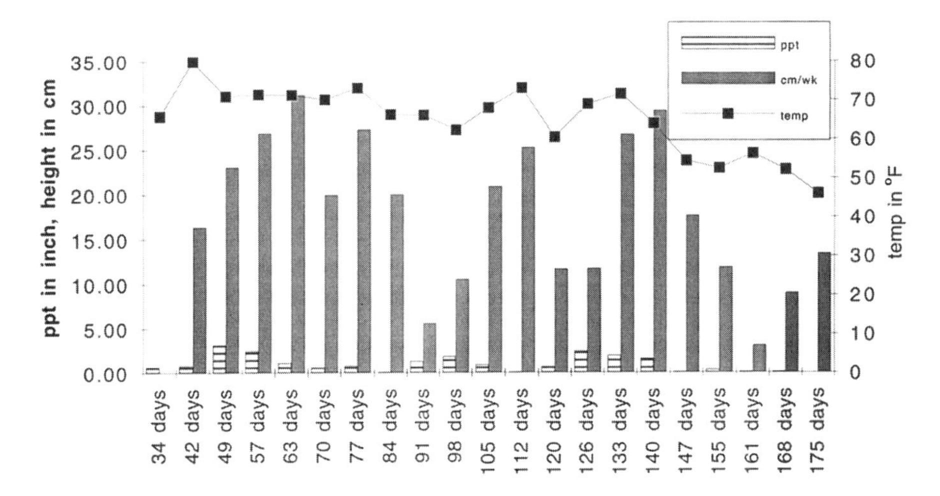

Figure 2.3 Weekly growth of kenaf in 1994.

At 84 and 108 dap, the bast fiber bundles comprised of primary and secondary phloem fibers tend to show more thickening and separation of primary and secondary phloem fibers become obvious. The secondary phloem fibers start thickening but with a somewhat weak middle lamella. At this stage of development, in addition to wall thickening, deposition of silica on the wall surface is seen (Figure 2.4-6). The fibers are long and broad mainly comprised of an S_2 layer (Figures 2.4-7 and 2.4-8), which is encrusted with amorphous silica.

At 112 dap (Figures 2.4-9 and 2.4-10), bast fiber bundles comprised of primary phloem fibers and secondary phloem fibers are thickened with prominent middle

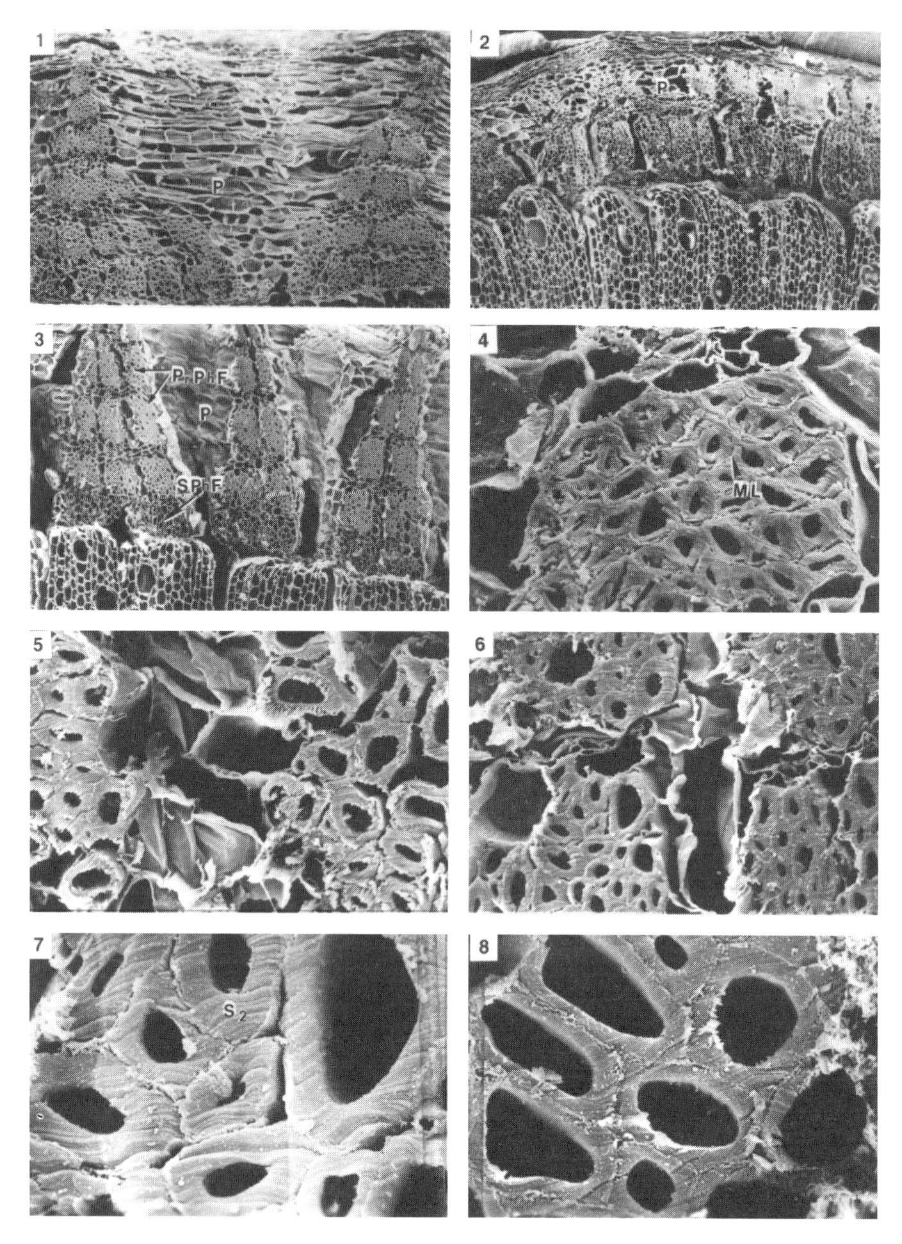

Figure 2.4 Scanning electron micrographs of kenaf cell walls. P = Parenchyma; PrPhF = Primary phloem fibers; SPhF = Secondary phloem fibers; ML = Middle lamella; and Si = Silica.

lamella formation. The cells are compact with thickened cell walls and decreased lumen width. The middle lamella is not well lignified at this stage of maturity. The fibers are long and broad with a well formed S_2 layer.

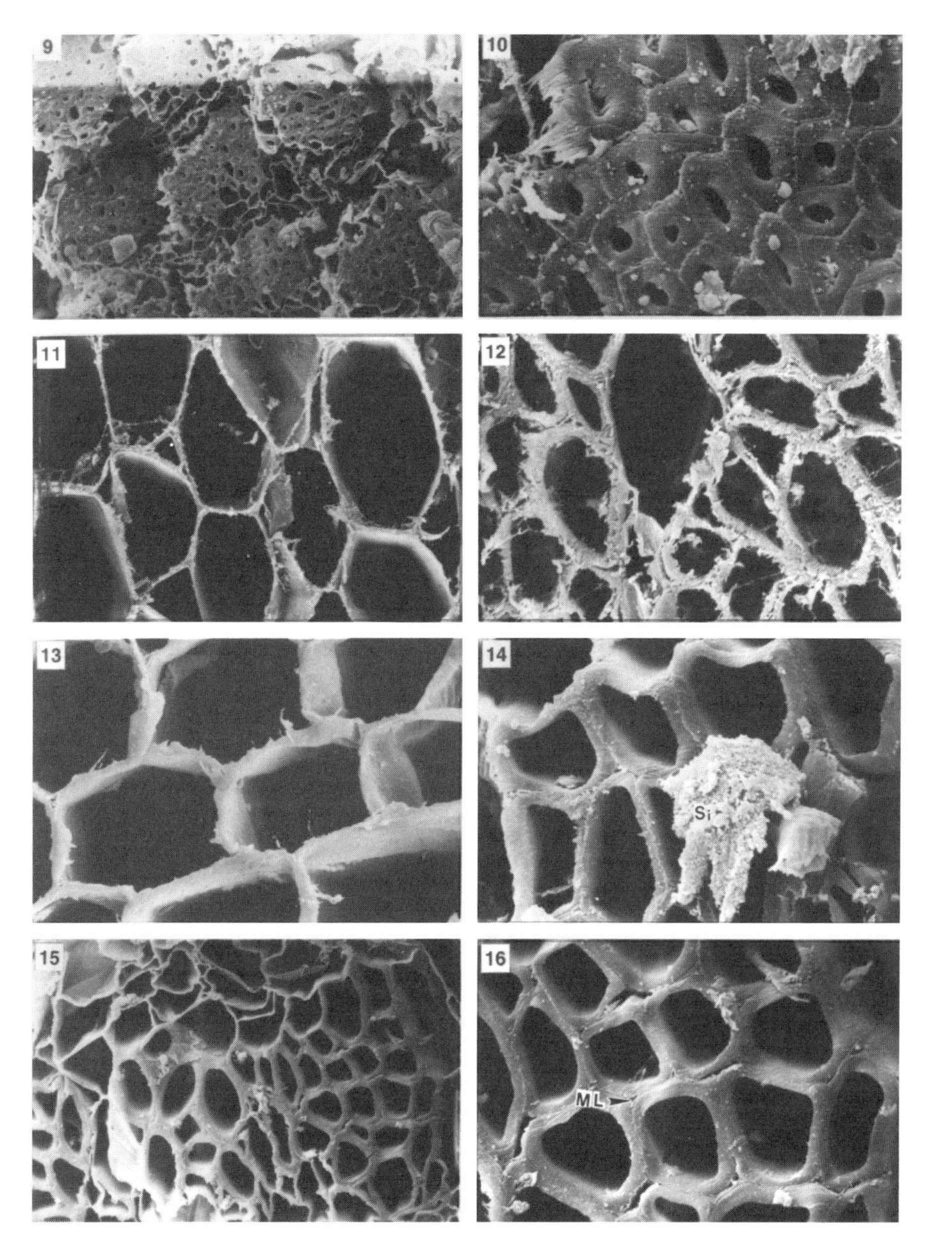

Figure 2.4 (continued)

Similar sequential development is seen in secondary phloem fibers. At 63 dap, there is little thickening of the fiber wall (Figure 2.4-11); however, the fibers thicken gradually with maturity from 73 dap to 112 dap (Figures 2.4-12 through 2.4-20).

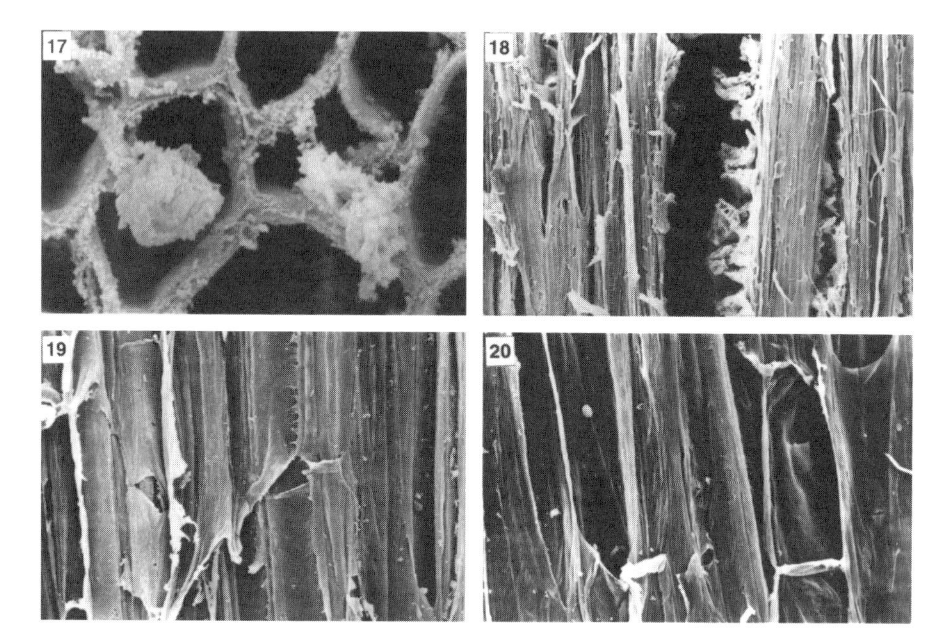

Figure 2.4 (continued)

REFERENCES

Bagby, M. O., Nelson, G. H., Helman, E. G., and Clark, T. F., Determination of lignin in non-wood plant fiber sources, *TAPPI*, 54, (11), pp. 1876–1878, 1971.

Chand, N. and Hashmi, S. A. R., Effect of plant age on structure and strength of sisal fibre, *Metals Materials and Processes*, 5(1), 51, 1993.

Chatterjee, H., Chemical characters of jute fibre at different stages of plant growth, *J. Sci. and Ind. Research*, 18C, 206, 1959.

Clark, T. F., Uhr, S. C., and Wolff, I. A., A search for new fiber crops. X. Effect of plant maturity and location of growth on kenaf composition and pulping characteristics, *TAPPI*, 50(11), 2261, 1967.

Clark, T. F. and Wolff, I. A., A search for new fiber crops. XI. Compositional characteristics of Illinois kenaf at several population densities and maturities, *TAPPI*, 52(11), 2606, 1969.

Fujii, Y., Azuma, J-I, Marchessault, R. H., Morin, F. G., and Aibara, S., Chemical composition change in bamboo accompanying its growth, *Holzforschung*, 47, 109, 1993.

Han, J. S., Kim, W., and Rowell, R. M., Chemical and physical properties of kenaf as a function of growth, the *International Kenaf Association Conference Proceedings*, Irving, TX, 1994.

Higuchi, T., Kawamura, I., and Ishikawa, H., On the formation of lignin in bamboo-shoot, *J. Japan For. Soc.*, 35, 258, 1953.

Higuchi, T., Kimura, N., and Kawamura, I., Differences of chemical properties of lignins of vascular bundles and of parenchyma cells of bamboo, *Mokuzai Gakkaishi*, 12, 173, 1966.

Higuchi, T and Shimada, M., Metabolism of phenylaline and tyrosine during lignification of bamboos. *Phytochemistry*, 8, 1183, 1969.

Higuchi, T., Shimada, M., and Ohashi, H., Role of 0-methyl-transferase in the lignification of bamboo, *Phytochemistry*, 8, 1183, 1967.

Migita, N., Chemical properties of bamboo, *Bull. Tokyo Univ. Forests*, 35, 139, 1947.

Mukherjee, A. C., Mukhopadhyay, A. K., and Mukhopadhyay, U., Surface characteristics of jute fibers at different stages of growth, *Textile Res. J.*, 56(9), 562, 1986.

Normura, T. and Yamada, T., X-ray analysis of tyrosine in growing stage of bamboo (*Phyllostacys edulis* A. & E. Riviere), *Japan Wood Res.*, 56, 21, 1974a.

Normura, T. and Yamada, T., Crystallinity change in growing stage of bamboo, *(Phyllostachys mitis)*, *Japan Wood Res.*, 57, 23, 1974b.

Ono, Y., Maturity of cotton fibers for spinning: IV. Relation between the maturity of cotton fibers and their chemical constituents, *Sen-i Gakkaishi*, 13, 785, 1957.

Ono, Y. and Sato, K., Maturity of cotton fibers for spinning: V. Relation between the maturity of cotton fibers and their reducing sugar content, *Sen-i Gakkaishi*, 13, 791, 1957.

Otoguro, C., Hikawa, Y., and Matsuno, O., Changes in physical and chemical compositions of edible burdock (*Arctium lappa* L.) in relation to varieties and growth, *Yamanashi-Ken Kogyo Gijutsu Senta Kenkyu Hokoku*, 5, 74, 1991.

Overbeke, M. van and Mazingue, G., Chemical study of several varieties of flax. *Bull. Inst. Textile France*, 13, 23, 1949.

Ozsanlav, V., British Textile Technology Group, Cheshire, UK, Personal communication, 1992.

Palo, R. T., Sunnerheim, K., and Theander, O., Seasonal variation of phenols, crude protein and cell wall content of birch (*Betula pendula* Roth.) in relation to ruminant *in vitro* digestibility, *Oecologia (Berlin)*, 65, 314, 1985.

Panshin, A. J. and de Zeeuw, C., *Textbook of Wood Technology*, McGraw-Hill, New York, 1980.

Phillips, M., Davidson, J., and Weihe, H. D., Studies of lignin in wheat straw with reference to lodging, *J. Agric. Res.* 43, 619, 1931.

Rowell, R.M., Ed., *The Chemistry of Solid Wood*, Advances in Chemistry Series No. 207, American Chemical Society, Washington, D.C., 1984.

Sjöström, E., *Wood Chemistry Fundamentals and Application*, Academic Press, New York, 1981.

Taniguchi, E., Crystalline region of cellulosic materials: XIV Variations of fine structure in *Pinus densiflora* and *Phyllostachys edulis* through their growth, *J. Japan Wood Research Soc.*, 2, 152, 1956.

Theander, O., Chemical characterization of some potential lignocellulosic raw materials in the Swedish agro-fiber project, *Production and utilization of lignocellulosics*, G.C. Galletti, ed., Elsevier Applied Science, London, UK, 49, 1991.

Usmanov, H. U., Physical chemistry of native cellulose, *J. Polymer Sci.*, 23, 831, 1957.

Usmanov, H. U. and Nikonovich, G. V., Electron microscopic studies of changes in cotton fiber structure during growth, *Uzbekh. Khim. Zhur.*, 3, 13, 1960.

Usmanov, H. U., Tillaev, R. S., and Mirsalikhov, M., Changes in the degree of polymerization of cotton fiber as a function of maturity, *Doklady Akad. Nauk Uzbek. S.S.R.*, 8, 17, 1958.

CHAPTER 3

Improvement of Fiber Crops Using Genetics and Biotechnology

Timothy La Farge, Sharon T. Friedman, and Charles G. Cook

CONTENTS

1. INTRODUCTION

Human beings have influenced the genetics and evolution of crop plants since long before recorded history. Since the times of Darwin and Mendel, the science of genetics and breeding has developed. The knowledge and application of genetic principles in crop improvement and biotechnology made the process of selection and breeding more efficient. Thus, the evolution of cultivated plants was guided by two processes: natural and artificial selection. In both of these processes, certain genetic patterns or mechanisms occur. Allard (1960) stated that the main patterns, or mechanisms, which prevail fall into three main categories: (1) Mendelian variation; (2) interspecific hybridization; and (3) polyploidy. Biotechnology, or genetic engineering, a more recent mechanism of selection and breeding developed since the discovery of DNA, will also be discussed.

1.1 Genetic Principles

There are two general sources of variability for any plant characteristic: genetics and environment. The challenge for a plant breeder is to evaluate the relative contribution of each of these sources of variability.

Another primary task of a breeder is to manage and use that portion of the variability which is due to heredity, and this portion is termed genetic variation. What a person sees in living plants in the field is called the phenotype. The phenotype is a combination of the plant's genetics, the environment, and the interaction between them. The genotype is the plant's genetic potential and can only be determined by well designed tests. In a general sense, heritability is that portion of the total variation for a given trait that is determined by the genes. The ratio of the genetic variance to the total or phenotypic variance is called heritability (h^2) and is shown in Equation 1.1, where V_G^2 represents the genetic variance, V_E^2 the variation due to environmental effects, and V_{GE}^2 the interaction between genotype and environment.

$$h^2 = \frac{V_G^2}{V_G^2 + V_E^2 + V_{GE}^2} \quad \textbf{(1.1)}$$

The heritability of a trait can be used to predict the amount of improvement in a trait that can be achieved through selection and breeding. Although classical Mendelian genetics defines a system of individual genes, in plant breeding each trait is controlled by a number of genes, each of which has a small effect. These are called quantitative traits. Traits that are controlled by a few genes, each of which has a large effect, are called qualitative traits.

However, since most traits of importance in commercial crops are quantitative, i.e., influenced by many genes having small effects, we cannot identify the specific genetic combinations that give rise to a certain value of a trait. Instead, we conduct experiments to determine what portion of the phenotypic variation is due to the genetic (or genotypic) potential of the plant, and then we are able to exert some genetic control.

A necessary first step in this process is to design a linear model that defines a phenotype for a particular trait of economic importance, such as yield. We might define yield as shown in Equation 1.2, where "μ" is the general population mean, "g" is the genotypic effect, "e" is the environmental effect, and "ge" is an interaction effect between the genotype and the environment. Each environmental and genotypic effect may add to or subtract from the population mean, μ.

$$P = \mu + g + e + (ge) \qquad \textbf{(1.2)}$$

If there are no genotype × environment interactions, i.e., when all genotypes perform consistently in all environments, then the interaction term (ge) will be zero. Since all measurements will be on phenotypes, we can only estimate the effect of a genotype in terms relative to other genotypes. For this system of genotypic evaluation to have practical utility, we must average all genotypes over a range of environments. If the relationship among the various genotypes being investigated is not constant, then the genotype × environment (ge) interaction will reflect this fact by a positive value (after Allard, 1960).

1.1.1 Heritability and Genetic Gain

We can create a model that is appropriate for any given investigation by partitioning some aspects of the environmental variation. Thus, we can divide "e," the environmental effect, into components such as replicates, locations, and years to reflect different components of environmental variation.

Even after we create this model, we must still set aside a composite error term for all remaining environmental effects, such as sampling errors among plants, sampling errors among plots in the same replicate, and errors of measurement. In this manner we create a linear model on which we base our experimental design. It is beyond the scope of this discussion to explore the details of a particular design, but suffice it to say that this allows us to make estimates of parameters in a given population so that we may estimate the genetic gain obtainable at a given selection intensity.

The genetic gain obtainable by selection among individuals or families in a base population will depend on three things: (1) the variability among genotypes (individuals or families) in the base population; (2) the amount of environmental noise, due to both the environmental (e) and interaction components of variability (ge); and (3) the selection differential, which may be defined as the proportion selected, e.g., q individuals (or families) as a portion of a total n individuals (or families), so that we select q/n individuals or families.

Because our population of phenotypes is assumed to be distributed normally, we standardize the selection differential by dividing it by the phenotypic standard deviation, Vp, and we call this the selection intensity, $i = (q/n)/Vp$ (after Falconer, 1960 and Becker, 1984).

One of several possible experimental designs (mentioned above) will be appropriate for any given investigation. After measurements of a sufficient number of individuals replicated over several environments (randomly or selectively obtained, depending on the assumptions or objectives of our investigation), we are able to obtain the components of genotypic and environmental variance which are essential for obtaining the heritabilities and genetic gains we are seeking. To review, we will first obtain the phenotypic, or total, variance (Equation 1.3).

$$V_P^2 = V_G^2 + V_E^2 + V_{GE}^2 \tag{1.3}$$

We then have the components of variance needed to obtain the heritability (Equation 1.4).

$$h^2 = \frac{V_G^2}{V_G^2 + V_E^2 + V_{GE}^2} = \frac{V_G^2}{V_P^2} \tag{1.4}$$

After consulting appropriate tables for selection intensities (i) under a normal curve (e.g., Becker, 1984), we would be able to estimate our genetic gain for any given selection intensity in Equation 1.5.

$$G = (i)(V_P)(h^2) \tag{1.5}$$

Thus, we may select, say, the five highest yielding families from a population of 100 families, and, after appropriate measurement of field trials and analysis of data, we can estimate the amount of genetic gain we have attained, either in absolute units of measure, or as genetic gain as a percentage of the population mean. If we have appropriate check lots representing a random selection from a natural parental population, they may be used as the basis for the absolute or relative genetic gain.

1.1.2 Genetic Strategies

The preceding paragraphs describe some of the essential principles involved in selection and breeding of individuals and families within a species. However, real genetic gains may be obtained by selection and breeding at several levels. We can evaluate, compare, and then select the best species (e.g., those species with the highest yields) for a given breeding zone. Such a zone combines the environmental factors (e.g., climate, soils, etc.) to which that species is best adapted.

Once we have determined a well-adapted species, we can evaluate and compare the best varieties for that breeding zone, or perhaps for different locations within that zone. We can also sometimes produce species hybrids which are better adapted, and produce higher yields in a given environment than either parent species. Finally, we can make crosses within each variety among pre-tested parents or families to obtain further gains in yield.

Each of these choices will depend on the circumstances and benefits obtainable. Actually, it is not usually a choice between one or the other. If a program is to be profitable, the optimum genetic gains will need to be attained. Such gains can be evaluated by standard economic analyses, such as cost-benefit ratios and internal rates of return.

2. APPLICATIONS OF GENETICS TO FIBER PRODUCTION

In recent decades there has been a growing interest in the development of species of annual crop plants having the potential to produce fiber. This interest in fiber production is directed at developing a range of products, including paper, composites, and cloth. Most research in the genetics of fiber producing plants has been performed on six species: (1) kenaf (*Hibiscus cannabinus* L.); (2) roselle *(H. sabdariffa* L. var. *altissima*); (3) jute (*Corchorus capsularis L.*); (4) jute (second species) *(C. olitorius* L.); (5) hemp (*Cannabis sativa* L.); and (6) sunn hemp *(Crotolaria juncea* L.). In this paper the genetics of fiber production will be limited to kenaf, jute, hemp, and sunn hemp.

2.1 Kenaf

Kenaf, a member of the *Hibiscus* (genus) section of *Furcaria*, is one of more than 50 species that occur in tropical and subtropical environments of every continent except Europe (Wilson, 1992). Only two of these species, kenaf (*H. cannabinus* L.) and roselle (*H. sabdariffa* L. var. *altissima*), have economic importance. Because of their capability to be utilized to produce pulp and paper, they have been investigated in the United States for over two decades as alternate sources of paper pulp.

Kenaf has a diploid chromosome number of n = 18, while roselle has a tetraploid number of n = 36. The other species of section *Furcaria* have haploid chromosome numbers ranging from n = 18–90. Of the 50 or so species of *Furcaria*, 11 other genomes have been identified and have much potential as sources of germplasm for improvement of kenaf by means of hybridization.

The importance of these other sources of germplasm lies in their resistance to root-knot nematode. Except for one wild strain from Tanzania, PI 292207, kenaf is very susceptible to nematodes. Exploitation of the moderately resistant Tanzanian strain has resulted in the release of four germplasm lines of kenaf having moderate resistance. But other species in section *Furcaria* have excellent resistance, and roselle strains vary in resistance from susceptible to resistant.

Three breeding strategies have been employed to improve nematode resistance in kenaf: (1) transfer of resistance from a wild, tetraploid species, *H. acetosella*; (2) doubling of the chromosome number of the *cannabinus* × *acetosella* hybrid to produce synthetic species; and (3) crossing kenaf and roselle and doubling the chromosome number of this hybrid to produce a synthetic polyploid kenaf × roselle species. The third approach proved best, since the hybrids had considerable growth and moderate nematode resistance. This synthetic species should show additional improvement when more resistant roselle parents are exploited.

2.1.1 Kenaf Variety Trials

While hybridization provides an avenue for the introduction of new genes for resistance to root-knot nematodes in kenaf, variety trials indicate that considerable genetic variation within this species offers many opportunities for genetic improvement in yield.

Neil and Kurtz (1994) conducted a series of variety trials which indicated a range of potential yields for 10 varieties tested at three locations in Mississippi representing three different soil types. There was also much variation in yield among the soil types, and there was much interaction in varietal performance on the different soil types. Hence, the authors could not recommend specific varieties for a soil type without more research on this problem. Similarly, Hovermale (1994) found interactions between kenaf yield among four varieties planted at six planting dates from April 1 to June 15 in a trial. These trials were repeated in 1990, 1991, 1992, and 1993, but only five dates were tested in 1990 and 1992. Results were not consistent from year to year. For example, roselle had higher yields than the three kenaf varieties in 1990 for four of the five planting dates but tended to have the lowest yields at most planting dates in 1991, 1992, and 1993. It appears that general conclusions as to the best overall varieties for Mississippi cannot be made from these studies at the present time. However, yield results from the 1990–1994 USDA–ARS Uniform Kenaf Variety test indicated that several entries had greater yield potential and stability across locations (Cook, unpublished). One of these entries, Tainung 2, is a released, commercially available variety. These findings are in agreement with the reports of Bhangoo et al. (1994) at Fresno, California and Hallmark et al. (1994) at Bossier City, Louisiana.

More recently a new kenaf cultivar with excellent yield potential and stability has been released by the USDA–ARS. This cultivar has also exhibited an excellent level of tolerance to the root-knot nematode/soil-borne fungi complex (Cook and Mullin, 1994). The ability to identify those cultivars with the greatest fiber yield potential, general adaptability, and disease/nematode resistance should help accelerate the progress of kenaf breeding programs (Figures 3.1 through 3.4).

2.1.2 Genetics of Yield Components in Kenaf

A strategy that is often useful in plant breeding is to break the yield of a plant species down into its yield components, i.e., those traits which contribute most to the inheritance of yield. If the heritability of each trait can be determined, then its genetic contribution to total yield can be estimated. When this is accomplished, a selection index for yield can be developed that may more precisely estimate yields.

Often for practical reasons it is more appropriate in designed field trials to calculate the general combining ability (gca) of each trait and determine its statistical significance. Sinha and Roy (1987) performed such an analysis for kenaf and found two out of 11 varieties which proved to be the best general combiners for those traits which contributed the most to fiber yield. However, no single parent possessed a better gca for all traits.

Several studies have shown that highly significant, positive correlations exist between plant height, stem diameter, and fiber yields (Sarma, 1967; Chakravarty, 1974; Adamson and Bagby, 1975). Since Lia (1991) has found that plant height and stem diameter of kenaf are of moderately high heritability, indirect selections of these yield components should be effective in advancing kenaf fiber yields.

Figure 3.1 Four kenaf genotypes, from left to right by row: (1) a cordate-leaf genotype; (2) a palmate-leaf genotype; (3) a cordate-leaf genotype; and (4) a photoperiod-insensitive (flowering palmate genotype). (Photo by C. G. Cook)

Figure 3.2 A comparison of cultivars: "Everglades 71" on the right and the root-knot nematode tolerant "SF459" on the left. The latter was released in 1994 by co-author Charles Cook of the USDA-ARS, Weslaco, TX. (Photo by C. G. Cook)

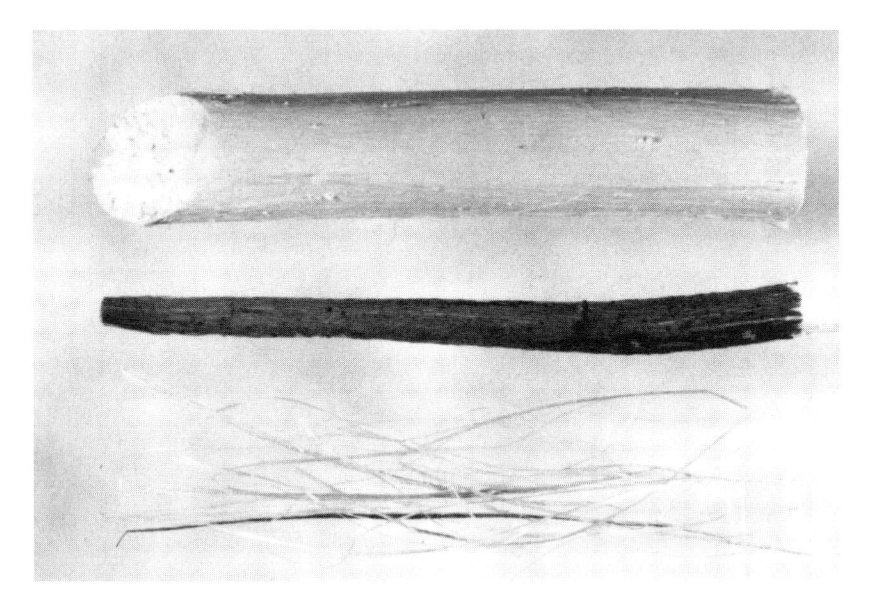

Figure 3.3 Kenaf fibers, from top to bottom: (1) short-fiber core; (2) the bark with bast; and (3) partially separated long-fiber bast. (Photo by C. G. Cook)

2.1.3 Genetic Divergence in Kenaf

Usually, before any breeding program is begun it is useful to determine the degree of genetic variability of a species and whether this variability is associated with geographic origin. If there are large differences for key adaptive traits across a species range, they are said to indicate a degree of genetic divergence or diversity.

One statistical measure of such diversity is the D^2 statistic, or genetic distance, originally devised by Mahalanobis (1936). Larger genetic distances suggest more distant relationships and hence potentially greater opportunities for heterosis (hybrid vigor) when wide crosses are made. Devi et al. (1991) used such an approach on 51 kenaf genotypes representing a range of ecogeographic origins. Their analysis resulted in eight groupings, or clusters, of genotypes which provided a basis for making wide-cross hybrids. The traits most useful for assembling these clusters and for determining genetic divergence were: (1) days to 50% flowering; (2) node number per plant; (3) fiber length; and (4) plant height. Three of the eight clusters (clusters III, VI and VII) showed the widest genetic divergence for most traits studied and therefore the greatest opportunity for maximizing the variation among the segregants. Thus, crossing genotypes from cluster III with those from cluster VI and VII would produce segregants with greater genetic diversity. Selection among such offspring for high yield and other desirable traits then provides an opportunity to obtain useful genetic gains.

Figure 3.4 A mature kenaf field. (Photo by C. G. Cook, USDA–ARS, Westlaco, TX))

3. JUTE

Jute is comprised of two cultivated species, *Corchorus capsularis* and *C. olitorius*, and both have a diploid chromosome number of 2n = 14. It is an important tropical crop that produces bast fibers and is second in importance only to cotton as a source of vegetable fibers. Ninety-five percent of the world production of jute fibers occurs in India, Bangladesh, China, Thailand, Nepal, and Indonesia (Saha and Sen, 1992). Although breeding research is ongoing, these species are very limited in their variability with respect to: (1) adaptability to different agronomic environments; (2) fiber quality; (3) fiber yield; and (4) susceptibility to some diseases and pests.

Jute has a growth habit which, like kenaf, lends itself to the breakdown of fiber yield into components. Thus, Sadaqat et al. (1989) evaluated genotypic and environmental components of variability and heritabilities of seven traits which contribute to fiber yield. They also calculated genotypic and phenotypic correlation coefficients among all traits, thus enabling the determination of correlated responses to selection of any one trait on fiber yield or any other trait. The three traits with the closest associations with fiber yield were, in order of declining association: (1) plant height; (2) number of nodes per plant; and (3) fiber length.

Much work has also been done on jute to induce mutations and breeding of resulting mutant genotypes. Geneticists sometimes use radiation to induce new mutations in plants for the purpose of inducing variability. While most induced mutants are deleterious, there is a chance that some mutations will be beneficial and can be selected. Mutants having severe deleterious effects at more than one locus (macromutants) are a common by-product of induced mutations in jute. Since they usually show poor agronomic performance and have severe deleterious pleiotropic (multiple phenotypic expressions of a single gene) effects, they are often rejected as unimportant or useless byproducts of mutation breeding. However, studies have shown that macromutants have much potential for providing useful genetic material when utilized in a cross breeding program in which subsequent deleterious mutants are eliminated and desirable offspring are retained.

Sinhamahapatra (1986) crossed five macromutants of jute (from both *C. olitorius* and *C. capsularis*) in a 9 × 9 diallel (a test design in which each parent is crossed with every other parent). The F1 (first generation) hybrids of these crosses showed, in almost all cases, strong heterosis for plant height, base diameter, days to flower, and fiber yield. The mutated locus showed no signs of deleterious effects when heterozygous (having different alleles or genes at one or more genetic loci) in the F1 generation. Further, the author found that induced polygenic variation was mostly additive, suggesting that selection for plant height in the segregation generation should be effective. Such induced mutation changes in the background genotypes should provide a broad genetic base in a breeding program.

Sometimes desirable traits have negative effects on each other under selection. A common problem in jute is the long duration of vegetative growth, which compels farmers to harvest the crop before maturity to release the land for other late season uses, resulting in loss of yield. So far, most induced early maturing mutants have reduced plant height and fiber yield. Sinhamahapatra and Rakshit (1986) radiated selfed seedlings and grew the treated and control progenies in the M2 generation (second generation after induced mutation). The M3 generation was grown and, subsequently, selection for earliness displayed significant drops in plant height, basal diameter, and fiber yield. They concluded that the possibility of selecting early lines with higher yield is unlikely. Selection for lateness, on the other hand, resulted in increases in basal diameter and fiber yield. It was suggested that early maturity and dwarfism might be due to close linkages of genes controlling earliness and plant height.

4. HEMP

The breeding of hemp (*C. sativa* L.) presents special problems because of its sexual polymorphism. Hemp is dioecious (male and female flowers occurring on separate plants), but breeders have developed monoecious (male and female flowers occurring separately on the same plant) forms. In dioecious forms of hemp the sexes flower at different times. The male hemp ripens 1 to $1^1/_2$ months earlier than female hemp. Hence, the male hemp is picked by hand and the female hemp is mechanically harvested later. In contrast, in monoecious hemp both sexual types mature simultaneously and, therefore, can be mechanically harvested at the same time (Migal, 1993).

The challenge for the breeder is to devise a breeding strategy that will result in a stable monoecious variety of hemp. Because of the very complex nature of the genetic structure, maintenance of stability of monoecious hemp is difficult. The offspring of open-pollinated populations tend to diverge to both monoecious and dioecious forms.

Monoecious hemp was first discovered in 1929–1933 when mutant monoecious plants appeared in a stand of dioecious hemp and were subsequently isolated and reproduced. Since then breeders have attempted a number of different breeding strategies to obtain stable monoecious varieties. Initial selection methods during the period 1937–1945 in Russia achieved a reduction in the frequency of staminate hemp (the dioecious male form) from 36–0.008%, a variety which is still the most stable for the monoecious trait. The two other strategies which have been tried with less success are inbreeding and the development of monoecious tetraploid hemp. Tetraploid hemp involves the doubling of the diploid chromosome number from 2n = 20–40. Tetraploid hemp has the greatest stability of the monoecious trait, while maintaining a very low frequency of staminate hemp (0.63–0.90%). However, tetraploid hemp tends to ripen late and has poor seed yield and viability. Further, it performs poorly with regard to certain other commercially desirable traits.

The net result of the inability of breeders to develop a stable variety of monoecious hemp is that workers must tediously cull by hand the staminate hemp plants before they begin to flower. Many years of research and breeding in Europe have not appreciably solved the problem of developing stable varieties of monoecious hemp (Migal, 1991, 1993). This fact, plus the notoriety of hemp (*Cannabinus sativa* L., marijuana) as an illegal drug, would seem to preclude this species as a source of nonwoody fiber production in the United States.

5. SUNN HEMP

Sunn hemp (*Crotalaria juncea* L.) is the sole member of the Leguminosea family that is grown commercially for the soft fiber produced in plant stems (Dempsey, 1975). Sunn hemp, which is indigenous to India, rates second only to kenaf as a raw fiber material for the manufacturing of paper and pulp. In addition, sunn hemp may be grown as a green manure and cover crop, and as a deterrent to nematodes (Breitenbach, 1958). The majority of the sunn hemp cultivars have resulted from selections at various research stations rather than by breeding. Past efforts to make interspecific crosses with other *Crotalaria* species have met with little success. Nevertheless, because of the lack of genetic exploitation in past research programs, improvements in yield and disease resistance through breeding within the species would appear to have great potential.

6. FOREST TREES

Although the emphasis in this chapter has been on annual fiber producing crop species, forest trees as a source of fiber should not be excluded. In the last 50 years, much work has been done in traits controlling growth, yield, and fiber quality in forest trees.

Figure 3.5 A fast-growing family of Virginia pine (in row on left) compared to a slow-growing family (in row on right) in a progeny test being grown to evaluate the genetic breeding values of parental clones. (Photo by T. LaFarge, USDA–Forest Service, Atlanta, GA)

Generally, forest tree improvement programs have concentrated on traits affecting growth and yield, stem straightness, and resistance to important diseases, such as fusiform rust in loblolly pine (*Pinus taeda* L.) caused by *Cronartium quercuum* (Figures 3.5– 3.10). However, Stonecypher and Zobel (1966) obtained genetic gains in loblolly pine for increases in specific gravity, and Zobel et al. (1978) significantly increased the specific gravity of juvenile wood in one generation. Some wood quality traits, such as specific gravity and tracheid length in pine species, are independently

Figure 3.6 A superior tree in an outstanding family in an older progeny test. Both parents of this tree have high scores for yield in several progeny tests. Scions will be cut from this tree and used to establish clones in a new second-generation clonal seed orchard. (Photo by T. LaFarge, USDA–Forest Service, Atlanta, GA)

inherited (Zobel and Talbert, 1984). Other traits, such as cell wall thickness and specific gravity, are strongly interdependent.

The relative independence of inheritance of some of the major wood properties means that it is possible to breed trees for certain specific wood properties. For example, trees can be bred for paper products that require long fibers, others for those products which need short fibers, those products for which high specific gravity is essential, or those which must have low specific gravity. The only problem with this diversity of product objectives is that it compounds the difficulties of selection and breeding in tree improvement programs. Separate populations in each species must be developed and maintained for each wood quality trait desired, even as those same populations are being bred for increased yield.

More recently some researchers in biotechnology are experimenting with lignin modification. Recently a gene has been identified in dicotyledonous angiosperms (hardwoods) which, if inserted in gymnosperms (e.g., conifers, pines, softwoods), could produce wood which is more easily digested in the pulping process. Therefore,

Figure 3.7 Harvest of genetically superior seed by means of a net retrieval system in a first-generation clonal seed orchard. Nylon netting is laid on the seed orchard floor for seed collection in the fall. At this point the seed is still mixed with pine straw. (Photo by T. LaFarge, USDA–Forest Service, Atlanta, GA)

Figure 3.8 The seed-straw separator. The netting is being rolled up onto the separator, where the seed will be separated from the pine straw. (Photo by T. LaFarge, USDA–Forest Service, Atlanta, GA)

Figure 3.9 The seed-straw separator. The straw and other trash are carried away from the seed, which is shaken through small openings into the seed tray underneath. (Photo by T. LaFarge, USDA–Forest Service, Atlanta, GA)

Figure 3.10 The seed-tray in the seed-straw separator. After the genetically-improved seed has been separated from the pine straw and other debris, it is collected in the seed-tray. (Photo by T. LaFarge, USDA–Forest Service, Atlanta, GA)

it has been argued that there are valid economic and environmental reasons to use genetic engineering to insert hardwood genes into conifers. Specifically, Malcolm Campbell (personal communication, 1995) has proposed inserting ferulate 5-hydroxylase (F5H), an enzyme necessary for synthesizing syringyl lignins found only in hardwoods, into conifers. However, some people are concerned about the adaptive effects on biosynthetic pathways of such modifications. Lignin strengthens plant tissues by protecting the cell walls from compression (Esau, 1965) and may be important in disease resistance mechanisms (Chapple et al., 1992). Hence, such work could have adverse effects on disease resistance mechanisms and even wood properties affecting strength. Environmental acceptance of such genetic engineering of forest trees could be adversely affected if trees having such modifications were placed in the forest without careful prior research to determine the potential effects on the ecosystems into which they are being inserted. Obviously, many questions remain to be answered. Nevertheless, the biotechnology of annual crops and forest trees is discussed in the following section.

7. THE POTENTIAL ROLE OF BIOTECHNOLOGY

Plant biotechnology has been defined as combined application of recombinant DNA (genetic engineering), cell and tissue culture, and plant breeding for developing new plant varieties and plant products (Kunimoto 1986). Certain genetic engineering techniques are the keys to any progress that may be made in this field.

In plants, genetic engineering is based on the regeneration or recovery of fully developed plants from single somatic cells into which have been inserted genes which are new to the host organism. Theoretically such genes may be isolated from any living source, such as animals, sexually incompatible plants, and microorganisms.

Four methods are available for transfer of such new genes: (1) microinjection; (2) electroporation; (3) *Agrobacterium*-mediated gene transfer; and (4) biolistics. Micro-injection consists of introducing genetic information into plant cells by means of a microscopic needle. However, this method has declined in use for several logistic reasons: (1) the needle tips are very fine and break easily, thereby clogging the cells; (2) transforming cells one at a time is time consuming and not commercially efficient; and (3) the incorporation of DNA into a cell genome, even after insertion, is very infrequent and uncertain (Gasser and Fraley, 1992).

Electroporation, which is electrically-mediated DNA uptake, involves the use of short, high-voltage impulses to produce pores in the protoplast membrane through which DNA molecules can enter. But even this method is limited to only a few crop species. The important grain crops, such as corn and wheat, respond poorly to this method (Gasser and Fraley, 1992).

Agrobacterium-mediated gene transfer (AT) is infection of host species by engineered *Agrobacterium*, a microorganism in the soil. Thus, this method of transfer can be used to insert a piece of DNA with new genes, i.e., novel DNA sequences, into susceptible host plants. In recent years, AT has included new ways of introducing *Agrobacterium* into plants that are not dependent on natural infection processes.

These methods include vacuum infiltration and soaking of seeds (David Harry, personal communication, 1994). *Agrobacterium*-mediated gene transfer has been utilized on cotton, many species of vegetables, and certain woody plants.

Biolistics is a recent form of gene transfer in which small metal particles (one or two microns in diameter) of tungsten or gold are coated with DNA and propelled into plant cells. One method of propulsion utilizes a "DNA particle gun," which consists of a .22 caliber blank cartridge as the propelling force (Gasser and Fraley, 1992). Other methods include electric discharge and pressurized gas.

Genetic engineering can be used to control precisely the function of a transferred gene. This control is possible because of the modular nature of genes. This modularity arises from the fact that genes that produce proteins comprise three regions. The first is a promoter region. Promoters can determine the location, level, and time of expression of novel genes. The recombinant gene can be controlled when a promoter sequence is isolated and a new gene is spliced to the promoter in a way similar to the gene from which the promoter sequence was obtained (Kunimoto, 1986).

The second region is a coding region which contains the information needed to determine the protein encoded by the gene. The third region is the polyadenylation (or poly-A) region, which codes the m-RNA transcript to terminate correctly (Gasser and Fraley, 1992).

Two other technologies, in addition to their own independent applications, are often included in gene transfer procedures: genome mapping and marker assisted breeding, also called marker breeding or marker aided selection (MAS). It should be added that these procedures also have a common interface with classical breeding. Neale and Harry (1994), discussing applications in forest trees, list three primary applications of genetic markers of genome maps: (1) basic studies of genomes; (2) measures of genetic diversity; and (3) mapping of quantitative trait loci (QTLs).

Although new technologies are evolving rapidly, there are basically two procedures available for obtaining genetic markers: (1) restriction fragment length polymorphisms (RFLPs); and (2) randomly amplified polymorphic DNA (RAPD). Two other important sources of markers are Short Sequence Repeats (SSRs) and Sequence Characterized Amplified Regions (SCARs) (Hans van Buijtenen, personal communication, 1995). RFLPs are formed by the presence or absence of certain restriction sites in DNA molecules. They are distributed throughout the genome and occur in all organisms (Hartl and Clark, 1989).

RFLPs are codominant markers, which means that they can be used to distinguish and identify multiple alleles. However, they require large amounts of DNA, they are technically complex to obtain, and the procedures needed to obtain them are often time consuming (Neale and Harry, 1994).

RAPD is an application of the polymerase chain reaction (PCR). It is a method of copying unlimited numbers of DNA sequences, hence the term "amplified." RAPDs require relatively little equipment, they can be produced quite quickly, and they require only small amounts of DNA. However, realistically, they are only useful in identifying dominant markers. Also, whereas RFLP technology is based on similarity of DNA sequences, RAPDs usually show little homology between DNA sequences. Also, RAPD fragments are classified by their molecular weight, so that fragments of similar size may not be homologous. In such cases false conclusions as to gene identity can be inferred.

Marker breeding is a method which has become prominent with the discovery of quantitative trait loci (QTLs). QTLs are genes affecting quantitative traits but which are themselves marked by genetic markers (David Harry, personal communication). Such Mendelian markers may take any form, including discrete segregating traits, isozyme markers, or RFLPs. They are linked to families of genes that may be located close together on the same chromosome. Identification of QTLs reduces the complexity of identifying the many genes associated with the control of any given quantitative trait. Typical QTLs would be loci controlling yield in crop breeding or forest tree breeding. The choice of which method to use in QTL investigations, i.e., between the slow and costly RFLP technique and the faster but often less precise RAPD method, is not always an easy one.

A half-sib approach to the mapping of quantitative traits has been used by Grattapaglia and Sederoff (1993) for mapping quantitative traits in hybrids of *Eucalyptus* spp., namely growth and wood quality. Likewise Wilcox et al. (1993) are conducting experiments to map height growth and resistance to fusiform rust (caused by *C. quercuum*) in loblolly pine (*Pinus taeda* L.).

Once gene transfer, by whatever method, has occurred, the cells successfully receiving the genes must be regenerated. One method of achieving this objective is by tissue culture. Tissue culture consists of numerous systems of micropropagation by which clonal propagation is obtained from explants of the original plant. The primary objective is to produce large numbers of genetically identical individuals. Such plantlets can be cultured from transgenic cells in leaves, roots, seeds, and tubers. The genetic engineering used to promote the desired genotypes in these tissues can even be targeted to specific cell types within complex tissues (Gasser and Fraley, 1992).

Tissue culture can be implemented to produce such plantlets by way of two developmental pathways: organogenesis and embryogenesis (Cheliak and Rogers, 1990). Organogenesis involves a process in which hormones are exogenously applied to meristematic tissues to initiate buds or other organ primordia. Somatic embryogenesis is a process in which nonmeristematic cells are cultured to result in a bipolar structure comprising shoots and roots at either end of a functional vascular tissue.

Recently, somatic embryogenesis has been utilized in an effort to provide a means of developing a jute hybrid between *C. olitorius* and *C. capsularis* (Saha and Sen, 1992). The authors isolated protoplasts from cotyledon, hypocotyl, and mesophyl cells of *C. capsularis* by a one-step enzyme digestion method. They were then successfully cultured on a special medium to form microcalli. The protoplasts created cultural conditions that achieved 34–78% plating efficiency, depending on which cell type was isolated. Hypocotyl protoplasts provided the highest plating efficiency. When the hypocotyl protoplasts were transferred to a regeneration medium, somatic embryos were developed at a high frequency. This appears to be an essential first step toward the production of appropriate somatic hybrids that can be developed into viable species hybrids and effective genetic improvement of jute as a fiber crop species.

8. THE ECONOMICS OF A BREEDING PROGRAM

Whatever methods are utilized in any attempt to attain genetic gains in a crop yield or its resistances to insects and diseases, it is important to estimate the potential costs and benefits or returns on investment. First, we can draw from an example of some basic production costs in a kenaf production project.

Fuller et al. (1991) have analyzed production costs of kenaf in Mississippi. They estimated costs at approximately $256 per acre. At an assumed price of $56 per dry ton of kenaf, and a yield of 6 tons per acre, returns would be about $80 per acre. But the authors warned that this $80 is not profit; it is returned to fixed costs on land, management, and farm overhead. A more careful analysis is required to fully evaluate a complete cost and return balance to an individual firm, or farm.

We can not resolve the profit questions raised by Fuller here, but we can use this example as a starting point for basic costs which might be incurred in a breeding program. Let us suppose an industrial or university cooperative breeding program had attained a 10% genetic gain for yield in kenaf in one year of breeding. The annual funding for this industrial/university program is $300,000 per year (there are three similar co-ops in the South in forest tree improvement). We wish to calculate the added value of one pound of genetically-improved seed.

To start, with fiber yields of six tons per acre and a genetic gain of 10%, the additional yield will be 0.1 × 6 tons per acre = 0.6 tons per acre. At $56.00 per ton, the added value per acre is 0.6 tons per acre × $56.00 per ton = $33.60. The number of acres needed to recover the breeding costs is $300,000/$33.60 = 8,929 acres. We can also calculate the added value of the seed as follows: eight pounds of seed per acre are needed for planting, and the added value of the improved variety is $33.60 per acre; therefore, the added value of the seed is $33.60/8 = $4.20 per pound.

These figures are, of course, somewhat simplistic and do not take into account the long-term investments needed to be considered in a breeding program. The costs of such a program may be higher or lower. However, it suggests that, if enough farmers or investors participated, the benefits could be well worth the investment.

REFERENCES

Adamson, W. C., and Bagby, M. O., Woody core fiber length, cellulose percentage, and yield components of kenaf, *Agronomy Journal*, 67, 57, 1975.

Allard, R. W., *Principles of Plant Breeding*, John Wiley & Sons, Inc., New York, 1960.

Becker, W. A., *Manual of Quantitative Genetics,* 4th Edition. Academic Enterprises, Pullman, Washington, D.C., 1984.

Bhangoo, M. S., Cook, C. G., Jacobson, T. A., and Gaberhiwet, A., Performance of kenaf varieties in the San Joaquin Valley, California, *California Agricultural Technology Institute*, Pub. 940802, 1994.

Breitenbach, C. A., Land management practices for kenaf production, *Proc. World Conference on Kenaf,* Havana, Cuba, 26, 1958.

Campbell, M. M., Phone conversation, February, 1995.

Chakravarty, K., Breeding for high fiber yield in mesta, *Indian Journal of Genetics and Plant Breeding*, 34, 937, 1974.

Chapple, C. C. S., Vogt, T., Ellis, B. E., and Somerville, C. R., An arabidopsis mutant defective in the general phenylpropanoid pathway, *The Plant Cell*, 4, 1413, 1992.

Cheliak, W. M., and Rogers, D. L., Integrating biotechnology into tree improvement programs, *Canadian Journal of Forest Research*, 20, 452, 1990.

Cook, C. G., and Mullin, B. A., Growth response of kenaf cultivar in root-knot nematode/soil-borne fungi infested soils, *Crop Science*, 34, 1455, 1994.

Dempsey, J. M., *Fiber Crops*, The University Press of Florida, Gainesville, FL, 1975.

Devi, B. S., Subramanyam, D., Kumar, P. V. R., Satyanarayana, G., and Krishnamurthy, B., Genetic divergence in kenaf, *Indian Journal of Genetics and Plant Breeding*, 51, 260, 1991.

Esau, Katherine, *Plant Anatomy,* John Wiley & Sons, Inc., New York, 1965.

Falconer, D. S., *Introduction to quantitative genetics,* The Ronald Press Company, New York, 1960.

Fuller, M. J., Brasher, B. B., and Little, D. E., Kenaf production costs in Mississippi: a preliminary assessment (Abstract), *Proc. Third International Kenaf Association Conference*, Tulsa, Oklahoma, 26, 1991.

Gasser, C. S., and Fraley, R. T., Transgenic crops, *Scientific American*, 266, 62, 1992.

Grattapaglia, D., and Sederoff, R., QTL mapping in Eucalyptus using pseudo-testcross RAPD maps, half and full-sib families, *Proc. of the 22nd Southern Forest Tree Improvement Conference,* Atlanta, Georgia, 14, 1993.

Hallmark, W. B., Brown, L. P., Viator, H. P., Habetz, R. J., Caldwell, W. D., and Cook, C. G., Kenaf: a new crop for Louisiana, *Louisiana Agriculture,* 37, 28, 1994.

Harry, D. E., Letter, November 16, 1994.

Hartl, D. L., and Clark, A. G., *Principles of Population Genetics*, Sinauer Associates, Inc., Sunderland, Massachusetts, 1989.

Hovermale, C. H., Kenaf variety by date of planting in Mississippi. In: *A Summary of Kenaf Production and Product Development Research, 1989–1993,* Office of Agricultural Communications, Division of Agriculture, Forestry, and Veterinary Medicine, Mississippi State University, 3, 1994.

Kunimoto, L., Commercial opportunities in plant biotechnology for the food industry, *Food Technology*, 40, 58, 1986.

Liu, W. J., Jute and kenaf breeding-principles and practice in China. Institute of Bast Fiber Crops,

Mahalanobis, P. C., On the generalized distance in statistics, *National Institute of Sciences of India*, 2, 49, 1936.

Migal, N. D., Genetic aspects of the evolution of sex in hemp, *Genetika* (Moskva), 27, 1561, 1991.

Migal, N. D., Genetic principles of the creation of Monoecious hemp, *Genetika* (Moskva), 29, 420, 1993.

Neale, D. B., and Harry, D. E., Genetic mapping in forest trees: RFLPs, RAPDs and beyond, *AgBiotech News and Information*, 6, 107N, 1994.

Neil, S. W., and Kurtz, M. E., *Agronomic research for kenaf crop production in Mississippi,* Office of Agricultural Communications, Division of Agriculture, Forestry, and Veterinary Medicine, Mississippi State University, 1, 1994.

Sadaqat, H. A., Khan, M. A., Bhatti, M. S., and Khan, A. A., Assessment of fibre yield components in jute (C. olitorius), *Journal of Agricultural Research*, 27, 281, 1989.

Saha, T., and Sen, S. K., Somatic embryogenesis in protoplast derived calli of cultivated jute, *Corchorus capsularis* L., *Plant Cell Reports*, 10, 633, 1992.

Sarma, M. S., *Breeding procedures for Hibiscus*, I.C.A.R. Technical Bulletin No. 11, New Delhi, 1967.

Sinha, M. K., and Roy, M. K. G., Genetics of yield and its components in mesta, *Indian Journal of Agricultural Sciences*, 57, 788, 1987.

Sinhamahapatra, S. P., Induced changes in the background genotype in relation to the mutated locus in jute, *Indian Journal of Genetics and Plant Breeding*, 46, 496, 1986.

Sinhamahapatra, S. P., and Rakshit, S. C., Effect of selection for days to flower in X-ray treated population of jute (*Corchorus capsularis* L.), *Plant Breeding*, 97, 352, 1986.

Stonecypher, R. W., and Zobel, B. J., Inheritance of specific gravity in five-year-old seedlings of loblolly pine, *TAPPI*, 49, 303, 1966.

van Buijtenen, J. P., Manuscript review, January 5, 1995.

Wilcox, P. L., Amerson, H. V., O'Malley, D., Carson, S., Carson, M. J., Kuhlman, G., and Sederoff, R. R., Fusiform rust – a model for marker assisted selection in loblolly pine, *Proc. of the 22nd Southern Forest Tree Improvement Conference*, Atlanta, Georgia, 174, 1993.

Wilson, M. D., The wild relatives of kenaf and roselle – rich sources of germplasm (Abstract), *Proc. Fourth Annual International Kenaf Association Conference*, Biloxi, Mississippi, 24, 1992.

Zobel, B. J., and Talbert, J. T., *Applied forest tree improvement*, John Wiley & Sons, New York, 1984.

Zobel, B. J., Jett, J. B., and Hutto, R., Improving wood density of short rotation of southern pine, *TAPPI*, 61, 41, 1978.

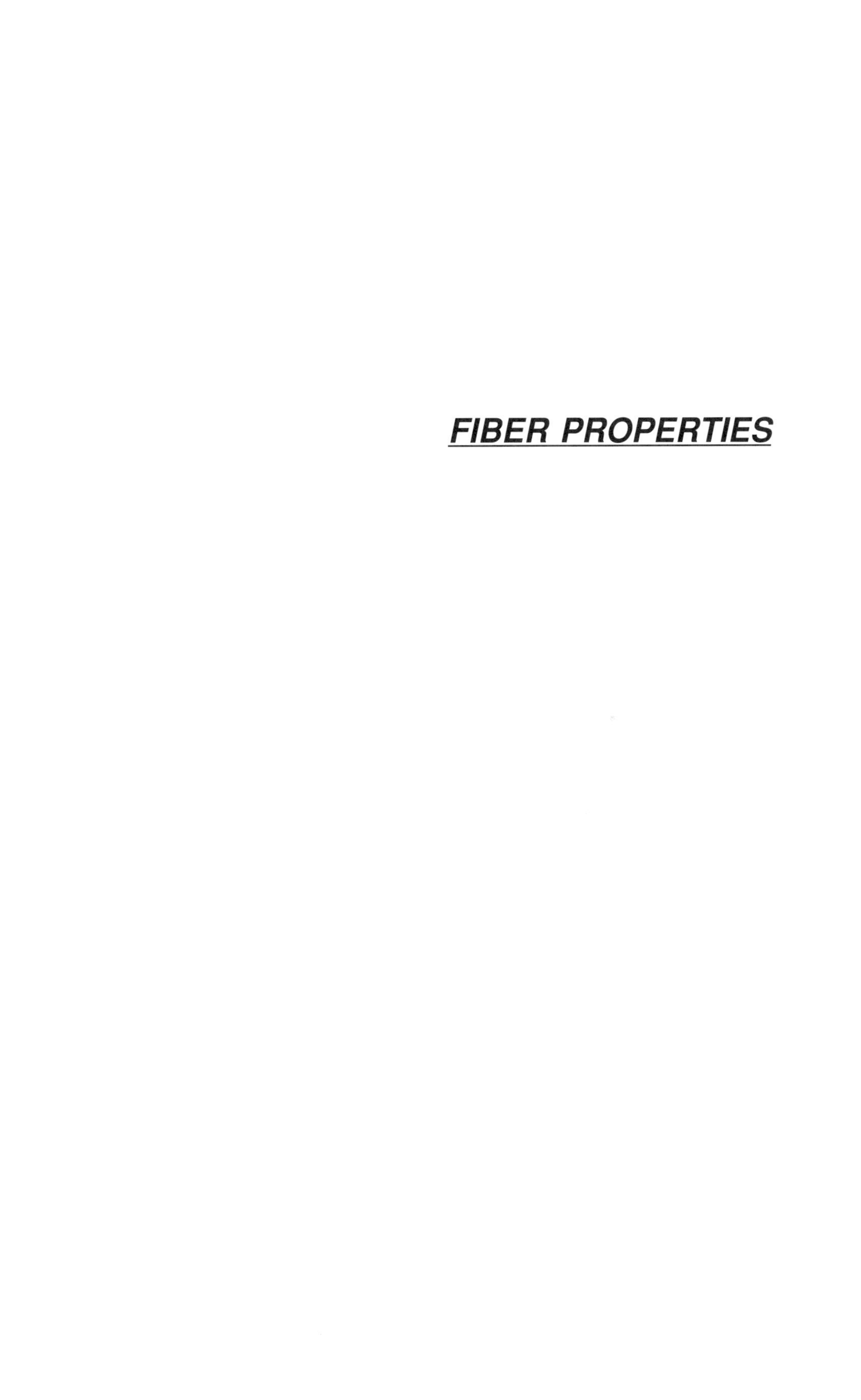

FIBER PROPERTIES

CHAPTER 4

Physical and Mechanical Properties of Agro-Based Fibers

Timothy G. Rials and Michael P. Wolcott

CONTENTS

1. INTRODUCTION

Vegetable fibers can be generally classified as bast, leaf, or seed-hair fibers, depending on their origin. In the plant, the bast and leaf fibers lend mechanical support to the plant's stem and leaf, respectively. In contrast, seed-hair fibers, such as cotton and milkweed, are attached to the plant's seeds and aid in wind dispersal. Historically, humans have used all these natural fibers for textiles because of their wide availability, long fiber geometries, chemical stability, and favorable mechanical properties. The coarse texture and high mechanical properties of many bast and leaf fibers have made them common cordage fibers for rope, twine, and string. Since the earliest written records in Egypt and China, vegetable fibers have been used to produce writing surfaces and paper.

With the increased development and availability of synthetic polymers, many of the natural fibers have been largely replaced. In addition, the large demand for printing surfaces spurred the change from parchment, to rag, to straw-paper, and finally to wood-fiber paper. These shifts from locally available vegetable fibers to world-wide synthetic and wood fiber markets have seriously impacted the rural farming communities of many countries. New restrictions on North American wood supply and the abundance of agricultural residues have spurred new interest in using agricultural fibers for both paper and composite materials.

In many cases, the similarities in the chemical composition (see Chapter 5) make agricultural fibers readily adaptable to traditional wood production processes. When exploring new product uses for agro-based fibers, the physical and mechanical properties often dictate the final use. In this case, it may be the long fibers and favorable mechanical properties that encourage new uses of these agro-based cellulosic fibers.

2. FIBER GEOMETRIES AND STRUCTURES

2.1 Bast and Core Fibers of Dicotyledons

Bast fibers exist in the inner bark or phloem of many dicotyledonous plants to provide structural rigidity to the stems. These fibers occur in bundles which run parallel to the stem between nodes. The fiber strands are composed of many smaller cells termed *ultimate fibers*. Just inside the phloem is a wood-like core material consisting of short, fine fibers. Bast fibers have been used for generations to produce textiles and are removed from the phloem by a controlled decay and separation process called retting. Several reviews of various retting processes exist (Young, 1993; McGovern, 1990; *American Fabrics Encyclopedia of Textiles*, 1960; Ramaswamy et al., 1994). In general, the core material is unused after the outer bast fibers are removed.

Characteristic geometries and dimensions for bast and core fibers are compiled in Table 4.1. In general, bast fiber strands are large with lengths in excess of 100 cm and widths less than 1 mm. The ultimate fibers comprising these strands range

Table 4.1 Dimensions of Bast and Core Fibers from Dicotyledons[1]

Type	Property	Flax	Hemp	Jute	Kenaf	Ramie	Cotton
Bast	Length (cm)	25–120	100–400	150–360	200–400	10–180	
Strd.	Width (mm)	0.04–0.6	0.5–5				
Bast	Length (mm)	4–69	5–55	0.7–6	2–11	60–250	
Ult.	Diameter (mm)	8–31	16	15–25	13–33	16–120	
Core	Length (mm)	0.2[2]	0.7[2]	1.06[3]	0.6[4]		0.6–0.8[4]
Ult.	Diameter (mm)			26[3]	30[4]		20–30[4]

[1] Young, 1993.
[2] Schafer and Simmonds, 1929.
[3] McGovern et al., 1983.
[4] Atchison, 1983b.

in size. For jute and kenaf, the ultimate fibers have a similar size to many coniferous wood species. However, flax, hemp, and ramie all have extremely long ultimate fibers that are only minimally larger in diameter to wood. These fibers typically have aspect ratios (length/diameter) on the order of 1,000.

The core material of kenaf, jute, hemp, flax, and cotton stalks have all been studied for an alternative paper fiber (Atchison and McGovern, 1983; McGovern et al., 1983; Schafer and Simmonds, 1929). Unlike the bast fibers, the woody core material of these dicotyledons is composed of relatively short fibers. In general, these core fibers are less than 1 mm long with aspect ratios of approximately 20.

2.2 Leaf Fibers from Monocotyledons

In many monocotyledons, long strand leaf fibers are embedded in parenchymatous tissue to provide mechanical support for the long, broad leaves. As in the bast, these composite fiber strands are composed of many smaller ultimate fibers. In general, leaf fibers are more coarse than bast fibers and are commonly used as cordage, mats, rugs, and carpet backings rather than clothing fabrics. Common examples of these materials are manila hemp (abaca) and sisal, both of which are used for twines and ropes worldwide. In addition, abaca and sisal have been both used historically for paper fibers.

Whereas the leaf fiber strands display the characteristically long length of bast fibers, the ultimate fibers comprising these strands are typically less than 12 mm long (Table 4.2). Like most of the plant fibers, the leaf ultimate fibers have typical diameters of approximately 30 mm.

2.3 Seed-Hair Fibers

Unlike the bast and leaf fibers, seed-hair fibers are single-celled. These fibers are attached to the seeds of certain plants for aid in wind-dispersal. One exception to this is coir, a fiber produced from the husks of coconuts. Like bast fibers, coir is produced by separation techniques after the husks undergo retting. Coir can produce

Table 4.2 Dimensions of Leaf Fibers from Monocotyledons[1]

Type	Property	Abaca	Henequin	Mauritius	Palm	Phormium	Sisal
Strd.	Length (cm)	365[2]	60–150[3,4]	124–210	30–60[3]	150–240	60–12
	Width (mm)	0.2–1[2]	0.1–0.5[3,4]		0.15–0.25[3]	0.1–0.5	0.1–0.5[2,5]
Ult.	Length (mm)	2–12	1.5–4	1.3–6		2–11	0.8–7
	Diameter (mm)	6–40	8.3–33	15–32	18[3]	5.25	8–48

[1] Cazaurang-Martinez et al., 1991.
[2] Young, 1993.
[3] Amin et al., 1986.
[4] Belmares et al., 1981.
[5] Cazaurang-Martinez et al., 1991.

long fiber strands, depending on the production process. The ultimate fibers are quite small: less than 1 mm long and 6 mm in diameter (Table 4.3).

Table 4.3 Seed-hair Fiber Dimensions[1]

Property	Coir	Cotton	Cotton Linters	Kapok	Milkweed
Length (mm)	0.2–1	10–50	3–7[2]	15–30	5–40[3]
Diameter (mm)	6.24	12–25	30[2]	10–30	6–38[3]

[1] Young, 1993.
[2] McGovern et al., 1983.
[3] Patry et al., 1993.

The more typical seed-hair fibers are all similar in morphology to cotton, with long lengths (ca. 20 mm) and small diameters (ca. 20 mm). Kapok and milkweed both have thin cell walls and large diameter lumens. Because kapok is difficult to spin, it is primarily used as loose fill for buoyancy or insulation purposes. Milkweed has been used only on a limited basis for fill; however, it has been studied for its papermaking and textile spinning qualities (Patry et al., 1993).

Short fibers that are left on the seed after processing are termed linters. Cotton linters are commercially available and are currently used as rag content in fine papers (McGovern et al., 1983). These waste fibers have similar diameters to the textile cotton, with much shorter lengths (less than 7 mm).

2.4 Cereal Straws

Historically, straws from wheat, rye, and rice were widely used as a pulp source for paper-making. Whereas this practice became extinct in North America and much of Europe by 1960, it is still practiced in southern and eastern Europe as well as many Asian, Mid-Eastern, and South American countries (Misra, 1983). Although corn stalks are widely available and have been studied as a pulp source, their commercial use has been limited (McGovern et al., 1983).

In general, cereal straw and corn stalk fibers have small diameters (ca. 13 mm) and average lengths (ca. 1.5 mm) producing a favorable aspect ratio of 110 (Table 4.4). Fibers from rice straws typically have smaller diameters (ca. 8 mm) with a long aspect ratio of 170.

Table 4.4 Cereal Straw Fiber Dimensions

Property	General[1]	Corn[2]	Rice[1]	Wheat[2]
Length (mm)	0.68–3.12	1.0–1.5	.65–3.48	1.5
Diameter (mm)	7–24	20	5–14	15

[1] Misra, 1983.
[2] Atchison, 1983b.

2.5 Other Grass Fibers

Many other grasses have been and are considered as a fiber source. Because of limited availability or processing difficulties, most of these fibers (except sugarcane bagasse) have never become widely used, however, they are often common in certain localities. These fibers (Table 4.5) include sugarcane bagasse, bamboo, esparto, and sabai grass.

Table 4.5 Other Grass Fiber Dimensions

Property	Sugarcane Bagasse[1]	Bamboo[2]	Esparto[3]	Sabai[3]
Length (mm)	2.8	2.7–4	1.1	0.5–4.9
Diameter (mm)	34.1	15	9	9–16

[1] Atchison, 1983a.
[2] Atchison, 1983b.
[3] McGovern et al., 1983.

Sugarcane bagasse is the most common and widely studied of these grass fibers. Sugarcane is grown as a source of sucrose in many tropical and sub-tropical countries on many continents. Bagasse is the residue remaining after the sugar has been extracted. Currently, this material is used for paper in India and Mexico and in boilers to generate process steam for the sugar production. Paper and material fiber provide the more economical use for this material (Atchison, 1992). The bagasse consists of pith, rind, and epidermis material. The rind consists of 50% of the dry bagasse weight and contains most of the usable fiber. In general, these fibers are similar to hardwood fibers averaging 1.7 mm in length and 20 mm in diameter, producing an aspect ratio of ca. 85.

Two grasses, esparto and sabai, are widely available in North Africa and India, respectively. Both grasses are locally common as a paper fiber, with sabai at one time comprising 20% of the pulped material in India. Compared to softwood fibers, these grass fibers have small diameters (ca. 9 mm) and average lengths (1–2 mm), producing fibers with large aspect ratios between 100 and 200.

3. MECHANICAL PROPERTIES

The long historical use of many agro-based natural fibers has afforded a wealth of literature regarding comparative mechanical and physical properties (Young, 1993; McGovern, 1983). The present literature primarily reflects the traditional textile uses. For textiles, strength and stiffness properties are evaluated in tension along the length of the fiber. As in wood, the orientation of the cellulose crystals along the fiber length likely results in anisotropic mechanical properties. Although no report of anisotropy is found in the literature, comparisons to wood would indicate that properties parallel to the fiber direction would be in excess of 100 times greater than those perpendicular (Mark, 1967).

3.1 Engineering Units

For textiles, strength and stiffness properties are reported using *tenacity* for a unit of measure. Tenacity can be expressed as a force per unit size area; Newtons/tex or grams/denier (Rebenfeld, 1990). By definition, a denier is the weight in grams of 9,000 m of fiber, whereas a tex is the weight in grams of 1,000 m of fiber. The use of tex or denier rather than units of area stems from the difficulty in measuring the cross-sectional area for small, natural fibers. Because natural plant fibers are cellular, the cross-sectional area for the cell wall must be determined from both the outside and inside fiber diameters. The fiber bundles of bast and leaf fibers are even more difficult to evaluate because the fiber is composed of many hollow, discontinuous ultimate fibers. The ratio of the fiber density to fiber denier provides an approximate measure of cross-sectional area. Because the density of natural fibers remains relatively constant at ca. 1.5 g/cm^3, the units of denier and tex render a relative measure of cross-sectional area with substantially reduced experimental difficulties. Tenacity can be converted to traditional engineering units of stress, by multiplying by the fiber density and a conversion constant (Billmeyer, 1984). This conversion has been used to develop much of the property data compiled for discussion in the remainder of Section 3.

3.2 Comparison of Fibers

Mechanical properties of both natural and synthetic fibers are presented for comparison in Tables 4.6 through 4.9. The synthetic fibers presented in Table 4.6 can be separated into the engineering fibers (glass, carbon, and aramid) and the non-engineering polymer fibers (polyester, rayon). In general, all of the natural fibers fall between the stiffness values for the engineering and non-engineering synthetic fibers. Across the range, the strength of natural fibers falls near those of the non-engineering synthetic fibers with elongations at failure similar to the engineering fibers.

As a group, the bast fibers (Table 4.7) are the stiffest of the natural fibers with a modulus in the range of 4 million psi, or nearly half that of glass fibers. Hemp is the stiffest natural fiber (ca. 4.3 million psi), although ramie has comparable strengths of nearly 120,000 psi. Although flax is relatively stiff (ca. 4.1 million psi), it is the weakest of the bast fibers with a strength of only 51,000 psi.

Leaf fibers (Table 4.8) average approximately half the stiffness of bast fibers with comparable and sometimes greater strength values. Abaca fibers are both the

Table 4.6 Mechanical and Physical Properties of Selected Synthetic Fibers[1]

Property	Glass	Carbon	Aramid	P'ester	V. Rayon	A. Rayon
Density (g/cm^3)	1.35	1.77	1.44	1.39	1.5	1.32
Moisture Regain (%)	—	—	4.5	0.4	12	6.4
Stiffness (10^6 psi)	10.52	33.96	17.96	0.53	0.33	0.09
Strength (10^3 psi)	400	520	405	142	138	24
Elongation (%)	3	1.5	2.5	20	25	35

[1] Billmeyer, 1984.

Table 4.7 Mechanical and Physical Properties of Bast and Core Fibers from Dicotyledons[1]

Property	Flax	Hemp	Jute	Kenaf	Ramie
Density (g/cm^3)	1.5	1.48	1.5	1.47	1.51
Moisture Regain (%)	12	12	13.7		6
Stiffness (10^6 psi)	4.13	4.29	3.8		3.28
Strength (10^3 psi)	51	119	84	232[2]	121
Elongation (%)	2.5	3.5	1.5	2.7	4

[1] Joseph, 1986.
[2] Mukherjee et al., 1992.

Table 4.8 Mechanical and Physical Properties of Leaf Fibers from Monocotyledons[1,2]

Property	Abaca	Henequin	Pineapple	Palm	Sisal
Density (g/cm^3)	1.35	1.5	1.44	1.5	1.45
Stiffness (10^6 psi)	2.86	1.98	2.23[3]	2.17	2.50
Strength (10^3 psi)	132	78	20[3]	84	76
Elongation (%)	3.5	4.7	0.88[3]	3.3	2.8

[1] Belmares et al., 1979.
[2] Satyanarayana et al., 1982.
[3] Samal and Bhuyan, 1994.

Table 4.9 Mechanical Properties of Seed-hair Fibers

Property	Coir[1]	Cotton[2]	Kapok[3]	Milkweed[4]
Stiffness (million psi)	1.03	1.19	2.83	
Strength (thousand psi)	31	80		34
Elongation (%)	16	5	1.2	

[1] Satyanarayana et al., 1982.
[2] Billmeyer, 1984.
[3] Young, 1993.
[4] Drean et al., 1993.

stiffest and strongest in this fiber grouping with stiffness and strength properties of 2.86 million and 132,000 psi, respectively. Sisal fibers are also relatively stiff (ca. 2.5 million psi) with fairly poor strength properties. Both palm and henequin have similar balanced strength and stiffness properties.Mechanical properties of seed-hair fibers (Table 4.9) are incomplete in the literature; however, what is present suggests that their properties are the poorest of the natural fibers. The reported stiffness values for coir and kapok vary greatly; both fibers have poor strength properties (ca. 30,000 psi). These values appear to be indicative of the remaining seed-hair fibers.

4. FACTORS INFLUENCING MECHANICAL PROPERTIES

Given the polymeric character of cellulosic fibers, their mechanical properties should be determined by the strength of their inter- and intra-molecular bonds, and the conformation of the polymer chains. From this ideal viewpoint, the theoretical modulus of cellulose fibers has been calculated to be 19.7 million psi (Zeronian, 1991), which leads to an estimated tensile strength of about 2 million psi, assuming the theoretical strength is about 10% of the fiber's elastic modulus. This is in excellent agreement with the experimental modulus of bleached ramie that has been determined from lattice extension studies to be 18.5 million psi. However, it is apparent from the information presented in the previous section that the experimental observations of lignocellulosic fiber strength and stiffness do not begin to approach this ideal. In reality, the mechanical property performance of these natural, lignocellulosic fibers are the result of complex interactions between a number of external variables and inherent structural parameters that originate at the molecular, macromolecular, and microscopic levels. Among the most important of these are degree of polymerization, crystallinity, and orientation of the cellulose; lignin and hemicellulose content; microscopic and molecular defects in the fiber; and, the presence of moisture or other introduced chemicals. While the following discussion cannot be considered comprehensive in its scope, it is intended to highlight the significance of these factors on the mechanical properties of agro-based fibers.

4.1 Fiber Structure Variables

4.1.1 Fiber Morphology and Structure

One of the more readily apparent factors contributing to diminished fiber strength is the morphology of the fiber since the presence of natural features (e.g., pits, nodes, etc.) may act as stress concentration points along the fiber axis (Figure 4.1). Additionally, defects such as reversals and convolutions that are common in cotton can further weaken fiber strength. Consequently, the experimentally determined strength of natural fibers tends to increase as the test length is decreased and the number of imperfections in the test area is reduced (Roy and Mukherjee, 1953; Meredith, 1956). Lahiri (1987), working with white and tossa jute fibers, found that while this general phenomenon held, no clear relationship between intrinsic strength and fiber imperfection could be established. It was suggested that in multi-cellular fiber strands like jute, sugar cane, or sisal, the "cementing" polymers of the middle lamella (i.e., lignin and hemicelluloses) introduce an additional site for defects and, therefore, an additional influence on intrinsic fiber strength.

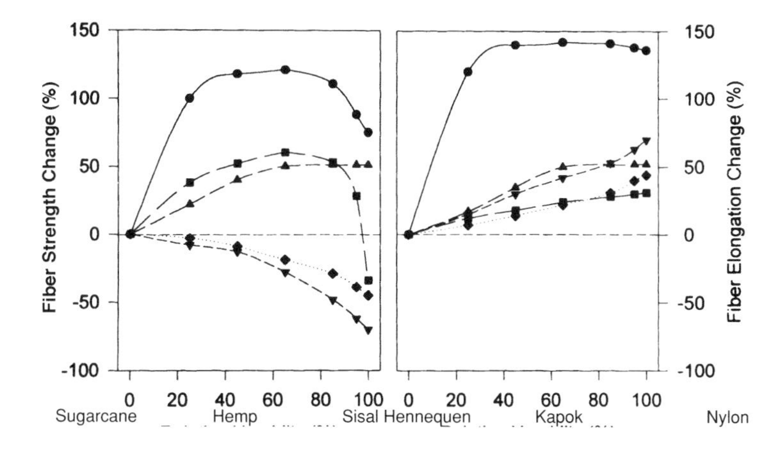

Figure 4.1 Photomicrographs of representative fibers from the different classifications. Magnifications are: Sugar cane (× 180), Hemp (× 270), Sisal (× 270), Kapok (× 180), and Nylon (× 360). (From Carpenter, C. H., Leney, L., Core, H. A., Cote, W. A. Jr., and Day, A. C., *Papermaking Fibers: A Photomicrographic Atlas of Woody, Non-woody, and Man-made Fibers Used in Paper-Making,* Technical Publ. No. 74, State University College of Forestry, Syracuse University, Syracuse, NY, 1963. With permission.)

4.1.2 Macromolecular Composition and Organization

It is critical to recognize that these agro-based fibers are macromolecular composite systems comprised of several polymeric constituents — most notably cellulose, hemicelluloses, and lignin. The polymer composition of the fibers varies widely in nature and is unquestionably a primary determinant of mechanical property performance. Unfortunately, these relationships remain ambiguous at best since no concerted research effort has addressed this daunting issue. Because the brittleness of many agro-based fibers presents processing problems, the influence of lignin content on the mechanical behavior of jute has been studied by Roy (1953). Figure 4.2 is adapted from this data and shows a gradual decrease in both the strength and stiffness of the fiber with lignin removal. The extensibility of the fiber was also found to decrease with lignin removal. The more substantial drop in strength at higher lignin removal was interpreted as being indicative of a portion that is intimately associated with the cellulose structure. This premise was further supported by observations of the fiber's wet strength in that a relatively constant 12% differential between the wet and dry state was found for lignin content's greater than 4%; however, with further lignin removal, the wet strength of the fiber was negligible. Similar experiments with sugarcane fiber (Collier et al., 1992) provide additional evidence of the significant contribution of lignin to fiber strength. Generally, as the lignin content decreased, the tensile and bending properties of the sugarcane fiber declined as might be anticipated. Observation of the failure surface by scanning

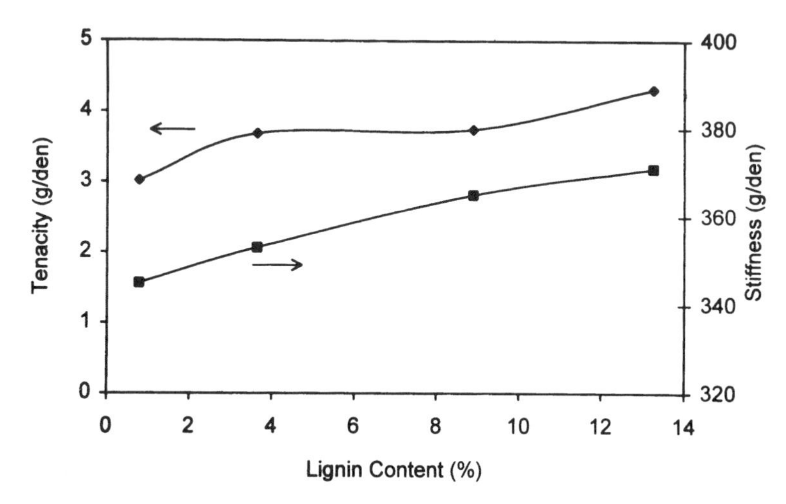

Figure 4.2 Variation in the strength and stiffness of jute fibers with lignin content. (From Roy, M. M., *Textile Institute Journal*, 44, T44, 1953. With permission.)

electron microscopy revealed that with decreasing lignin content, the ultimate fibers debonded from the matrix and then failed as they were required to carry the load. The relatively mild treatments used did not allow complete removal of lignin. No insight could be gained regarding fiber strength at very low lignin compositions.

It is worth noting that very little research has addressed the actual macromolecular organization of the cell wall and the fiber, as a whole. The widely held view is similar to that of wood fiber morphology (Fengel and Wegener, 1984) where the amorphous hemicelluloses and lignin polymers are concentrated in the middle lamella where they act as a matrix for the cellulosic fiber. It is pointed out by Roy (1953), however, that it is difficult to rationalize the relatively high strength and stiffness of jute based solely on its cellulose crystallinity, which is very low (relative to the other vegetable fibers). He suggests that a higher level of organization may be present in this "amorphous" matrix and cites several reports that suggest the occurrence of lignin-cellulose bonds, and cocrystallization between cellulose and xylan. Although cocrystallization of two polymers is extremely rare, experimental evidence of a lignin-carbohydrate-complex in wood (Sjöström, 1981) is widely accepted. Surprisingly, there were no recent reports found on the phase morphology of agricultural fibers and no consideration of the potential impact morphology might have on the fibers' mechanical properties.

4.1.3 Cellulose Crystallinity and Fine Structure

It is extremely difficult, if not risky, to directly compare the strength and stiffness characteristics of agro-based fibers because isolation and purification processes can significantly alter the crystalline structure of the cellulose polymer. These preparation methods routinely involve the use of water or mild alkaline solutions that strongly interact with the carbohydrate components, swelling the fiber and increasing molecular mobility. As a consequence, the relative ratio of the crystalline to amorphous phase can be altered as well as the orientation of the crystallites, thereby modifying the mechanical properties of the fiber. A detailed discussion of the macromolecular and supramolecular structure of cellulose is beyond the scope of this chapter, and the reader is referred to any of several excellent reviews on this subject (Zeronian, 1991; Nevell and Zeronian, 1985).

In comparing the properties of different textile fibers, several general guidelines have been established (Meredith, 1956; Zeronian, 1991). The "critical" degree of polymerization (DP) of cellulose has been determined to be about 700 repeat units, and very little change in mechanical properties occurs above this point. Since the DP of native cellulose is much higher than 700, this factor is relatively insignificant except under extreme fiber preparation conditions. Also, there has been very little success in correlating the degree of crystallinity found in different types of raw fiber with mechanical properties. This may be due in part to the experimental challenge presented in determining absolute crystallinity of cellulose, since it is reasonable to assume that crystallinity plays a significant role. The orientation of the crystallites clearly affects the mechanical properties of the agro-based fibers. Fibers like hemp, jute, flax, and ramie have their cellulosic chains oriented nearly parallel to the fiber axis (in the range 7–12°), and they exhibit high strength. Cotton fibers, with appreciably lower tensile strengths, are characterized by spiraled cellulose chains that are oriented approximately 30° to the fiber axis.

Ray (1969) has shown that the degree of crystallinity (DC) in jute and mesta fibers varies consistently with the moisture content and fiber purity. In all of the fiber systems studied, the crystalline component decreased as the moisture sorption increased, although the extent of the decrease became smaller with fiber purification. Furthermore, the degree of crystallinity increased as the lignin and hemicellulose polymers were removed. For the tossa jute, the DC increased from a low of 42% for the wet, delignified fiber to 66% for the dry α-cellulose. It is interesting to note, too, that DC values of 66, 67, and 69% were found for the tossa jute, white jute, and mesta fibers, respectively. These results indicate that the non-cellulosic materials may interfere with crystallization; however, upon the removal of the amorphous materials, the crystalline structure of these fibers is comparable to ramie and cotton.

The relationships between fine structure and mechanical properties have also been investigated for various agro-based fibers (Ray, 1973; Ray et al., 1976). In this work, the moisture-swollen fibers were dried using either direct heating or a solvent exchange technique. As seen in Table 4.10, the degree of crystallinity of the fibers was consistently lower when moisture was removed by the indirect, solvent exchange method. At the same time, the crystallite orientation also increased. Furthermore, the difference in direct and indirect moisture removal methods was greatest for mesta when compared to the jute and roselle fibers. These observations were interpreted to indicate that moisture tends to disrupt the crystalline structure of the cellulose, since the benzene displaces the moisture without strongly interacting with polymer, thereby retaining the reduced order of the swollen fiber. Interestingly, the strength of all the indirectly dried fibers is lower than that in the heat dried fibers. This leads to the general conclusion that a drop in crystallinity more strongly influences the fiber's mechanical properties than does an improvement in crystallite orientation.

Table 4.10 Effect of Drying Method on the Crystal Structure and Mechanical Properties of Various Agricultural Fibers

Fiber/ Drying Method	Crystal Structure			Mechanical Properties	
	Crystallinity (%)	Crystallite Orientation	Density (g/cm³)	Tenacity (g/tex)	Torsional Rigidity (dynes/cm², × 10¹⁰)
Direct Heat					
Ramie	74	7° 15'			
Mesta	64.8	11° 44'	1.48	19.9	0.30
Jute (Tossa)	59.3	10° 16'	1.478	18.7	0.38
Jute (White)	57.5	10° 08'	1.475	15.7	0.36
Roselle	34.9	9° 53'	1.455	9.3	0.29
Solvent Exchange					
Ramie	61.6	7° 08'			
Mesta	52.4	9° 24'	1.477	18.6	0.21
Jute (Tossa)	52.4	9° 38'	1.471	16.8	0.27
Jute (White)	51.5	9° 28'	1.457	14.3	0.20
Roselle	29.4	9° 32'	1.445	8.0	0.18

Reference: Ray, 1969.

This is somewhat contradictory to the observations on raw fibers, although the increase in orientation of these fibers is relatively small.

Stronger swelling agents, like alkaline solutions, are commonly employed in the preparation of textile fibers. Mild treatments are similar to water in that they swell the fiber, allowing the release of internal strains and improved orientation of the crystallites. As such, small improvements in strength and elongation have been demonstrated for jute fibers (Sikdar et al., 1993). The process of mercerization involves treating textile fibers with more concentrated alkaline solutions (greater than 18 wt.%) that not only swell the fiber, but also result in a conversion of the crystal structure from cellulose I to cellulose II. The extent of conversion to cellulose II can vary considerably for different vegetable fibers as reported by Barkakaty (1976) resulting in some contradictory literature reports. Cheek and Russel (1989) studied the effect of mercerization on the mechanical properties of ramie, flax, and cotton yarns with and without tensioning the fibers. A significant drop in strength was observed for both flax and ramie in the absence of tension, while the ultimate elongation increased 4–5 times the control. In contrast, the tension-mercerized fibers increased significantly in their strength. The authors suggest that this behavior is the result of decreased crystalline orientation that arises as a consequence of unrestrained lateral swelling of the fibrils. This behavior is contrasted by cotton fibers that exhibit higher strength after slack mercerization, and even greater improvements when tension is applied (Warwicker, 1966; Rajagopalan et al., 1975). This further evidences the significant role of molecular order on mechanical properties of agro-based fibers.

4.2 External Variables

4.2.1 Moisture

Lignocellulosic fibers are highly hygroscopic materials. At a relative humidity of 65%, the equilibrium moisture content typically ranges from 8% to 12%, although it is dependent on a number of morphological and structural parameters. As such, water can and must be considered as an integral, and significant, part of the molecular structure. The action of water on fiber properties is generally characterized as that of an external plasticizer that lowers the glass transition temperature (T_g) of the fiber's polymeric constituents. This would normally be expected to reduce the strength and stiffness of the fiber; however, the situation is complicated somewhat since the increased chain mobility can alter the crystalline structure and morphology of the cellulosic component as discussed above. Consequently, literature reports on the variation of mechanical properties with moisture are somewhat contradictory.

As seen in Figure 4.3, when the moisture content of regenerated cellulose fibers increases, the breaking strength decreases and the elongation increases. This is consistent with the premise that water serves to disrupt the bonding between polymer chains, increasing their mobility and facilitating chain slippage. The natural vegetable fibers — cotton, jute, and flax — exhibit very different behavior. Although the breaking extension increases with moisture content of the fiber, the breaking strength also increases from the dry condition to a relative humidity of 65% (corresponding to a moisture content of about 10%). Fiber stiffness also increased at low moisture contents. This anti-plasticizing effect is explained by the authors as a result of the release of internal strains within the fiber micro-structure that is afforded by the increased chain flexibility (Meredith, 1956; Roy, 1953; Zeronian, 1991). The result is a more uniform stress distribution throughout the fiber and a corresponding increase in its load bearing capacity. Furthermore, this behavior may be reflective of a disruption in crystalline suprastructure that has been demonstrated for jute fibers (Ray, 1969). It is worth noting that this unusual increase in strength at low moisture contents has also been reported for hemp (Meredith, 1956; Mark, 1967). The strength of other textile fibers including rayon, protein and synthetic fibers declines as they absorb water.

Interestingly, as the relative humidity is raised above 65% the behavior among the natural fibers is quite different. The breaking strength of cotton reaches a plateau and changes very little over the relative humidity range 65%–100%; however, a slight drop in strength is observed in jute, and the flax fiber exhibits a dramatic decline in its load bearing capacity. The ultimate elongation increased slightly for all three fiber types. This is again consistent with the concept that water serves as a plasticizer for the polymer constituents of the fiber. The difference presumably lies in the fact that both the jute and flax are lignocellulosic fibers containing both lignin and hemicellulose, and they are further plasticized by the increased moisture sorption. Experiments with wood fiber and isolated wood polymers have indicated that the glass transition temperature of lignin reaches a plateau of about 50°–90° C at moisture contents of 10% and above (Kelley et al., 1987). Hemicelluloses, as noted by numerous researchers (Roy, 1953; Zeronian and Menefee, 1976), are more

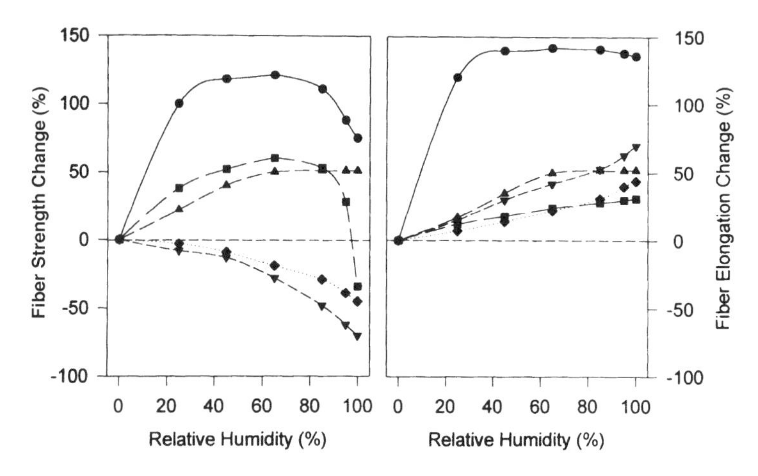

Figure 4.3 Percent change in strength (left) and elongation (right) with relative humidity for rayon[a] (▼), acetate[a] (◆), cotton[a] (▲), jute[b] (●), and flax[b] (■). (References: [a] Meredith, 1956; [b] Roy and Mukherjee, 1953). (From [a]Meredith, R., *Mechanical Properties of Textile Fibers*, Interscience Publishers, Inc., New York, 1956, chaps. VI and VII. With permission. [b]Roy, M. M., and Mukherjee, R. R., *Textile Institute Journal*, 44, T36, 1953. With permission.)

strongly affected by moisture, and their T_g is depressed below ambient temperatures at moisture contents less than 20%. Consequently, the significant drop in the breaking strength of jute and flax probably reflects the continued plasticization and softening of the hemicellulose component, and to a lesser extent, lignin.

There are inconsistencies regarding the variation of fiber mechanical properties with moisture in the literature. While Mark (1967) substantiates the increase in tensile strength at low moisture contents for several natural textile fibers, he makes note that this trend does not hold for jute. Recent work with coir fibers (Kulkarni et al., 1983) found that, although elongation increased with moisture content, fiber tenacity decreased from 1.99 g/denier at a moisture content of 10.5% to 1.64 g/denier for the dry fiber. It is difficult to rationalize these contrasting observations; however, differences could result if mechanical properties of the fiber were determined during the desorption, rather than the adsorption, cycle. Additionally, it is not clear if the experiments were run on bundles or ultimates, which could also influence the observations.

4.2.2 Chemical Modification

As discussed above, water serves as an external plasticizer for vegetable fibers, increasing the mobility and plasticity of the fiber's individual polymeric constituents. A similar, more permanent, effect can be achieved by covalently bonding molecules onto the fiber (see Chapter 11). The hydroxyl-rich character of agro-based fibers provides a convenient mechanism for property modification through a variety of chemical reactions, as has been demonstrated with wood (Rowell, 1991) and cotton (Ishizu, 1991). This approach has not been extensively investigated for these alternative vegetable fibers, although some reports on the effect of chemical modification on thermal properties (Varma et al., 1988a; Pandey et al., 1993; Basak et al., 1993) and moisture absorption (Varma et al., 1988b; Andersson and Tillman, 1989) are available. It is likely that research on chemical modification of agro-based fibers will intensify in the near future as interest in alternative applications like polymer reinforcement expands.

Preliminary investigations have appeared in the literature on the effect of chemical modification on the mechanical properties of various fiber types. The limited modification of jute (Andersson and Tillman, 1989) and sunn hemp (Chand et al., 1987) with acetic anhydride was shown to be an effective means of improving moisture resistance and dimensional stability for these fibers. The effect of acetylation on mechanical properties was negligible under appropriate conditions. A substantial decline in strength was reported for jute at higher acetic anhydride concentration, presumably as a consequence of cellulose degradation by acetic acid. The modification of kenaf with succinic anhydride (Rowell et al., 1994) was shown to impart some flow to the fiber, suggesting increased plasticization. It was apparent, however, that the modification is extremely sensitive to reaction conditions. Varma et al. (1988b) investigated the influence of several chemical treatments including tolylene diisocyanate, isopropyl triisostearoyl titanate, and γ-amino-propyl trimethoxysilane on the behavior of jute fibers. At relatively low levels of modification, these treatments were shown to be highly effective in reducing hyroscopicity without degrading the mechanical performance of the fiber. Interestingly, a decline in both the initial modulus and tenacity of almost 50% was observed when the fibers were coated with a small amount (16%) of an unsaturated polyester or vinyl ester resin. The reason for this different behavior is not clear.

Graft polymerization holds considerable promise as a simple and versatile approach to property enhancement of vegetable fibers. Several reports have appeared in the recent literature investigating the graft polymerization of acrylonitrile on jute (Ghosh and Ganguly, 1994) and pineapple leaf (Samal and Bhuyan, 1994). The moisture regain was dramatically reduced as the degree of grafting increased for all of the material systems. In the case of the acrylonitrile-grafted pineapple leaf fibers, strength and elongation increased only slightly over the percent grafting range investigated (0–90%). Similarly, mechanical properties varied only slightly with degree of grafting of the jute fibers, although some dependence on the fiber source was detected. Generally, defatted jute fiber was much more sensitive to the degree of grafting than bleached jute.

Sisal fibers were modified using a constant 10:90 comonomer mixture of styrene and ethyl acrylate (El-Naggar et al., 1992) initiated with gamma radiation. The tenacity of the grafted fibers decreased from approximately 22–2 g/tex as the degree of grafting reached 130%. This difference in behavior can be attributed in part to the use of a γ-radiation initiator system that deteriorated the sisal fiber, as evidenced by its decline in strength and elongation with increased radiation dosage. It is surprising that despite the decrease in elongation of the sisal fiber, the elongation of the highly grafted fiber (degree of grafting = 130%) was twice that of the ungrafted control. This may indicate the establishment of polymer bridges between microfibrils. Barkakaty and Robson (1979) and Habbibudowla (1981) also reported diminished strength for sisal fibers grafted with methyl methacrylate, methyl acrylate, or ethyl acrylate. Although the synthetic polymers vary widely and consistently in their properties, no correlation could be established between the type of polymer graft and either fiber strength or elongation. This is explained by the authors as a consequence of polymer distribution throughout the fiber structure. These limited results indicate the complexity of polymer grafting on vegetable fibers, as well as the versatility that it affords. Considerable research is needed in this area in order to develop a better understanding of the complex relationships between morphology and properties of polymer-grafted vegetable fibers.

5. CONCLUDING REMARKS

> In point of fact, no two of the many tests reported on fibers are comparable, for three reasons. The first is the variation in test conditions, the second is the fact that each investigator has been seeking to ascertain a different relationship, and the third is the individuality of molecular architecture. These three factors are not mutually exclusive; each testing program has to be designed to accommodate the other two factors.
> Of the three, the last factor is by far the most important as it is the least understood and appreciated. But upon the molecular conformation rests the pattern of stress distribution under load, and this is what ultimately determines every mechanical property of a wood fiber. (Mark, 1967)

Although these statements were made by Mark more than 25 years before this publication, they are equally appropriate to the current literature on mechanical properties of agro-based fibers. Historically, the evaluation of mechanical and physical properties for any material is determined with end uses in mind. It is no surprise, then, that during times of changing uses, much of this data seems inadequate and conflicting. This presentation of the data attempts to describe the trends in properties for comparative purposes. The mechanical properties have been converted to engineering values to reflect the current interest in using agro-based fibers for engineered composites. However, the work is by no means definitive.

A great need is present to evaluate a wide variety of these materials under standard conditions to produce statistical distributions for mechanical properties over representative growing ranges. Many of these fibers must be tested in both fiber-strand and ultimate fiber forms to contribute to different potential uses. In addition, several fundamental differences appear to exist between many of the fiber-types, especially when compared to wood. These differences are especially apparent when assessing influences of moisture on strength and stiffness. Lastly, any research efforts to establish such a comprehensive database should be conducted in concert with evaluations of chemical constituents and polymer morphology, since these will fundamentally govern the mechanical behavior.

REFERENCES

American Fabrics Encyclopedia of Textiles, Prentice Hall, Inc., Englewood Cliffs, 1960.

Amin, M. B., Maadhah, A. G., and Usmani, A. M., Natural vegetable fibers: a status report, in *Renewable-Resource Materials: New Polymer Sources*, Carraher, C. E., and Sperling, L. H., Eds., Plenum Press, New York, 1986, 29.

Andersson, M., and Tillman, A. -E., Acetylation of jute: Effects on strength, rot resistance, and hydrophobicity, *Journal of Applied Polymer Science*, 37, 3437, 1989.

Atchison, J. E., Making the right choices for successful bagasse newsprint production: Part I, *TAPPI*, 75(12), 63, 1992.

Atchison, J. E., Bagasse, in *Pulp and Paper Manufacture, Vol. 3, Secondary Fibers and Agro-based Pulping*, Hamilton, F., and Leopold, B., Eds., TAPPI Press, Atlanta, 1983a.

Atchison, J. E., Data on agro-based plant fibers, in *Pulp and Paper Manufacture, Vol. 3, Secondary Fibers and Agro-based Pulping*, Hamilton, F., and Leopold, B., Eds., TAPPI Press, Atlanta, 1983b.

Atchison, J. E., and McGovern, J. N., History of paper and the importance of agro-based plant fibers, in *Pulp and Paper Manufacture, Vol. 3, Secondary Fibers and Agro-based Pulping*, Hamilton, F., and Leopold, B., Eds., TAPPI Press, Atlanta, 1983.

Barkakaty, B. C., Some structural aspects of sisal fibers, *Journal of Applied Polymer Science*, 20, 2921, 1976.

Barkakaty, B. C., and Robson, A., Polymer deposition in sisal fibers: A structural investigation, *Journal of Applied Polymer Science*, 24, 269, 1979.

Basak, R. K., Saha, S. G., Sarkar, A. K., Saha, M., Das, N. N., and Mukherjee, A. K., Thermal properties of jute constituents and flame retardant jute fabrics, *Textile Research Journal*, 63, 658, 1993.

Belmares, H., Barrera, A., Castillo, E., Verheugen, E., Monjaras, M., Patfoort, G. A., and Bucquoye, M. E. N., New composite materials from natural hard fibers, *Industrial Engineering and Chemistry: Product Research and Development*, 20, 555, 1981.

Belmares, H., Castillo, J. E., and Barrea, A., Natural hard fibers of the North American continent. Statistical correlations of physical and mechanical properties of lechuguilla (*Agave lechuguilla*) fibers, *Textile Research Journal*, 49(11), 619, 1979.

Billmeyer, F. W., Jr., *Textbook of Polymer Science*, 3rd ed., Wiley-Interscience, New York, 1984.

Carpenter, C. H., Leney, L., Core, H. A., Cote, W. A. Jr., and Day, A. C., *Papermaking Fibers: A photomicrographic atlas of woody, non-woody, and man-made fibers used in paper-making*, Technical Publication No. 74, State University College of Forestry at Syracuse, Syracuse, New York, 1963.

Cazaurang-Martinez, M. N., Herrera-Franco, P. J., Gonzalez-Chi, P. I., and Aguilar-Vega, M., Physical and mechanical properties of Henequen fibers, *Journal of Applied Polymer Science*, 43, 749, 1991.

Chand, N., Verna, S., and Khazanchi, A. C., Acetylation of sunhemp fibre and its characteristics, *Indian Journal of Textile Research*, 12(3), 167, 1987.

Cheek, L., and Russel, L., Mercerization of ramie: comparison with flax and cotton, Part I: Effects on physical, mechanical, and accessibility characteristics, *Textile Research Institute*, 59, 478, 1989.

Collier, B. J., Collier, J. R., Agarwal, P., and Lo, Y.-W., Extraction and evaluation of fibers from sugarcane, *Textile Research Journal*, 62, 741, 1992.

Drean, J. Y., Patry, J. L., Lombard, G. F., and Weltrowski, M., Mechanical characterization and behavior in spinning processing of milkweed fibers, *Textile Research Journal*, 63, 443, 1993.

El-Naggar, A. M., El-Hosamy, M. B., Zahran, A. H., and Zohdy, M. H., Surface morphology/mechanical/dyeability properties of radiation-grafted sisal fibers, *American Dyestuff Reporter*, 81, 40, 1992.

Fengel, D., and Wegener, F., *Wood: Chemistry, Ultrastructure, Reactions*, Walter de Gruyter, New York, 1984, Chap. 2.

Ghosh, P., and Ganguly, P. K., Polyacrylonitrile (PAN)-grafted jute fibres: some physical and chemical properties and morphology, *Journal of Applied Polymer Science*, 52, 77, 1994.

Habbibuddowla, M., Grafting of methyl methacrylate onto KPM (Karnafuli Paper Mill) rayon and jute fiber: Effect on properties of grafted fibers, in *ACS Symposium Series No. 187*, Hon, D. N.-S., Ed., American Chemical Society, Washington, D.C., 1984.

Ishizu, A., Chemical modification of cellulose, in *Wood and Cellulosic Chemistry*, Hon, D. N.-S., and Shiraishi, N., Eds., Marcel Dekker, New York, 1991, Chap. 16.

Joseph, M. L., *Textile Science Introduction*, 5th ed. Holt, Reinhart and Winston, New York, 1986, 63-80.

Kelley, S. S., Rials, T. G., and Glasser, W. G., Relaxation behavior of the amorphous components of wood, *Journal of Materials Science*, 22, 617, 1987.

Kulkarni, A. G., Cheriyan, K. A., Satyanarayana, K. G., and Rohatgi, P. K., Studies on moisture sorption in coir fibers (*Cocos Nucifera* L.), *Journal of Applied Polymer Science*, 28, 625, 1983.

Lahiri, A., Bundle tenacity of jute fibres at different test lengths, *Indian Journal of Textile Research*, 12(3), 149, 1987.

Mark, R. E., *Cell Wall Mechanics of Tracheids*. Yale University Press, New Haven, 1967.

McGovern, J. N. Vegetable fibers, in *Polymers: Fibers and Textiles, A Compendium, Encyclopedia Reprint Series*, Kroschwitz, J. I., Ed., John Wiley & Sons, New York, 1990, Chap. 11.

McGovern, J. N., Coffelt, D. E., Hurter, A. M., Ahuja, N. K., and Weidermann, A., Other fibers, in *Pulp and Paper Manufacture, Vol. 3, Secondary Fibers and Agro-based Pulping*, Hamilton, F., and Leopold, B., Eds., TAPPI Press, Atlanta, 1983, Chap. IX.

Meredith, R., *Mechanical Properties of Textile Fibers*, Interscience Publishers, Inc., New York, 1956, Chaps. VI and VII.

Misra, D. K., Cereal straw, in *Pulp and Paper Manufacture, Vol. 3, Secondary Fibers and Agro-based Pulping*, Hamilton, F., and Leopold, B., Eds., TAPPI Press, Atlanta, 1983, Chap. VI.

Mukherjee, A. C., Bandyopadhyay, S. K., Mukhopadhyay, A. K., and Mukhopadhyay, U., Effect of ethylenediamine treatment on jute fibre, *Indian Journal of Fibre and Textile Research*, 17, 80, 1992.

Nevell, T. P., and Zeronian, S. H., *Cellulose Chemistry and Its Applications*, Ellis Horwood, Chichester/Halsted, New York, 1985.

Pandey, S. N., Day, A., and Mathew, M. D., Thermal analysis of chemically treated jute fibers, *Textile Research Journal*, 63(3), 143, 1993.

Patry, J. J., Gerard, L. F., and Weltrowski, M., Mechanical characterization and behavior in spinning processing of milkweed fibers, *Textile Research Journal*, 63(8), 443, 1993.

Rajagopalan, A., Venkatesh, G. M., and Dweltz, N. E., Differential response of cottons to slack mercerization, *Textile Research Journal*, 45, 409, 1975.

Ramaswamy, G. N., Ruff, C. G., and Boyd, C. R., Effect of bacterial and chemical retting on kenaf fiber quality, *Textile Research Journal*, 64, 305, 1994.

Ray, P. K., On the degree of crystallinity in jute and mesta fibers in different states of purifications and moisture conditions, *Journal of Applied Polymer Science*, 13, 2593, 1969.

Ray, P. K., The effect of methods of drying on the fine structure, density, and some mechanical properties of jute and allied fibers, *Journal of Applied Polymer Science*, 17, 951, 1973.

Ray, P. K., Chakravarty, A. C., and Bandyopadhyay, S. B., Fine structure and mechanical properties of jute differently dried after retting, *Journal of Applied Polymer Science*, 20, 1765, 1976.

Rebenfeld, L., Fibers, in *Polymers: Fibers and Textiles, A Compendium*, Kroschwitz, J. I., Ed., John Wiley & Sons, New York, 1990, 219.

Rowell, R. M., Chemical modification of wood, in *Wood and Cellulosic Chemistry*, Hon, D. N. -S., and Shiraishi, N., Eds., Marcel Dekker, New York, 1991, 15.

Rowell, R. M., Rials, T. G., and O'Dell, J., Chemical modification of agro-fiber for thermoplasticization, in *Proceedings of the Second Pacific Rim Bio-Based Composites Symposium*, Steiner, P. R., Ed., Wood Science Department, Faculty of Forestry, University of British Columbia, Vancouver, 144, 1994.

Roy, M. M., and Mukherjee, R. R., Mechanical properties of jute. I, *Textile Institute Journal*, 44, T36, 1953.

Roy, M. M., Mechanical properties of jute. II. The study of chemically treated fibers, *Textile Institute Journal*, 44, T44, 1953.

Samal, R. K., and Bhuyan, B. L., Chemical modification of lignocellulosic fibers 1. Functionality changes and graft copolymerization of acrylonitrile onto pineapple leaf fibers; their characterization and behavior, *Journal of Applied Polymer Science*, 52, 1675, 1994.

Satyanarayana, K. G., Pillai, C. K. S., Sukumarab, K., Pillai, S. G. K., Rohatgi, P. K., and Vijayan, K., Structure property studies of fibers from various parts of the coconut tree, *Journal of Materials Science*, 17(8), 2453, 1982.

Schafer, E. R., and Simmonds, F. A., Physical and chemical characteristics of hemp stalks and of seed flax straw, *Industrial and Engineering Chemistry*, 21(12), 1241, 1929.

Sikdar, B., Mukhopadhyay, A. K., and Mitra, B. C., Action of weak alkali on jute, *Indian Journal of Fibre and Textile Research*, 18, 139, 1993.

Sjöström, E., *Wood Chemistry, Fundamentals and Applications*, Academic Press, New York, 1981, 79.

Varma, I. K., Anantha Krishnan, S. R., and Krishnamoorthy, S., Effect of chemical treatment on mechanical properties and moisture regain of jute fibers, *Textile Research Journal*, 58(9), 537, 1988a.

Varma, I. K., Anantha Krishnan, S. R., and Krishnamoorthy, S., Effect of chemical treatment on thermal behavior of jute fibers, *Textile Research Journal*, 58(8), 486, 1988b.

Varma, D. S., Varma, M., and Varma, I. K., Coir fibers. 3. Effect of resin treatment on properties of fibers and composites, *Industrial Engineering and Chemistry: Product Research and Development*, 25, 282, 1986.

Warwicker, J. O., (1966). Effect of chemical reagents on the fine structure of cellulose, part III: Action of caustic soda on cotton and ramie, *Journal of Polymer Science*, Part A-2, 4, 571, 1966.

Young, R. A., *Kirk-Othmer Encyclopedia of Chemical Technology*, 4th edition, Volume 10, John Wiley and Sons, New York, 1993.

Zeronian, S. H., The mechanical properties of cotton fibers, *Journal of Applied Polymer Science: Applied Polymer Symposium*, 47, 445, 1991.

Zeronian, S. H., and Menefee, E., Thermally-induced changes in the mechanical properties of ramie and chemically-modified ramie, *Applied Polymer Symposium*, 28, 869, 1976.

CHAPTER 5

Chemical Composition of Fibers

James S. Han and Jeffrey S. Rowell

CONTENTS

1. CHEMISTRY

The major chemical component of a living tree is water, but on a dry weight basis, all plant cell walls consist mainly of sugar-based polymers (carbohydrates) that are combined with lignin with lesser amounts of extractives, protein, starch, and inorganics. The chemical components are distributed throughout the cell wall, which is composed of primary and secondary wall layers. Chemical composition varies from plant to plant, and within different parts of the same plant. Chemical composition also varies within plants from different geographic locations, ages, climate, and soil conditions.

There are hundreds of reports on the chemical composition of plant material. In reviewing this vast amount of data, it becomes apparent that the analytical procedures used, in many cases, are different from lab to lab, and a complete description of what procedure was used in the analysis is not clear. For example, many reports do not describe if the samples were pre-extracted with some solvents before analysis. Others do not follow a published procedure, so comparison of data is not possible.

This chapter will present a general description of the chemistry of plant components followed by suggested analytical procedures that could be used by all future laboratories so that consistent, comparable results may be obtained. The final section will be a listing of the chemical components of many different types of plants.

1.1 Carbohydrates

1.1.1 Holocellulose

The carbohydrate portion of the vast majority of plants is composed of cellulose and hemicellulose polymers with minor amounts of other sugar polymers such as starch and pectins. Table 5.1 shows the chemical analysis of the major components of plant fibers. The combination of cellulose and the hemicelluloses are called holocellulose and usually accounts for 65–70 percent of the plant dry weight. These polymers are made up of simple sugars, mainly, D-glucose, D-mannose, D-galactose, D-xylose, L-arabinose, D-glucuronic acid, and lesser amounts of other sugars such as L-rhamnose and D-fucose. Table 5.2 shows the sugar content of different plant holocelluloses. These polymers are rich in hydroxyl groups, which are responsible for moisture sorption through hydrogen bonding.

1.1.2 Cellulose

Cellulose is the most abundant organic chemical on the face of the earth. It is a glucan polymer of D-glucopyranose units, which are linked together by β-(1–4)-glucosidic bonds (Figure 5.1). Actually the building block for cellulose is cellobiose, since the repeating unit in cellulose is a two-sugar unit. The number of glucose units in a cellulose molecule is referred to as the degree of polymerization (DP), and the average DP for plant cellulose ranges from a low of about 50 for a sulfite pulp to approximately 600, depending on the determination method used (Stamm, 1964).

Table 5.1 Chemical Composition of Some Common Fibers (% of total)

Type of Fiber	Cellulose	Lignin	Pentosan	Ash	Silica
Stalk fiber					
Rice	28–48	12–16	23–28	15–20	9–14
Wheat	29–51	16–21	26–32	4.5–9	3–7
Barley	31–45	14–15	24–29	5–7	3–6
Oat	31–48	16–19	27–38	6–8	4–6.5
Rye	33–50	16–19	27–30	2–5	0.5–4
Cane fiber					
Bagasse	32–48	19–24	27–32	1.5–5	0.7–3.5
Bamboo	26–43	21–31	15–26	1.7–5	0.7
Grass fiber					
Esparto	33–38	17–19	27–32	6–8	—
Sabai	—	22	24	6	—
Reed fiber					
Phragmites communis	44–46	22–24	20	3	2
Bast fiber					
Seed flax	43–47	21–23	24–26	5	—
Kenaf	44–57	15–19	22–23	2–5	—
Jute	45–63	21–26	18–21	0.5–2	—
Hemp	57–77	9–13	14–17	0.8	—
Ramie	87–91	—	5–8	—	—
Core fiber					
Kenaf	37–49	15–21	18–24	2–4	—
Jute	41–48	21–24	18–22	0.8	—
Leaf fiber					
Abaca (Manila)	56–63	7–9	15–17	3	—
Sisal (agave)	47–62	7–9	21–24	0.6–1	—
Seed hull fiber					
Cotton	85–90	0.7–1.6	1–3	0.8–2	—
Wood fiber					
Coniferous	40–45	26–34	7–14	<1	—
Deciduous	38–49	23–30	19–26	<1	—

Table 5.2 Sugar Content of Selected Plant Holocelluloses

	Sugars Present, %				
Fiber	Glucose	Xylose	Galactose	Arabinose	Mannose
Cotton	92.0	—	—	—	—
Southern Pine	49.0	5.4	2.4	—	19.2
Aspen	53.3	18.5	1.0	—	1.4
Bamboo	52.0	21.7	—	0.8	—
Bagasse	47.4	27.6	—	1.7	—
Kenaf	47.2	17.7	1.4	0.9	1.4
Jute	63.8	13.1	1.2	—	0.6
Penny Wort	39.0	3.5	2.8	0.8	2.9
Water Hyacinth	37.2	8.7	5.0	11.4	1.4

This would mean an approximate molecular weight for cellulose ranging from about 10,000–150,000.

Cellulose molecules are randomly oriented and have a tendency to form intra- and intermolecular hydrogen bonds. As the packing density of cellulose increases,

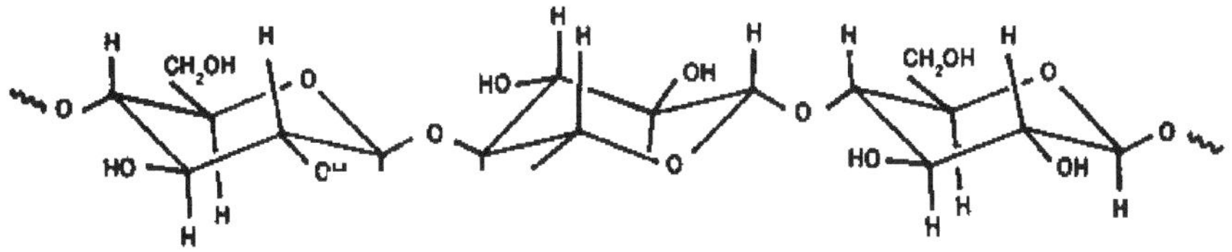

Figure 5.1 Partial structure of cellulose.

crystalline regions are formed. Most plant-derived cellulose is highly crystalline and may contain as much as 80% crystalline regions. The remaining portion has a lower packing density and is referred to as amorphous cellulose. Table 5.1 shows the range of average cellulose contents for a wide variety of plant types (Atchison, 1983). On a dry weight basis, most plants consist of approximately 45–50% cellulose. This can vary from a high (cotton) of almost 90% to a low of about 30% for stalk fibers. The cellulose content of many different types of plants is listed in the table at the end of this chapter.

1.1.3 Hemicelluloses

In general, the hemicellulose fraction of plants consists of a collection of polysaccharide polymers with a DP lower than cellulose and containing mainly the sugars D-xylopyranose, D-glucopyranose, D-galactopyranose, L-arabinofuranose, D-mannopyranose, and D-glucopyranosyluronic acid with minor amounts of other sugars (Figure 5.2). They usually contain a backbone consisting of one repeating sugar unit linked β-(1–4) with branch points (1–2), (1–3), and/or (1–6).

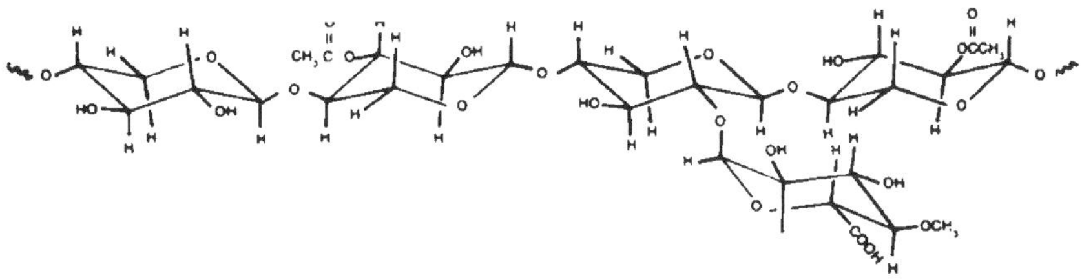

Figure 5.2 Partial structure of glucuronoxylan, a hardwood hemicellulose.

Hemicelluloses usually consist of more than one type of sugar unit and are sometimes referred to by the sugars they contain, for example, galactoglucomannan, arabinoglucuronoxylan, arabinogalactan, glucuronoxylan, glucomannan, etc. The hemicelluloses also contain acetyl and methyl substituted groups.

The hemicelluloses from bamboo consist of a backbone polymer of D-xylopyranose, linked β-(1-4) with an average of every eight xylose units containing a side chain of d-glucuronic acid attached glycosidically to the 2-position of the xylose sugar (Bhargava, 1987). The hemicelluloses from kenaf also contain a backbone polymer of D-xylopyranose with side chains of D-galactose and L-arabinose (Cunningham et al., 1986).

One of the main hemicelluloses from softwoods contains a backbone polymer of D-galactose, D-glucose, and D-mannose (Sjöström, 1981). The galactoglucomannan is the principal hemicellulose (ca. 20%), with a linear or possibly slightly branched chain with β-(1–4) linkages. Glucose and mannose make up the backbone polymer with branches containing galactose. There are two fractions of these polymers differing by their galactose content. The low galactose fraction has a ratio of galactose:glucose:mannose of about 0.1:1:4, while the high galactose fraction has a ratio of 1:1:3. The D-galactopyranose units are linked as a single-unit side chain by a (1–6) bond. The 2- and 3-positions of the backbone polymer have acetyl groups substituted an average of 3–4 hexose units. Another major hemicellulose polymer in softwoods (5–10%) is an arabinoglucuronoxylan consisting of a backbone of β-(1–4) xylopyranose units with (1–2) branches of D-glucopyranosyluronic acid on the average of every 2–10 xylose units and the (1–3) branches of l-arabinofuranose on the average of every 1.3 xylose units.

The major hemicellulose from hardwoods contains a backbone of D-xylose units linked β-(1–4) with acetyl groups at C-2 or C-3 of the xylose units, on an average of 7 acetyls per 10 xylose units (Sjöström, 1981). The xylan is substituted with side chains of 4-*0*-methylglucuronic acid units linked to the xylan backbone through a link (1–2) with an average frequency of approximately 1 uronic acid group per 10 xylose units. This class of hemicelluloses is usually referred to as glucuronoxylans. Hardwoods also contain 2–5% of a glucomannan composed of β-D-glucopyranose and β-D-mannopyranose units linked (1–4). The glucose:mannose ratio varies between 1:2 and 1:1, depending on the wood species.

The major hemicellulose from kenaf is similar to a hardwood xylan (Duckart et al., 1988). It has a backbone of β-(1–4) D-xylopyranose with side chains of 4-*0*-methylglucuronic acid linked (1–2) with an average frequency of 1 uronic acid group per 13 xylose units. There are terminal rhamnose and arabinose units linked (1–3), but the nature of the glycosidic linkage is unknown. The major hemicellulose from bamboo is composed of a backbone of β-(1–4) D-xylopyranose residues with an average of every eighth xylose unit containing a side chain of D-glucuronic acid, attached glycosidically to the 2-position of the xylose unit (Bhargave, 1987).

The detailed structures of most plant hemicelluloses have not been determined. Only the ratio of sugars these polysaccharides contain have been studied. Table 5.3 shows the sugar analysis of the two major hemicelluloses from several types of plant stalks (Jones et al., 1979). The table at the end of this chapter lists the sugars present in a wide variety of plant sources.

1.1.4 Pentosans

Part of the hemicellulose fraction consists of pentose sugars, mainly D-xylose and L-arabinose. The polymers containing these five carbon sugars are referred to as pentosans. Identification of this fraction in a plant material has been important

Table 5.3 Sugar Content of Selected Plant Stalk Hemicelluloses

	Sugars Present %				
Fiber	**Glucose**	**Xylose**	**Galactose**	**Arabinose**	**Mannose**
Corn					
Hemicellulose A	10.9	70.5	—	4.3	—
Hemicellulose B	12.1	43.8	3.6	9.7	1.0
Bagasse					
Hemicellulose A	6.1	60.9	trace	3.5	trace
Hemicellulose B	13.9	33.0	3.6	9.0	2.7
Sunflower					
Hemicellulose A	19.0	5.8	—	0.9	21.8
Hemicellulose B	11.8	24.5	3.6	3.4	12.6

to indicate its potential utilization for furan-type chemicals. It is therefore common to see tables of chemical composition data include pentosan content.

1.2 Lignin

Lignins are amorphous, highly complex, mainly aromatic, polymers of phenylpropane units. Lignins can be classified in several ways but they are usually divided according to their structural elements (Sjöström, 1981). All plant lignins consist mainly of three basic building blocks of guaiacyl, syringyl, and *p*-hydroxyphenyl moieties, although other aromatic type units also exist in many different types of plants (Figure 5.3). There is a wide variation of structures within different plant species. The phenylpropane can be substituted at the α, β, and γ positions into various combinations linked together both by ether and carbon to carbon linkages (see Figure 5.4).

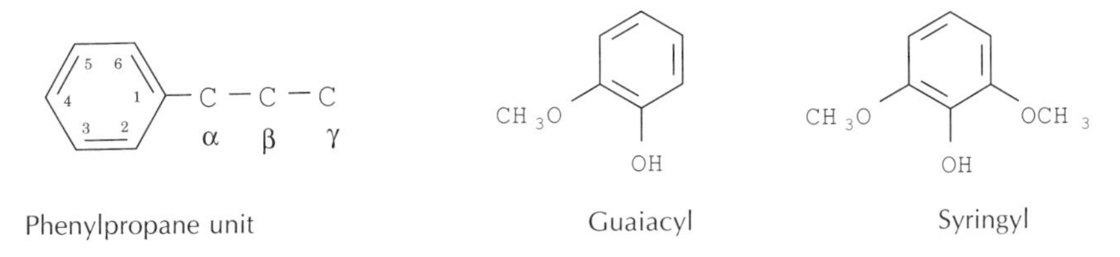

Figure 5.3 Building blocks of lignin.

Lignins from softwoods are mainly a polymerization product of coniferyl alcohol and are called "guaiacyl lignin." Hardwood lignins are mainly "syringyl-guaiacyl lignin" as they are a copolymer of coniferyl and sinapyl alcohols. The ratio of these two varies in different lignins from a ratio of 4:1 to 1:2.

Lignins found in plants contain significant amounts of constituents other than guaiacyl- and syringyl-propane units (Sarkanen and Ludwig, 1971). Lignin from corn contains vanillin and syringaldehyde units along with substantial amounts of *p*-hydroxybenzaldehyde. Bamboo lignin is a mixed dehydration polymer of coniferyl, sinapyl, and *p*-coumaryl alcohols (Bhargava, 1987). A recent study showed

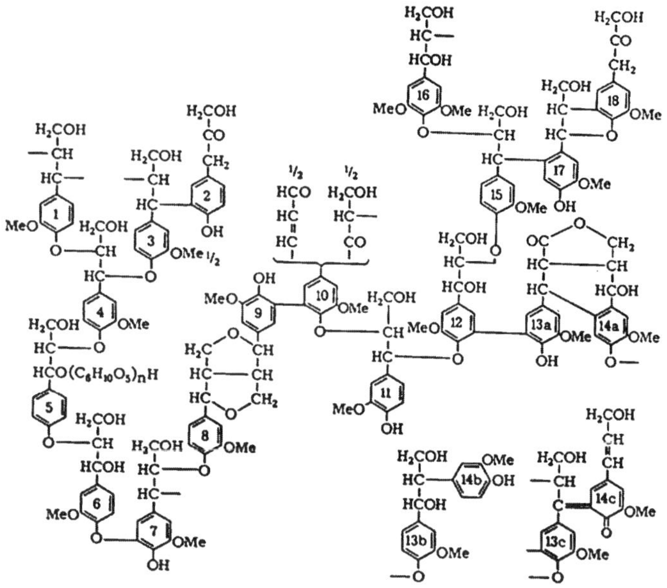

Figure 5.4 Partial structure of one type of lignin.

that the lignin from kenaf contains a very high level of syringyl functionality (Ralph et al., 1995).

Lignin in wood is distributed throughout the secondary cell wall with the highest concentration in the middle lamella. Because of the difference in the volume of middle lamella to secondary cell wall, about 70% of the lignin is located in the secondary wall.

The function of lignin in plants is as an encrusting agent in the cellulose/hemicellulose matrix. It is often referred to as the plant cell wall adhesive. Both lignin and extractives in plants reduce the digestibility of grasses to animals (Jung et al., 1993). Ligins are also associated with the hemicelluloses, in some cases forming lignin-carbohydrate complexes that are resistant to hydrolysis even under pulping conditions.

1.3 Inorganics

The inorganic content of a plant is usually referred to as ash content, which is an approximate measure of the mineral salts and other inorganic matter in the fiber after combustion at a temperature of 575 ± 25°C. The inorganic content can be quite high in plants containing large amounts of silica.

1.4 Proteins

Proteins are polymers of amino acids that are normally high in concentration in young growing cells but can also be found in some plants in high concentration throughout their life cycle. Proteins include enzymes and toxins as well as those involved in wound responses and pathogen resistance (Iiyama et al., 1993). Pathogen resistance proteins are related to the structural proteins that are thought to provide the framework, in addition to the microfibrillar phase, onto and around which the various non-cellulosic polysaccharides are arranged.

Three classes of structural proteins have been identified and classified on the basis of their repeating amino acid sequences (Iiyama et al., 1993). These three are: hydroxyproline-rich glycoproteins, the glycine-rich proteins, and the proline-rich proteins. The hydroxyproline-rich glycoproteins are usually associated with L-arabinose and D-galactose. The glycine-rich and the proline-rich proteins lack *N*-glycosylation sites in their primary sequence.

In wood, the protein content of the cell is usually less than 1% but can be much higher in grasses. The protein content is often reported as part of the lignin content if the laboratory personnel are not aware of its presence in the plant tissue when doing a lignin determination, since both protein and lignin are isolated in the sulfuric acid procedure.

1.5 Extractives

The extractives are a group of cell wall chemicals mainly consisting of fats, fatty acids, fatty alcohols, phenols, terpenes, steroids, resin acids, rosin, waxes, etc. These chemicals exist as monomers, dimers, and polymers. They derive their name as chemicals that are removed by one of several extraction procedures.

2. SAMPLING PROCEDURE

In reporting the chemical content of a plant, it is very important to report as much information about the samples as possible. Since the chemical content of a given species may vary depending upon the growing conditions, harvesting times of the year, etc., it is critical to report these conditions along with the chemical analysis. It is also important to report the exact analytical conditions and procedures used. This way, it may be possible to reproduce the results by other workers in different laboratories. *Without this information, it is not possible to compare data from different laboratories.*

The following information should accompany each chemical analysis:

(1) Source of the plant
 (a) Place of the growth
 (b) Year of the growth
 (c) Age of the plant
 (d) Condition of the soil and fertilizers applied
(2) Sampling
 (a) Different anatomical parts
 (b) Degree of biological deterioration if any
 (c) Sample size
 (d) Drying method applied
 (e) Time of year the sample was taken

(3) Analytical procedure used
(4) Reporting technique

All of the above mentioned criteria could contribute in one way or another toward variations in chemical analyses. Every criterion is as important as the other.

3. ANALYTICAL PROCEDURES

3.1 Extraction

3.1.1 Scope and Summary

Plant materials = Extractives + holocellulose + lignin + inorganics (ash)

This method describes a procedure for extraction of non-wood fiber for further analysis, such as holocellulose, hemicellulose, cellulose, and lignin analysis.

Neutral solvents, water, toluene or ethanol, or combinations of solvents are employed to remove extractives in agro-based fibers. However, other solvents ranging from diethyl ether to 1% NaOH, etc. could be applied according to the nature of extractives.

3.1.2 Sample Preparation

It is highly recommended to have a fresh sample. If not, keep the sample in a refrigerator to avoid fungal attack. Peel off the bark from the stem and separate the sample into component parts. Dry samples are oven dried for 24 hours (usually at 105°C) before milling. Wet samples can be milled while frozen in order to prevent oxidation or other undesirable chemical reactions. Samples are ground to pass 40 mesh (0.40 mm) using a Wiley Mill.

3.1.3 Apparatus

- Buchner funnel
- Extraction thimbles, ASTM 170-220 or Pyrex ™ 33950-MEC E or-MC.
- Extraction apparatus, extraction flask, 500 mL, Soxhlet extraction tube
- Heating device, heating mantle or equivalent
- Boiling chips, glass beads, boilers or any inert granules for taming boiling action
- Chemical fume hood
- Vacuum oven

3.1.4 Reagents and Materials

- Ethanol (ethyl alcohol), 200 proof ethanol
- Toluene, reagent grade
- Toluene–ethanol mixture, mix one volume of ethanol and two volumes of toluene

3.1.5 Procedures

Weigh 2–3 g of sample into several covered (yet ventilated) preweighed extraction thimbles. Place the thimbles in a vacuum oven not hotter than 45°C for 24 h, or to a constant weight. Cool the thimbles in a desiccator for one h and weigh. Then, place the thimbles in Soxhlet extraction units. Place 200 mL of the toluene–ethanol mixture* in a 500 mL round bottom flask with several boiling chips to prevent bumping. Carry out the extraction in a well ventilated chemical fume hood for 24 h, keeping the liquid boiling so that siphoning from the extractor is no less than four times per h. After extraction with the toluene:ethanol mixture, take the thimbles out of the extractors, drain the excess solvent, and wash the samples with ethanol. Place them in the vacuum oven overnight at temperatures not exceeding 45°C for 24 h. When dry, remove them to a desiccator for one h and weigh. Generally, the extraction is complete at this stage; however, the extractability depends upon the matrix of the sample and the nature of extractives. Second and third extractions using a different polarity of solvents may be necessary. Browning (1967) suggests 4 h of successive extraction with 95% alcohol. TAPPI Standard T 264 (TAPPI, 1988) designates two successive extractions, 4 h with ethanol, followed with distilled water for 1 hour. Pettersen et al. (1991) extracted pine samples with acetone/water, followed by the toluene–ethanol mixture.

3.2 Ash Content

3.2.1 Scope

The ash content of fiber is defined as the residue remaining after ignition at 575° ± 25° C (1067° ± 5°F) for 3 h, or longer if necessary to burn off all the carbon. It is a measure of mineral salts in the fiber, but it is not necessarily quantitatively equal to them. Fiber, like wood and pulp, is ashed at a lower temperature than paper (925°C) to minimize the volatilization of inorganic compounds.

3.2.2 Sample Preparation

Obtain a representative sample of the fiber, preferably ground to pass a 40-mesh screen. Weigh, to 5 mg or less, a specimen of about 5 g of moisture-free fiber for ashing, preferably in duplicate. If the moisture in the sample is not known, determine it by drying a corresponding specimen to constant weight in a vacuum oven at 105 ± 3°C.

3.2.3 Apparatus

- Crucible. A platinum crucible or dish with lid or cover is recommended. If platinum is not available, silica may be used.

* OSHA Standard for occupational exposure to Benzene is 29CFR 1910.1028 which became effective as of 12/10/87. Benzene is an OSHA regulated chemical and no longer used due to the health hazard.

- Analytical balance with a sensitivity to 0.1 mg.
- Electric muffle furnace adjusted to maintain a temperature of 575 ± 25°C.

3.2.4 Procedure

Carefully clean the empty crucible and cover, and ignite them to constant weight in a muffle furnace at 575 ± 25°C. After ignition, cool slightly and place in a desiccator. When cooled to room temperature, weigh the crucible and cover on the analytical balance.

Place all, or as much as practicable, of the weighed specimen in the crucible. Burn the fiber directly over a low flame of a Bunsen burner (or preferably on the hearth of the furnace) until it is well carbonized, taking care not to blow portions of the ash from the crucible. If a sample tends to flare up or lose ash during charring, the crucible should be covered, or at least partially covered during this step. If the crucible is too small to hold the entire specimen, gently burn the portion added and add more as the flame subsides. Continue heating with the burner only as long as the residue burns with a flame. Place the crucible in the furnace at 575 ± 25°C for a period of at least 3 h, or longer if needed, to burn off all the carbon.

When ignition is complete, as indicated by the absence of black particles, remove the crucible from the furnace, replace the cover and allow the crucible to cool somewhat. Then place in a desiccator and cool to room temperature. Reweigh the ash and calculate the percentage based on the moisture-free weight of the fiber.

3.2.5 Report

Report the ash as a percentage of the moisture-free fiber to two significant figures, or to only one significant figure if the ash is less than 0.1%.

3.2.6 Precision

The results of duplicate determinations should be suspect if they differ by more than 0.5 mg.

Additional Information

1. Since the ignition temperature affects the weight of the ash, only values obtained at 575 ± 25°C should be reported as being in accordance with this method.
2. In this procedure, the temperature of ignition has been specified at 575 ± 25°C, the same as given in TAPPI Standard T 211, "Ash in pulp."
3. Similar Method: Australia, APPITA, P 3m.
   Related Methods: ASTM D 1102; Canadian, C.P.P.A., G-10.
4. Porcelain crucibles can also be used in most cases for the determination of ash.
5. Special precautions are required in the use of platinum crucibles; a list of rules to follow is given by Pierce and Haenish (1948).
6. If the fiber ash is to be analyzed to determine its various constituents, wet ashing is recommended by Phifer (1957).
7. Data on the volatility of some ash constituents of wood pulp are reported by Bethge and Troeng (1958), Grove, Jones, and Mathews (1961) and Phifer and Maginnes

(1960). They report significant losses in sodium, calciums, irons and copper at temperatures of more than 600°C.

3.3 Preparation of Holocellulose (Chlorite Holocellulose)

3.3.1 Scope

Holocellulose is defined as a water-insoluble carbohydrate fraction of plant materials. According to Browning (1967) there are three ways of preparing holocellulose and their modified methods: (1) Chlorination method (also ASTM Standard D1104); (2) Modified chlorination methods; and (3) Chlorine dioxide and chlorite methods. The standard purity of holocellulose is checked following lignin analysis.

3.3.2 Sample Preparation

The sample should be extractive and moisture free and prepared after Procedure 3.1. If Procedure 3.1 is skipped for some reason, the weight of the extractives should be accounted for in the calculation of holocellulose.

3.3.3 Apparatus

- Buchner funnel
- 250 mL Erlenmeyer flasks
- 25 mL Erlenmeyer flasks
- Water bath
- Filter paper
- Chemical fume hood

3.3.4 Reagents

- Acetic acid, reagent grade
- Sodium chlorite, $NaClO_2$, technical grade, 80%

3.3.5 Procedure

To 2.5 g of sample, add 80 mL of hot distilled water, 0.5 mL acetic acid, and 1 g of sodium chlorite in a 250 mL Erlenmeyer flask. An optional 25 mL Erlenmeyer flask is inverted in the neck of the reaction flask. The mixture is heated on a water bath at 70°C. After 60 min, 0.5 mL of acetic acid and 1 g of sodium chlorite are added. After each succeeding hour, fresh portions of 0.5 mL acetic acid and 1 g sodium chlorite are added with shaking. The delignification process degrades some of the polysaccharides, and the application of excess chloriting should be avoided. Continued reaction will remove more lignin but hemicellulose will also be lost (Rowell, 1980).

Addition of 0.5 mL acetic acid and 1 g of sodium chlorite is repeated until the fibers are completely separated from lignin. It usually requires 6 h of chloriting, and the sample can be left without further addition of acetic acid and sodium chlorite in the water bath overnight. At the end of 24 h of reaction, cool the sample and filter the holocellulose on filter paper using a Buchner funnel until the yellow color (the color of holocellulose is white) and the odor of chlorine dioxide are removed. If the weight of the holocellulose is desired, filter the holocellulose on a tarred fritted disc glass thimble, wash with acetone, vacuum-oven dry at 105°C for 24 h, place in a desiccator for an hour and weigh. The holocellulose should not contain any lignin and the lignin content of holocellulose should be determined and subtracted from the weight of the prepared holocellulose.

3.4 Preparation of α-Cellulose (Determination of Hemicellulose)

3.4.1 Scope

The preparation of α-cellulose is a continuous procedure from Procedure 3.3.5 in pursuit of the ultimately pure form of fiber. The terms; α-cellulose, β-cellulose, γ-cellulose, cellulose, cellulose I, cellulose II, cellulose III, cellulose IV, cellulose V are defined in ASTM 1695-77. The term hemicellulose was introduced by Schulze [Schulze, *E. Ber.*, 24, 2274(1891)] and defined as the cell-wall components that are readily hydrolyzed by hot dilute mineral acids, hot dilute alkalies, or cold 5% sodium hydroxide.

3.4.2 Principle of Method

Extractive-free, lignin-free holocellulose is treated with sodium hydroxide and then with acetic acid, with the residue defined as α-cellulose. Thus the last fraction gives the hemicellulose content.

3.4.3 Apparatus

A *thermostat* or other constant-temperature device will be required that will maintain a temperature of 20 ± 0.1°C in a container large enough to hold a row of at least three 250 mL beakers kept in an upright position at all times.

Filtering Crucibles of Alundum™ or fritted glass thimbles of medium porosity.

3.4.4 Reagents

- Sodium hydroxide solution, NaOH, 17.5%, and 8.3%
- Acetic acid, 10%, mix one part by weight of glacial acetic acid with nine parts of distilled water.

3.4.5 Procedure

- Weigh out about 2 g of vacuum-oven dried holocellulose and place into a 250-mL glass beaker provided with a glass cover. Measure 25 mL of 17.5% NaOH solution in a graduated cylinder, and maintain at 20°C.

- Add 10 mL of 17.5% NaOH solution to the holocellulose in the 250-mL beaker, cover with a watch glass, and maintain at 20°C in the water bath. Manipulate the holocellulose lightly with a glass rod with the flat end so that all of the specimen becomes soaked with the NaOH solution.
- After 2 min, manipulate the specimen with the glass rod by pressing and stirring until the particles are separated from one another. After the addition of the first portion of 17.5% NaOH solution to the specimen, at 5 min intervals, add 5 mL more of the NaOH solution and thoroughly stir the mixture with the glass rod, until the NaOH is gone.
- Allow the mixture to stand at 20°C for 30 min, making the total time for NaOH treatment 45 min.
- Add 33 mL of distilled water at 20°C to the mixture. Thoroughly mix the contents of the beaker and allow to stand at 20°C for 1 h before filtering.
- Filter the cellulose with the aid of suction into the tarred, alkali-resistant Alundum or fritted-glass crucible of medium porosity.
- Transfer all the holocellulose residue to the crucible, and wash with 100 mL of 8.3% NaOH solution at 20°C. After the NaOH wash solution has passed through the residue in the crucible, continue the washing at 20°C with distilled water, making certain that all particles have been transferred from the 250-mL beaker to the crucible. Washing the sample in the crucible is facilitated by releasing the suction, filling the crucible to within 6 mm of the top with water, carefully breaking up the cellulose mat with a glass rod to separate any lumps present, and again applying suction. Repeat this step twice. The combined filtrate at this stage of the procedure may be set aside for the determination of β-cellulose.

Pour 15 mL of 10% acetic acid (at room temperature) into the crucible, drawing the acid into the cellulose by suction but, while the cellulose is still covered with acid, release the suction. Subject the cellulose to the acid treatment for 3 min from the time the suction is released, then apply suction to draw off the acetic acid. Without releasing the suction, fill the crucible almost to the top with distilled water at 20°C and allow to drain completely. Repeat the washing until the cellulose residue is free of acid as indicated by litmus paper. Give the cellulose a final washing by drawing, by suction, an additional 250 mL of distilled water through the cellulose in the crucible. Dry the crucible on the bottom and sides with a cloth and then, together with the weighing bottle in which the sample was originally weighed, place it overnight in a vacuum oven to dry at 100–105°C. Cool the crucible and weighing bottle in a desiccator for 1 h before weighing.

3.4.6 Calculation and Report

Calculate the percentage of α-cellulose on the basis of the oven-dry holocellulose sample:

$$\alpha\text{-cellulose, percent} = (W2/WI) \times 100$$

W2 = weight of the oven-dry α-cellulose residue
W1 = weight of the original oven-dry holocellulose sample.

3.5 Preparation of Klason Lignin

3.5.1 Scope

Klason lignin gives a quantitative measure of lignin and is not suitable for the study of lignin structures and some other lignins such as cellulolytic enzyme lignin, or Björkman (milled wood lignin) should be prepared (Sjöström, 1981) for the study of lignin structure. About 10–15% of Klason lignin of non-wood sources could be protein, and the protein content should be subtracted from the Klason lignin value applying the Kjeldahl procedure (Procedure 3.6). This procedure is a modified version of TAPPI T222 acid-insoluble lignin in wood and pulp (TAPPI, T-222, 1988). The lignin isolated using this procedure is also called sulfuric acid lignin.

3.5.2 Apparatus

- Autoclave
- Buchner funnel
- 100 mL centrifuge tube, Pyrex™ 8240
- Desiccator
- Glass rods
- Water bath
- 60 mL syringe
- Glass fiber filter paper, Whatman Cat. No. 1827-021, 934-AH
- Glass microfiber filter, 2.1 cm

3.5.3 Reagents

Sulfuric acid, H_2SO_4, 72% and 4% by volume
Fucose, 24.125% in 4% H_2SO_4 [w/w]

3.5.4 Procedure

Prepare samples by Procedure 3.1.5 and dry the sample at 45°C in a vacuum oven overnight. Accurately weigh out approximately 200 mg of ground vacuum-dried sample into a 100 mL centrifuge tube. To the sample in a 100 mL centrifuge tube, add 1 mL of 72% (w/w) H_2SO_4 for each 100 mg of sample. Stir and disperse the mixture thoroughly with a glass rod twice, then incubate the tubes in a water bath at 30°C for 60 min. Add 56 mL of distilled water (use a 60-mL syringe). This results in a 4% solution for the secondary hydrolysis. Add 1 mL fucose internal standard (this procedure is required only if five sugars are to be analyzed by HPLC as a part of the analysis).

Autoclave at 121°C, 15 psi, for 60 min. Remove the samples from the autoclave and filter off the lignin, with glass fiber filters (filters were rinsed into crucibles, dried and tarred) in crucibles using suction, keeping the solution hot. Wash the residue thoroughly with hot water and dry at 105°C overnight. Move to a desiccator, and let it sit 1 h and weigh to five places. Calculate Klason lignin content from weights.

3.5.5 Interference by Protein in Klason Lignin Determinations

Condensation reactions involving protein can cause artificially high Klason lignin measurements when tissues containing significant protein contents are analyzed, as in the case of non-wood fibers.

The Forest Products Laboratory (FPL) conducted a study (1994) on protein content as a function of growth on kenaf. A trend is apparent: although mature kenaf contains less protein, a greater percentage of this protein is condensed by acid hydrolysis than that of the younger kenaf. As a result, the positive interference from protein remains significant even in the less proteinaceous mature samples. It is reasonable to assume that the same proteins are condensed in samples harvested at either sample time. The ratio of structural protein to total protein increases with increasing maturity. A final note in this regard: hot acid detergent (Goering and Van Soest, 1970) extracted hay gave a protein of 4.3% as compared to 18.5% for raw hay. The initial impression might be that positive interference from protein is thereby substantially reduced. However, structural proteins are the most likely candidates to be resistant to extraction. Thus, if structural proteins do tend to condense under acid hydrolysis conditions, the outcome for hot acid detergent extracted materials may be similar to that of the more mature kenaf samples.

3.6 Nitrogen Content

3.6.1 Protein Determination by Kjeldahl Method

3.6.1.1 Scope

An FPL study on kenaf showed that by separate Kjeldahl (1883) analysis, 32% of nitrogen was found in Klason lignin and 68% in the hydrolysate of acid hydrolysis. Investigators at the USDA, Dairy Forage Laboratory developed (Goering and Soer, 1970) the acid-detergent lignin procedure where the detergent removed the protein and other acid-soluble materials that would interfere with the lignin determination. Further study is desired in this area.

This Kjeldahl method was modified at FPL (Moore and Johnson, 1967) in 1967, and further modification was achieved in 1993 for use in determination of the amine and amide nitrogen content in nonwood fibers. The organic compound is digested with concentrated sulfuric acid, which converts combined nitrogen into ammonium sulfate. The solution is then made alkaline. The liberated ammonia is then distilled, and the amount determined by titration with standard acid. It is directly applicable to amines and amides but not to nitro-, azo-, and azoxy-compounds. These latter compounds must be reduced (Zn–Hg amalgam and acid or salicylic acid, sodium thiosulfate and acid) before the Kjeldahl treatment. The protein content is then obtained by multiplying the percent nitrogen in an aliquot of fiber by an empirical factor of 6.25.

3.6.1.2 Sample Preparation

Prepare samples by Procedure 3.1 and place in a vacuum oven at 45°C overnight. Place the sample in a desiccator prior to actual chemical analysis. A quality control sample of DL-Norvaline (%N = 11.96%) and a blank sample should be carried out through this entire procedure.

3.6.1.3 Apparatus

- Burette, 10, 25, or 50 mL
- Desiccator
- Erlenmeyer flask, 250 mL
- Micro–Kjeldahl digestion apparatus
- Micro–Kjeldahl digestion rack, Labconco 7053-S10
- Heating element, Labconco 7053-S10
- Kjeldahl flasks, 30 mL, 100 mL, or 125 mL, Pyrex™ and Kimax
- Micro–Kjeldahl distilling apparatus, Thomas Scientific
- Micro–Kjeldahl distilling unit, 7052-J10
- Distilling unit, 7052-J20
- Steam generator, ASTM, 7052-J30
- Immersion heater, ASTM, 7052-J40

Note: See ASTM E 147 for detailed dimensions of the apparatus.

3.6.1.4 Reagents

- Boric acid, H_3BO_3
- Copper sulfate, $CuSO_4 \cdot 5H_2O$
- Hydrochloric acid, HCl, 0.01 N
- Mixed indicator (Place 200 mg of bromocresol green and 40 mg of methyl red in 100 mL volumetric flask. Dissolve and fill up to mark with 95% ethanol.)
- DL-Norvaline, 99%, Aldrich 85,162-0
- Potassium sulfate, K_2SO_4
- Sodium carbonate, $NaCO_3$
- Sodium hydroxide, NaOH
- Sulfuric acid, H_2SO_4
- Sodium thiosulfate, anhydrous, $Na_2S_2O_3$
- Mercury (II) oxide, red, HgO

3.6.1.5 Procedure

Digestion: Weigh approximately 100 mg of sample to the nearest 0.1 mg into 30 mL Kjeldahl flask. Add 5 g of K_2SO_4 per gram of sample and 250 mg $CuSO_4 \cdot 5H_2O$ per gram of sample to each flask. Next, add 10 mL of conc. H_2SO_4 per gram of sample. Place specimens on low heat at first and cook until all black carbon has disappeared and the solution appears tint in color. Kenaf fiber requires about 2 h for complete digestion, while 10 mg of DL-Norvaline should be fully digested within 1 h.

Note: the weight of sample should be adjusted depending upon the nitrogen content followed by the size of Kjeldahl flask. More sulfuric acid may be needed and distilled water may be added to rinse the sample. For nitro-, azo- or azoxy-compounds: 1 mL 5% salicylic acid in H_2SO_4 and wait 30 min, add 100 mg $Na_2S_2O_3$ wait 10–15 min and proceed with digestion. For amine and amide compounds: skip the step above and start with 650 mg of K_2SO_4, 16 mg of HgO, 1 mL of H_2SO_4 proceed with digestion.

***Distillation and Titration*:** Close the upper stopcock (sample stopcock), open the lower (vacuum) stopcock and pull distilled water, from a large beaker submerged to the condenser tip, by suction and close the lower stopcock. Open the upper stopcock and fill the still with distilled water. Repeat this process until approximately 1–2 liters of water have been washed through the entire system. The lower drain spout is connected to an aspirator via a water trap, and waste water is removed after the rinsing. Add 5 mL of 4% boric acid and 5 drops of the mixed indicator to a 250 mL Erlenmeyer flask. Dilute with 20 mL of distilled water (the solution should be green) and submerge the tip of the condenser in the solution. Open the upper stopcock and quantitatively transfer the digested sample from the Kjeldahl flask to the still. Also rinse the filling cup to insure the complete transfer (Caution: when rinsing, the flask will become hot and sulfuric acid fumes may be emitted). Close the upper stopcock and fill the cup with 28 mL of 40% NaOH. If the filling cup cannot hold the full volume of NaOH, open the stopcock slightly and transfer the remaining NaOH to the cup. Close the stopcock immediately once the NaOH has completely drained. Replace the rubber stopper and plug in the heating coil.

Distill until the volume in the Erlenmeyer flask has doubled. The solution should be blue in color. Lower the flask and rinse the condenser tip. Remove the rubber stopper and turn off the heating coil. Allow the sample to cool. Titrate the distillate from blue to a green endpoint with the standardized 0.01N HCl solution. Calculate the percent of nitrogen in the sample as follows:

$$\%N = \frac{\text{mL HCl (sample} - \text{blank)} \times \text{(normality of HCl) } 14.0067 \times 100}{\text{sample weight in mg}}$$

$$\%\ \text{Protein} = 6.25 \times \%N$$

3.7 Determination of Methoxyl Groups

3.7.1 Scope

Methoxyl groups ($-OCH_3$) are present in the lignin and polysaccharide portions of plants. Methoxyl groups occur in lignin and lignin derivatives as side chains of aromatic phenylpropanes and in the polysaccharides mainly as methoxy-uronic acids. Methoxyl content is determined by some modification of the original method of Zeisel (1886) and an instrumental method such as gas chromatography, HPLC or infrared. This is a modified version of ASTM, D1166-84, Standard Test Method for Methoxyl Groups in Wood and Related Materials (ASTM, 1960), TAPPI T 2 wd-72 Methoxyl Group in Wood, and T 209 wd-79 Methoxyl Content of Pulp and Wood (TAPPI, 1988). Related materials can be found in the following references: Friedrich (1927), Clark (1929), Vieböck and Schwappach (1930), Peniston and Hibbert (1939), Bailey (1942), and Samsel (1942).

3.7.2 Principle of Method

In the original method of Zeisel (1886), the methyl iodide was absorbed in an alcoholic solution of silver nitrate. The solution was diluted with water, acidified with nitric acid, and boiled. The silver iodide was removed by filtration, washed, and weighed in the manner usual for halide determinations. A volumetric modification is based on absorption of the methyl iodide in a known volume of standard silver nitrate solution and titration of the unused silver nitrate with standard potassium thiocyanate solution (ferric alum indicator solution). In this procedure, the methyl iodide is collected in an acetic acid solution of potassium acetate containing bromine.

$$CH_3I + Br_2 \rightarrow CH_3Br + IBr$$

$$IBr + 2Br_2 + 3H_2O \rightarrow HIO_3 + 5HBr$$

The excess bromine is destroyed by addition of acid, and the iodate equivalent of the original methoxyl content is determined by titration with sodium thiosulfate of the iodine liberated in the reaction

$$HIO_3 + 5HI \rightarrow 3I_2 + 3H_2O$$

One methoxyl group is equivalent to six atoms of iodine and, consequently, a favorable analytical factor is obtained.

3.7.3 Sample Preparation

The sample is dried, ground, and extracted accordingly prior to the actual analysis taking place.

3.7.4 Apparatus

- Reaction flask
- Heat source–A micro burner
- Vertical air-cooled condenser
- Scrubber
- Absorption vessels

3.7.5 Reagents

Bromine, liquid
Cadmium sulfate solution—dissolve 67.2 g of $CdSO_4$–$4H_2O$ in 1 L of water.
Carbon dioxide gas
Formic acid, 90%
Hydroiodic acid
Phenol
Potassium acetate solution in acetic acid—anhydrous potassium acetate (100g) is dissolved in 1 L of glacial acetic acid
Potassium iodide solution—dissolve 100 g of KI in water and dilute to 1 L
Sodium acetate solution—dissolve 415 g of sodium acetate trihydrate in water and dilute to 1 L
Sodium thiosulfate solution (0.1N)—Dissolve 25 g of $Na_2S_2O_3{\cdot}H_2O$ in 200 mL of water and dilute to 1 L
Starch indicator solution (10 g/L)
Sulfuric acid—Mix one volume of H_2SO_4 (sp gr 1.84) with nine volumes of water

3.7.6 Procedure

Weigh the sample (about 100 mg of fiber or 50 mg of lignin) accurately in a gelatin capsule and place with the capsule in the reaction flask. Place in the reaction flask 15 mL of HI, 7 g of phenol, and a boiling tube. Place in the scrubber a mixture of equal volumes of $CdSO_4$ solution and $Na_2S_2O_3$. The volume of solution should be adjusted so that the inlet tube of the scrubber is covered to a depth of about 4 mm. Adjust the flow of CO_2 to about 60 bubbles per minute through the scrubber. Bring the contents of the flask to reaction temperature. Adjust the rate of heating so that the vapors of the boiling HI rise about 100 mm into the condenser. Heat the flask at reaction temperature for 30–45 min, or longer if necessary, to remove methoxyl-containing substances or other interfering substances that usually are present in the reagents.

Let the distilling flask cool below 100°C. In the meantime, add to 20 mL of the potassium acetate solution about 0.6 mL of bromine, and mix. Add approximately 15 mL of the mixture to the first receiver and 5 mL to the second, and attach the receiver to the apparatus. Seal the ground-glass joint with a small drop of water from a glass rod.

Remove the distilling flask and introduce the test specimen. Immediately reconnect the flask and seal the ground-glass joint with a drop of molten phenol from a glass rod. Bring the contents of the flask to reaction temperature while passing a uniform stream of CO_2 through the apparatus. Adjust the rate of heating so that the vapors of the boiling HI rise about 100 mL into the condenser. Continue the heating for a time sufficient to complete the reaction and sweep out the apparatus. Usually, not more than 50 min are required.

Wash the contents of both receivers into a 250-mL Erlenmeyer flask that contains 15 mL of sodium acetate solution. Dilute with water to approximately 125 mL, and add 6 drops of formic acid. Rotate the flask until the color of the bromine is discharged, then add 12 more drops of formic acid and allow the solution to stand for 1 to 2 min. Add 10 mL of KI solution and 10 mL of H_2SO_4, and titrate the liberated iodine with $Na_2S_2O_3$ solution, adding 1 mL of starch indicator solution just before the end point is reached, continuing the titration to the disappearance of the blue color.

3.7.7 Calculation and Report

$$\text{Methoxyl, \%} = (VN \times 31.030 \times 100)/(G \times 1000 \times 6) = (VN/G) \times 0.517$$

V = milliliters of $Na_2S_2O_3$ solution required for the titration,
N = normality of $Na_2S_2O_3$ solution, and
g = grams of moisture-free sample.

3.8 Determination of Acetyl by Gas-Liquid Chromatography (GLC)

3.8.1 Scope

The aliphatic acyl groups in woods and grasses are acetyl and formyl groups which are combined as *O*-acyl groups with the polysaccharide portion. There are basically three ways of determinations: (1) Acid hydrolysis—sample is hydrolyzed to form acetic acid; (2) Saponification—acetyl groups are split from polysaccharides with hot alkaline solution and acidified to form acetic acid; and (3) Transesterification—sample is treated with methanol in acid or alkaline solution to form methyl acetate; acetic acid and methyl acetate are analyzed by gas chromatography.

This procedure is an application of saponification.

$$CH_3COOR + 3NaOH \rightarrow 3CH_3COONa$$

$$CH_3COONa + H^+ \rightarrow CH_3COOH$$

3.8.2 Sample Preparation

Weigh an oven-dried sample in a long handled weighing tube and transfer it to an acetyl digestion flask and add four boiling chips. Pipette 2 mL 1 *N* NaOH solution to wash down the neck of the flask. Connect the reaction flask to a water cooled reflux condenser. Reflux for 1 h, heating the flask in a phosphoric acid bath with a Bunsen burner. Pipette 1 mL of propionic acid (internal standard) into a 10-mL volumetric flask. Quantitatively transfer the liquid from the reaction flask to the volumetric. Wash the reaction flask and the solid residue with several portions of distilled water. Add 0.2 mL of 85% phosphoric acid and make to volume with distilled water. This solution may be filtered through a small plug of glass wool to remove solid particles.

Analyze the sample by GLC and determine the average ratio. Milligrams of acetic acid are determined from the calibration curve.

3.8.3 Reagents

Internal Standard Stock Solution: Weigh 25.18 g of 99+% propionic acid in 500 mL volumetric flask, make to volume with 2% formic acid. Internal Standard Solution: Pipette 10 mL stock solution into a 200-mL volumetric flask; make to volume with distilled water.

Acetic Acid Standard Solution: Weigh 100 mg 99.7% glacial acetic acid into a 100-mL volumetric flask; make to volume with distilled water.

NaOH Solution 1*N*: Weigh 4 grams NaOH; dissolve in 100 mL distilled water.

3.8.4 Gas Chromatography

Column: Supelco 60/80 Carbopack C/0.3% carbowax 20M/0.1% H_3PO_4 – 3 ft-1/4 inch O. D. and 4 mm I. D.; Oven temperature 120°C; Injection port 150°C; F.I.D. 175°C; Nitrogen 20 mL/min.

The ratio of the area is determined by dividing the peak area of the acetic acid by the area of the propionic acid (internal standard).

The average of the ratios is used to determine mg/mL of acetic acid from the calibration curve.

Preparation of a Calibration Curve: Pipette 1, 2, 4, 6, and 8 mL of standard acetic acid solution into 10-mL volumetric flasks. Pipette 1 mL of propionic acid internal standard into each sample, then add 0.2 mL 85% phosphoric acid. Make to volume with distilled water. Analyze each solution six times by GLC. Calculate the ratios by dividing the peak area of the acetic acid by the peak area of the propionic acid (internal standard). Plot the average ratios against milligrams per milliliter of acetic acid. The results may be reported as acetic acid:

$$\text{mg/mL} \frac{\text{Acetic acid found} \times 10\text{ mL}}{\text{sample weight (mg)}} \times 100 = \%\text{Acetic acid or as Acetyl}$$

$$\text{mg/mL} \frac{\text{found} \times 10\text{ mL} \times 0.7172}{\text{sample weight (mg)}} \times 100 = \%\text{Acetyl}$$

4. CHEMICAL PROPERTIES OF FIBERS

The following table is a compilation of the chemical composition of some non-wood and wood materials. The data include different anatomical parts of the non-wood plants such as bast fibers, cortex, etc. It should be known that this data was collected from different times in different places using a variety of analytical procedures. The lignin contents of non-wood materials are generally lower than those of woods, but the pentosan and extractive contents are higher.

According to an FPL study on kenaf, the lignin content increases whereas the extractive content decreases as a function of growth. Extractive content of cores could be as much as twice of the bast fibers within the given plant. The extractives content could be twice that of at the top of the plant than at the bottom of the plant in both core and fiber. Details of variations based on growth were seen in Chapter 2. The general sources of variations on chemical compositions can be outlined as follows.

(1) Location–a growing season for an annual plant could be anywhere between 80 to 200 days. The height of kenaf could reach from 1 m to 3 m per year.
(2) Cultivars–different varieties of species. Tainung grows tallest.
(3) Conditions, types of soil, fertilizer applied, moisture, temperature, etc.
(4) Sampling procedure–top of the plant vs. bottom.
(5) Analytical procedure

Table 5.4 Chemical Properties of Agro-based Fibers

[a] Acetyl	[h] Holocellulose	[p] Polyuronide
[c] Cellulose	[i] Insoluble ash	[s] Silica
[cc] Crude cellulose	[k] Pectins	[t] Total ash
[d] Monoethanoamine procedure	[m] Hemicellulose	[u] Uronic acid
[e] Extractives	[o] Protein	[w] Wax

Fibers		Variety/or	Cross				Solubility					
Botanical name	Common name	Place of growth	& Bevan Cellulose	α-Cellulose	Lignin	Pento-sans	Alcohol Benzene	Hot Water	1% NaOH	Ash	Others	Ref.
Monocotyledoneae												
Agavaceae												
Agave sisalana	Sisal	India	55-73	43-56	8-9	21-24	—	—	—	.6-1	<1[s]	[17]
Agave sisalana	Sisal	—	—	63.9	8.6	17.9	—	—	—	0.7	4.6[a]	[25]
Agave sisalana	Sisal	—	73.1[c]	—	11.0	—	—	—	—	—	.9[k], 1.6[e]	[19]
Musa textilis	Abaca or	—	70.1[c]	—	5.7	—	—	—	—	—	.6[k], 1.8[e]	[19]
Musa textilis	Manila hemp	—	78.0	61.0	9.0	17.0	—	—	—	1.0	1.0[s]	[17]
Phormium tenax	Phormium	—	71.3[c]	—	—	—	—	—	—	—	—	[19]
Yucca schidigera (leaves)	—	CA	33.2[d]	22.3	—	—	15.3	—	51.4	—	—	[22]
Bromeliaceae												
Ananas comosus	Pineapple	Leaf	—	69.5	4.4	17.8	—	—	—	0.9	2.7[a]	[25]
Ananas comosus	Pineapple	—	81[c]	—	12.0	—	—	—	—	—	—	[26]
Cyperaceae												
Bulboslylis capilleris	—	MD	26.5[d]	7.8	—	—	5.9	—	43.3	—	—	[22]
Cyperus esculentus	Nut sedge	MD	38.2[d]	23.1	—	—	14.9	—	54.2	—	—	[24]
Cyperus filiculmis	—	MD	39.0[d]	24.9	—	—	12.0	—	52.2	—	—	[24]
Cyperus papyrus	Papyrus	Pith	21.0[h]	39.6	19.7	—	—	—	—	4.6	4.1[a], 10.7[e]	[20]
Cyperus papyrus	Papyrus	India	24.8[h]	40.1	16.1	—	—	—	—	5.2	4.7[a], 9.2[e]	[20]
Scirpus rubricosus	—	MD	44.5[d]	30.4	—	—	7.4	—	43.9	—	—	[22]
Scirpus americanus	—	CA	36.2[d]	22.3	—	—	12.1	—	59.5	—	—	[24]
Scirpus americanus	—	CA	37.2[d]	23.5	—	—	9.6	—	58.6	—	—	[24]
Scirpus paludosus	—	CA	35.3[d]	23.8	—	—	15.3	—	62.7	—	—	[24]

Table 5.4 Chemical Properties of Agro-based Fibers

Fibers		Variety/or	Cross				Solubility					
Botanical name	Common name	Place of growth	& Bevan Cellulose	α-Cellulose	Lignin	Pento-sans	Alcohol Benzene	Hot Water	1% NaOH	Ash	Others	Ref.
Scirpus sp.	—	NE	40.8[d]	27.0	—	—	9.5	—	47.5	—	—	[24]
Gramineae												
Agropyron elongatum	Wheatgrass	NE	35.4[d]	21.4	—	—	14.3	—	54.2	—	—	[22]
Agropyron intermedium	Wheatgrass	NE	44.1[d]	26.2	—	—	12.2	—	51.2	—	—	[24]
Agrostis sp.	Bent grass	MD	30.4[d]	14.4	—	—	5.7	—	56.0	—	—	[22]
Andropogon barbinodis	Beard grass	TX	46.8[d]	27.7	—	—	5.6	—	52.4	—	—	[22]
Andropogon gerardi	—	KS	52.2[d]	33.4	—	—	5.7	—	39.9	—	—	[23]
Andropogon hallii	—	KS	52.7[d]	33 5	—	—	5.8	—	39.1	—	—	[23]
Andropogon intermedius	—	KS	50.0[d]	31.5	—	—	5.7	—	42.8	—	—	[23]
Andropogon virginicus	—	TX	52.9[d]	33.1	—	—	4.5	—	40.2	—	—	[22]
Aristida wrighii	—	TX	41.3[d]	25.9	—	—	6.3	—	0.9	—	—	[22]
Arundinaria alpina	Bamboo	Ethiopia	43.8[d]	30.4	—	—	3.2	—	25.4	—	—	[22]
Arundinaria longifolia	Bamboo	AZ	60.3[d]	39.3	—	—	3.9	—	27.4	—	—	[22]
Arundinaria tecta	Bamboo	MD	46.1[d]	26.4	—	—	7.9	—	43.0	—	—	[24]
Arundo donax	Addar grass	MD	46.0[d]	29.3	—	—	10.2	—	43.8	—	—	[23]
Avena barbata	Oat	AZ	41.0[d]	27.1	—	—	12.5	—	54.2	—	—	[22]
Avena sativa	Oat	American	71.0[h]	35.9	16.1	26.2	—	14.9	9.7	—	2.4[p]	[20]
Avena sativa	Oat	—	—	39.4	17.5	27.1	4.4	15.3	—	7.2[t]	—	[20]
Avena sativa	Oat	Holland	44-53	31-37	16-19	27-38	—	—	—	6-8	4-7[s]	[17]
Avena sativa	Oat	India	—	37.1	15.5	24.9	—	13.8	—	2.9[i]	1.3[a], 1.1[u]	[20]
Bambusa vulgaris	Bamboo	Brazil	55.2[d]	36.3	—	—	5.0	—	25.8	—	—	[23]
Botanical names not given	Bamboo	—	49-62	32-44	19-24	27-32	—	—	—	1.5-5	.7-3[s]	[17]
"	Bamboo	Dowga	57-66	26-43	21-31	15-26	—	—	—	1.7-5	1.5-3[s]	[17]
"	Bamboo	Medar	68.8[h]	—	24.6	17.2	5.5	6.5	24.9	2.7	1.9[s]	[5]
"	Bamboo	Chiva	67.3[h]	—	26.6	17.7	4.3	5.8	24.0	4.1	2.5[s]	[5]
"	Bamboo	—	72.0[h]	—	27.5	17.5	2.9	4.9	25.4	3.9	2.3[s]	[5]
Bothriochloa intermidia	Bluestem	OK	46.7[d]	8.5	—	—	9.9	—	46.5	—	—	[22]
Bouteloua curtipendula	Gramagrass	TX	40.3[d]	23.0	—	—	6.4	—	52.4	—	—	[22]

Bromus rigidus	Bromgrass	MD	40.4[d]	24.8	—	—	5.8	—	49.3	—	—	[22]
Cenchrus myosuroides	—	AZ	45.5[d]	30.4	—	—	9.3	—	43.0	—	—	[22]
Cymbopogon validus	Oilgrass	S. Africa	46.0[d]	29.3	—	—	12.6	—	44.0	—	—	[22]
Cynodon dactylon	Bermuda grass	GA	41.3[d]	24.5	—	—	9. 9	—	51.4	—	—	[22]
Dactylis glomerata	Orchard grass	MD	47.0[d]	31.5	—	—	8.5	—	43.8	—	—	[24]
Digitaria sanguinalis	Crabgrass	MD	34.0[d]	16.7	—	—	8.4	—	46.1	—	—	[22]
Echinochloa crusgalli	Barnyard grass	MD	43.9[d]	27.5	—	—	9.2	—	48.5	—	—	[24]
Echinochloa pyramidalis	—	S. Africa	49.1[d]	30.6	—	—	9.0	—	41.4	—	—	[22]
Eleusine indica	Wiregrass	NC	29.0[d]	8.0	—	—	5.3	—	35.0	—	—	[24]
Elymus canadensis	Wild rye	ND	42.9[d]	27.0	—	—	15.2	—	51.2	—	—	[24]
Elymus giganteus	Wild rye	NE	47.9[d]	32.1	—	—	9.2	—	44.9	—	—	[24]
Elymus sp.	Wild rye	CA	48.3[d]	30.8	—	—	14.2	—	50.0	—	—	[24]
Eragrostis chloromelas	Love grass	AZ	39.9[d]	94.6	—	—	11.5	—	49.0	—	—	[24]
Eragrostis curvula	Love grass	OK	44.7[d]	28.2	—	—	7.5	—	44.3	—	—	[24]
Eragrostis curvula	Love grass	OK	40.8[d]	25.4	—	—	9.3	—	50.2	—	—	[24]
Erianthus ravennae, stalks	Ravenna grass	IL	52.6[d]	35.1	—	—	4.7	—	38.8	—	—	[22]
Guadua amplexifolia, base	—	Mexico	58.2[d]	40.6	—	—	5.8	—	32.2	—	—	[23]
Guadua amplexifolia, middle	—	Mexico	60.9[d]	42.8	—	—	3.7	—	28.5	—	—	[23]
Guadua amplexifolia, top	—	Mexico	61.6[d]	43.0	—	—	4.1	—	29.5	—	—	[23]
Guadua angustifolia	—	Ecuador	52.1[d]	29.8	—	—	3.6	—	37.3	—	—	[23]
Gynerium sagittatum	Wild cane	Honduras	58.4[d]	38.5	—	—	3.2	—	29.9	—	—	[23]
Holcus lanatus	Velvet grass	MD	43.3[d]	29.1	—	—	7.4	—	50.0	—	—	[24]
Hordeum vulgare	Barley	Europe	73.3[h]	36.4	16.6	26.7	—	12.2	—	9.6[t]	2.7[p]	[20]
Hordeum vulgare	Barley	America	47-48	31-34	16-19	24-29	—	—	—	5-7	3-6[s]	[17]
Hordeum vulgare	Barley	whole stalk	—	33.8	14.5	14.7	4.7	16.1	—	6.4[t]	—	[20]
Hyparrhenia hirta	—	S. Africa	54.6[d]	34.1	—	—	8.7	—	37.8	—	—	[22]
Ischaemum arcuatum	—	S. Africa	48.0[d]	30. 8	—	—	11.0	—	42.9	—	—	[22]
Lolium multiflorum	Ryegrass	Spain	49.1[d]	34.5	—	—	7.7	—	42.5	—	—	[23]
Lygeum spartum	—	MD	39.3[d]	24.4	—	—	11.8	—	46.5	—	—	[22]
Melica mulica	Melic grass	MD	52.3[d]	35 5	—	—	6.7	—	39.8	—	—	[23]
Miscanthus sinensis	—	MD	45.5[d]	26.6	—	—	4.0	—	37.8	—	—	[22]
Muhlenbergia rigens	—	CA	39.5[d]	22.8	—	—	5.0	—	47.8	—	—	[22]

Table 5.4 Chemical Properties of Agro-based Fibers

Fibers		Variety/or	Cross				Solubility					
		Place of	& Bevan	α-		Pento-	Alcohol	Hot	1%			
Botanical name	Common name	growth	Cellulose	Cellulose	Lignin	sans	Benzene	Water	NaOH	Ash	Others	Ref.
Oryza sativa	Rice	Egypti	—	36.2	11.9	24.5	4.6	13.3	—	16.1[t]	—	[20]
Oryza sativa	Rice	Sri Lanka	—	28.1	12.5	26.5	5.1	16.1	—	—	—	[20]
Oryza sativa	Rice	—	—	—	12.0	21.0	2.1	17.7	—	24.8	14.3[s]	[20]
Oryza sativa	Rice	India	43-49	28-36	12-16	23-28	—	—	—	15-20	9-14[s]	[17]
Oryza sativa	Rice	—	53.5[c]	—	25.5	21.0	—	10-14	—	12.0	—	[27]
Oxytenanthera abyssinica	—	Ethiopia	56.7[d]	39.8	—	—	5.4	—	29.5	—	—	[22]
Panicum antidotale	Panic grass	TX	39.3[d]	24.4	—	—	11.8	—	43.3	—	—	[22]
Panicum deustum	Panic grass	S. Africa	39.8[d]	24.3	—	—	11.8	—	50.8	—	—	[22]
Panicum subjunceum	Panic grass	Uruguay	42.0[d]	27.4	—	—	0.4	—	46.1	—	—	[22]
Panicum virgatum	Panic grass	KS	51.6[d]	32.4	—	—	5.9	—	39.4	—	—	[22]
Paspalum arechavaletae	—	Uruguay	45.2[d]	31.1	—	—	5.4	—	47.5	—	—	[22]
Paspalum exaltatum	—	Uruguay	44.8[d]	31.2	—	—	6.4	—	47.7	—	—	[22]
Paspalum haumanii	—	Uruguay	41.9[d]	29.4	—	—	7.4	—	48.8	—	—	[22]
Paspalum quadrifarium	—	Uruguay	46.3[d]	30.5	—	—	4.4	—	43.8	—	—	[22]
Pennisetum macrourum	—	GA	46.2[d]	30.2	—	—	14.1	—	46.1	—	—	[23]
Pennisetum spicatum	—	MI	40.1[d]	26.O	—	—	16.1	—	54.4	—	—	[23]
Pennisetum typhoides	—	S. Africa	48.4[d]	31.4	—	—	6.9	—	43.5	—	—	[22]
Pennisetum typhoides	—	IL	49.2[d]	31.7	—	—	7.3	—	45.5	—	—	[22]
Phleum pratense	—	MD	36.5[d]	18.1	—	—	15.4	—	54.3	—	—	[22]
Phragmites communis	Reeds	China	59.6[c]	—	14.7	18.2	—	—	—	2.1	—	[32]
Phragmites communis	Reeds	Romania	47.2[c]	—	22.9	26.6	—	—	—	2.5	—	[32]
Phragmites communis	Reeds	U.S.S.R.	36.4[c]	—	—	38.8	—	—	—	10.4	—	[32]
Phragmites communis	Reeds	Germany	33.3[c]	—	35.8	16.8	—	—	—	5.7	—	[32]
Phragmites communis	Reeds	Italy	49.1[c]	—	15.7	23.7	—	—	—	6.5	—	[32]
Phragmites communis	Reeds	—	57.0	45.0	22.0	20.0	—	—	—	3.0	2.0[s]	[17]
Phragmites communis	Reeds	NE	46.1[d]	25.1	—	—	8.6	—	46.0	—	—	[22]
Phyllostachys angusta	Bamboo	GA	56.1[d]	35.6	—	—	7.2	—	26.9	—	—	[24]

Phyllostachys aureosuslcata	Bamboo	GA	57.2[d]	36.3	—	—	5.0	—	27.1	—	—	[24]
Phyllostachys bamusoides	Bamboo	GA	56.1[d]	36.3	—	—	6.3	—	26.2	—	—	[24]
Phyllostachys bamusoides *cv.* castillon	Bamboo	GA	57.6[d]	35.6	—	—	6.1	—	26.8	—	—	[24]
Phyllostachys bamusoides *cv.* slender crookstem	Bamboo	GA	56.4[d]	35.7	—	—	7.3	—	25.4	—	—	[24]
Phyllostachys bamusoides	Bamboo	GA	55.4[d]	34.2	—	—	9.2	—	28.0	—	—	[24]
cv. white crookstem	Bamboo	GA	58.0[d]	35.5	—	—	5.8	—	23.6	—	—	[24]
Phyllostachys bambusoides (base)	Bamboo	GA	58.0[d]	35.5	—	—	5.8	—	23.6	—	—	[23]
Phyllostachys bambusoides (middle)	Bamboo	GA	57.7[d]	36.6	—	—	5.9	—	22.5	—	—	[23]
Phyllostachys bissetii	Bamboo	GA	58.0[d]	35.0	—	—	9.4	—	29.1	—	—	[24]
Phyllostachys congesta	Bamboo	GA	57.0[d]	35.6	—	—	6.1	—	27.6	—	—	[24]
Phyllostachys decora	Bamboo	GA	58.7[d]	35.0	—	—	6.5	—	24.2	—	—	[24]
Phyllostachys dulcis	Bamboo	GA	52.9[d]	32.9	—	—	12.6	—	32.1	—	—	[24]
Phyllostachys flexuosa	Bamboo	GA	57.0[d]	35.5	—	—	6.1	—	26.1	—	—	[24]
Phyllostachys lithophila	Bamboo	Taiwan	61.4[d]	39.0	—	—	4.0	—	22.6	—	—	[22]
Phyllostachys makinoi	Bamboo	Taiwan	59.2[d]	38.7	—	—	2. 9	—	23.6	—	—	[22]
Phyllostachys meyeri	Bamboo	GA	57.8[d]	36.9	—	—	5.8	—	25.5	—	—	[24]
Phyllostachys nidularia *cv.* smooth sheath	Bamboo	GA	53.1[d]	33.0	—	—	13.4	—	31.3	—	—	[24]
Phyllostachys nigra	Bamboo	GA	57.2[d]	36.4	—	—	6.0	—	29.5	—	—	[24]
Phyllostachys nigra *cv.* bory	Bamboo	GA	58.1[d]	36.7	—	—	6.2	—	26.9	—	—	[24]
Phyllostachys nuda	Bamboo	GA	56.6[d]	36.1	—	—	5.3	—	25.6	—	—	[22]
Phyllostachys pubescens	Bamboo	GA	57.2[d]	34.1	—	—	6.4	—	28.7	—	—	[24]
Phyllostachys purpurata	Bamboo	GA	53.4[d]	35.7	—	—	8.5	—	30.9	—	—	[24]
Phyllostachys rubromarginata	Bamboo	GA	56.1[d]	35.6	—	—	8.4	—	27.4	—	—	[24]
Phyllostachys viridiglaucescens	Bamboo	GA	57.1[d]	34.9	—	—	8.3	—	28.6	—	—	[24]

Table 5.4 Chemical Properties of Agro-based Fibers

Fibers		Variety/or	Cross				Solubility					
		Place of	& Bevan	α-		Pento-	Alcohol	Hot	1%			
Botanical name	Common name	growth	Cellulose	Cellulose	Lignin	sans	Benzene	Water	NaOH	Ash	Others	Ref.
Phyllostachys viridis	Bamboo	GA	54.5[d]	34.1	—	—	9.6	—	29.8	—	—	[24]
Phyllostachys vivax	Bamboo	Israel	56.7[d]	36.0	—	—	7.8	—	25.9	—	—	[24]
Saccharum biflorum		S. Africa	48.9[d]	32.3	—	—	6.8	—	43.8	—	—	[22]
Saccharum officinanum	Sugar cane or	FL	47.1[d]	29.9	—	—	10.2	—	41.5	—	—	[22]
Saccharum officinanum	Bagasse	HI	49.4	31.6	17.8	27.6	8.0	12.9	41.0	1.8	—	[3]
Saccharum officinanum	"	LA	53.2	32.4	20.8	30.3	3.1	4.0	32.2	3.3	—	[3]
Saccharum officinanum	"	FL	53.5	—	18.9	30.0	6.0	8.8	35.9	2.4	—	[16]
Saccharum officinanum	"	HI	52.0	—	18.1	27.9	10.8	11.2	39.9	2.2	—	[16]
Saccharum officinanum	"	HI	50.2	—	21.3	27.7	3.2	5.7	33.9	5.4	—	[16]
Saccharum officinanum	"	PR	55.0	—	19.3	31.3	3.6	4.0	31.3	2.6	—	[16]
Saccharum officinanum	"	Mexico	50.9	—	18.1	29.6	5.4	8.0	27.3	3.9	—	[16]
Saccharum officinanum	"	Philippines	46.0	—	22.4	29.9	2.3	7.6	40.1	4.9	—	[16]
Saccharum officinanum	"	India	56.8	—	22.3	31.8	3.0	2.8	31.1	2.3	—	[16]
Saccharum officinanum	"		43[c]	—	23.0	27.5	—	25-30	—	2-4	—	[27]
Saccharum sp.	"	FL	49.0[d]	31.0	—	—	21.8	—	45.5	—	—	[24]
Secale cereale	Rye	America	50-54	33-35	16-19	27-30	—	—	—	2-5	.5-4[s]	[17]
Secale cereale	Rye	Europe	—	37.6	19.0	30.5	3.2	9.4	—	4.3t	—	[20]
Secale cereale	Rye		75.5[h]	39.3	17.8	27.9	—	8.8	—	3.8	3.5[p]	[20]
Secale cereale	Rye	Europe	—	37.2	17.6	25.7	2.9	12.0	—	1.2i	2.1[a], 1.0[u]	[20]
Setaria italica	Foxtail millet	NE	41.0[d]	25.3	—	—	2.6	—	52.6	—	—	[22]
Setaria sphacelata	—	S. Africa	48.1[d]	30.9	—	—	10.4	—	42.4	—	—	[22]
Setaria verticillata	—	MD	39.4[d]	17.5	—	—	8.5	—	43.0	—	—	[22]
Semiarundinaria munielae	Bamboo	Holland	60.7[d]	35.6	—	—	4.5	—	28.6	—	—	[22]
Sorghum almum	Sorghum	(mean)	46.2[cc]	29.1	15.8	23.0	14.3	—	47.5	5.7	—	[9]
Sorghum caffrorum	Sorghum		46.2[cc]	28.8	14.7	27.0	12.2	—	48.3	5.4	—	[9]
Sorghum caudatum	Sorghum		47.2[cc]	29.7	16.0	27.4	8.1	—	47.0	6.2	—	[9]
Sorghum drummondii	Sorghum		44.3[cc]	27.5	—	—	13.7	—	51.0	—	—	[9]
Sorghum nervosum	Sorghum		44.5[cc]	27.9	13.6	26.8	9.6	—	51.0	5.8	—	[9]

Sorghum subglabrecens	Sorghum		47.0[cc]	29.9	15.0	24.7	11.1	—	46.7	5.0	—	[9]
Sorghum technicum	Sorghum		44.9[cc]	28.0	—	—	10.8	—	52.8	—	—	[9]
Sorghum	Sorghum	All others	48.4[cc]	30.1	14.6	27.4	11.2	—	46.9	5.4	—	[9]
Sorghum	Sorghum		43.7[cc]	27.0	—	—	14.4	—	50.0	—	—	[9]
Sorghum durra	Sorghum		46.9[cc]	29.9	15.0	28.0	10.9	—	48.5	5.5	—	[9]
Sorghum helpense	Sorghum		45.8[cc]	29.3	12.1	26.5	9.2	—	46.4	9.6	—	[9]
Sorghum sudanese	Sorghum	Holland	44.5[cc]	28.0	13.8	—	12.2	—	49.2	6.3	—	[9]
Sorghum almum	—	IL	54.9[d]	36.3	—	—	12.7	—	40.8	—	—	[23]
Sorghum vulgare (grain)	—	TX	46.0[d]	29.4	—	—	11.9	—	49.6	—	—	[23]
Sorghum vulgare (broom corn)	—	TX	52.1[d]	32.5	—	—	8.2	—	44.3	—	—	[23]
Sorghum halepense (forage)	—	MD	52.0[d]	33.2	—	—	74	—	41.7	—	—	[23]
Sorghastrum nutans	Wood grass	NE	50.5[d]	31.4	—	—	9.0	—	43.5	—	—	[22]
Spartina cynosuroides	Marsh grass	VA	39.8[d]	25.0	—	—	9.0	—	50.1	—	—	[22]
Spartina pectinata	Marsh grass	NE	42.0[d]	27.3	—	—	13.1	—	50.4	—	—	[22]
Sporobolus cryptandrus	Rush grass	OK	43.6[d]	28.9	—	—	8.6	—	49.8	—	—	[22]
Sporobolus fimbriatus	Rush grass	S. Africa	44.6	26.3	—	—	8.6	—	46.0	—	—	[22]
Stipa coronata	Esparto	CA	44.6[d]	26.4	—	—	8.0	—	48.2	—	—	[23]
Stipa speciosa	Esparto	CA	45.8[d]	27.8	—	—	7.8	—	44.4	—	—	[23]
Stipa splendens	Esparto	NE	42.2[d]	27.2	—	—	10.6	—	46.5	—	—	[23]
Stipa tenacissima	Esparto	Spain	60.5[d]	47.8	—	—	4.2	—	33.1	—	—	[23]
Stipa viridula	Esparto	ND	47.2[d]	29.4	—	—	8.6	—	43.5	—	—	[23]
Stipa sp.	Esparto	—	50-54*	33-38	17-19	27-32	—	—	—	6-8	2-3[s]	[17]
Stipa sp.	Esparto	—	60.0[cc]	48.0	14-16	22-28	4.2	—	33.0	2.8-7.7	—	[10]
Trispeacum dactyloides	—	MD	39.6[d]	25.5	—	—	14.7	—	49.0	—	—	[23]
Triticum sp.	Wheat	America	73.7*	34.7	16.3	26.8	—	12.6	—	9.1	3.1[p]	[20]
Triticum sp.	Wheat	—	—	39.9	16.7	28.2	3.7	7.4	—	6.6[t]	—	[20]
Triticum sp.	Wheat	Holland	49-54	29-35	16-21	16-21	—	—	—	4-9	3—7[s]	[17]
Triticum sp.	Wheat	India	—	33.3	15.6	24.6	7.8	12.5	—	7.8[i]	2.0[a], 1.0[u]	[20]
Triticum sp.	Wheat	—	51.5[c]	—	21.5	23.5	—	14-16.5	—	15-18	—	[27]
Zea mays	Corn	IL	50.4[d]	34.2	—	—	12.1	—	46.7	—	—	[23]
Zea mays, stalk	Corn	Baroni	45.5[c]	—	16.7	27.1	7.0	14.9	47.6	7.0	6.9[s]	[20]

Table 5.4 Chemical Properties of Agro-based Fibers

Fibers		Variety/or	Cross				Solubility					
		Place of	& Bevan	α-		Pento-	Alcohol	Hot	1%			
Botanical name	Common name	growth	Cellulose	Cellulose	Lignin	sans	Benzene	Water	NaOH	Ash	Others	Ref.
Zea mays	Corn	—	35.4[c]	—	34.0	—	—	—	—	3.6	—	[20]
Zea mays, stalk	Corn	—	50.0[cc]	34.0	5.0	20.0	—	—	—	—	—	[7]
Zea mays, corncobs	Corn	Timell	40.0[cc]	32.0	14.0	41.0	—	—	—	—	—	[7]
Zea mays	Corn	Europe	46.5[c]	—	14.0	—	—	—	—	1.2	—	[20]
Juncaceae												
Juncus acutus	—	CA	36.6[d]	25.4	—	—	16.2	—	54.8	—	—	[24]
Juncus xiphioides	—	CA	43.5[d]	30.0	—	—	11.2	—	48.1	—	—	[24]
Liliaceae												
Asparagus officinalis	—	MD	43.3[d]	25.2	—	—	14.0	—	42.0	—	—	[24]
Musaceae												
Musa sp.	Banana	India	—	61.5	9.7	14.9	—	—	—	4.8	2.8[a]	[25]
Musa sp.	Banana	India	46.7[c]	—	14.0	18.0	—	12-20	—	10.0	—	[27]
Musa sp.	Banana	India	88.7[h]	61.5	9.7	14.9	1.4	28.6	26.0	4.8	1.6[o], 1.6[k]	[15]
Palmae												
Cocos nucifera	Coconut coir	—	60.2	29.8	34.8	19.3	—	10.2	24.4	3.9	0.6[s], 3.4[e]	[11]
Cocos nucifera	Coconut coir	India	43[c]	—	45.0	—	—	—	—	—	—	[26]
Cocos nucifera, pith	Coconut coir	Phillipines	35[c]	—	25.2	7.5	—	3.5-6	—	10.2	—	[27]
Cocos nucifera, husk dust	Coconut coir		44.4[h]	—	29.3	16.9	6.7	16.3	45.2	3.3	0.4[s]	[13]
Cocos nucifera	Coconut coir		51.4[h]	—	39.2	11.0	2.0	2.4	44.4	4.9	0.5[s]	[13]
Cocos nucifera, trunk	Coconut coir		62.3[h]	—	30.9	9.6	1.7	2.4	25.5	2.7	0.9[s]	[13]
Cocos nucifera, petiole	Coconut coir		66.7[h]	—	25.1	22.9	2.6	2.8	22.9	2.8	0.2[s]	[13]
Cocos nucifera, leaves	Coconut coir		64.4[h]	—	16.2	18.6	6.6	8.1	28.6	4.7	2.2[s]	[13]
Cocos nucifera	Coconut coir		47.1[h]	—	27.7	11.6	7.6	9.5	47.2	8.1	4.3[s]	[13]
Sabal texana	Mexican palm	TX	25.1[d]	14.9			12.3		44.6			[22]
Papavearceae												
Papaver sp.	Poppy straw	America	78.0[h]	42.2	20.0	27.6	1.7	19.1	—	—	—	[6]
Restionaceae												

Wildenowia striata	—	S. Africa	44.9[d]	23.1	—	—	8.4	—	44.7	—	—	[22]
Typhaceae												
Typha angustifolia	—	CA	40.7[d]	28.1	—	—	9.4	—	48.4	—	—	[22]
Dicotyledoneae												
Amaranthaceae												
Amaranthus graecizans	Tumbleweed	MD	39.0[d]	25.3	—	—	5.8	—	43.2	—	—	[22]
Amaranthus hybridus	Pigweed	NE	38.4[d]	24.9	—	—	11.4	—	50.0	—	—	[22]
Amaranthus palmeri	—	MD	36.9[d]	24.3	—	—	10.4	—	48.7	—	—	[24]
Amaranthus retroflexus	Redweed	NC	41.2[d]	26.8	—	—	7.0	—	46.4	—	—	[24]
Apcynaceae												
Apocynum cannabinum	Indian hemp	MD	45.3[d]	29.5	—	—	9.9	—	36.8	—	—	[23]
Asclepiadaceae												
Asclepias syriaca	—	NE	46.1[d]	31.2	—	—	8.2	—	39.4	—	—	[23]
Asclepias tuberosa	Butterfly weed	NJ	40.8[d]	27.2	—	—	11.5	—	44.5	—	—	[24]
Asclepias incarnata	—	MD	44.5[d]	29.9	—	—	11.2	—	38.6	—	—	[24]
Caprifoliaceae												
Sambucus canadensis	American elder	NE	44.8[d]	27.9	—	—	9.5	—	31.5	—	—	[23]
Sambucus canadensis	American elder	MD	45.9[d]	28.3	—	—	5.5	—	30.2			[22]
Bombacaceae												
Ceiba petandra	Kapok	—	64.0[c]	—	13.0	—	—	—	—	—	23.0[k]	[19]
Boraginaceae												
Echium vulare	Blueweed	CT	36.6[d]	20.0	—	—	3.2	—	44.2	—	—	[24]
Lithostpermum arvense	—	MD	33.8[d]	21.5	—	—	10.2	—	53.6	—	—	[24]
Campanulaceae												
Lobelia cardinalis	Cardinal flower	VA	44.1[d]	28.3	—	—	11.2	—	49.9	—	—	[22]
Specularia perfoliata	—	MD	37.2[d]	22.6	—	—	10.7	—	18.1	—	—	[22]

Table 5.4 Chemical Properties of Agro-based Fibers

Fibers		Variety/or	Cross				Solubility					
Botanical name	Common name	Place of growth	& Bevan Cellulose	α-Cellulose	Lignin	Pento-sans	Alcohol Benzene	Hot Water	1% NaOH	Ash	Others	Ref.
Capparidaceae												
Cleome spinosa	—	MD	46.5[d]	29.9	—	—	6.9	—	39.9	—	—	[24]
Cleome serrulata	Stinking clover	NE	53.2[d]	33.2	—	—	4.4	—	30.6	—	—	[24]
Caryophyllaceae												
Agroustemma githago	Corn cockle	MD	44.1[d]	28.3	—	—	7.6	—	41.5	—	—	[24]
Arenaria serpyllifolia	Thyme—eaves	MD	32.0[d]	17.6	—	—	7.8	—	43.7	—	—	[22]
Dianthus ameria	—	MD	38.8[d]	23.8	—	—	11.7	—	47.6	—	—	[24]
Saponaria officinalis	Bouncing bet	CT	53.9[d]	35.3	—	—	12.3	—	52.9	—	—	[22]
Silene antirrhina	Sleepy catchfly	MD	39.2[d]	23.1	—	—	11.0	—	49.2	—	—	[22]
Silene noctiflola	Sticky cockle	CT	41.0[d]	24.3	—	—	3.9	—	42.7	—	—	[22]
Chenopodiaceae												
Atriplex mulleri	Saltbush	GA	45.4[d]	27.2	—	—	6.2	—	29.4	—	—	[24]
Atripex patula var. hastuta	Saltbush	MD	41.3d	25.5	—	—	8.9	—	45.5	—	—	[22]
Chenopodium ambrosioides	Mexican tea	VA	49.5d	31.1	—	—	3.6	—	33.9	—	—	[24]
Chenopodium album	Pigweed	NC	44.1d	27.7	—	—	7.6	—	37.0	—	—	[24]
Kochia scopria	Belvedere	KS	51.8d	32.8	21.4	—	2.2	—	29.9	—	—	[24]
Spinacea oleracea	Spinach	MD	35.3d	21.4	—	—	5.5	—	50.7	—	—	[22]
Cistaceae												
Lechea maritima	Leggett	DE	34.8d	20.0	—	—	6.7	—	38.9	—	—	[22]
Compositae												
Achilea millefolium	Common yarrow	CA	44.2d	28.0	—	—	8.0	—	29.7	—	—	[22]
Agoseris apargiodes	Mountain dandlion	CA	44.2d	28.0	—	—	6.7	—	35.9	—	—	[22]
Ambrosia artemisifolia	Common ragweed	CT	44.9d	28.0	—	—	8.6	—	39.1	—	v	[23]
Ambrosia psilostachya	—	KS	40.8d	25.2	—	—	7.0	—	43.4	—	—	[24]
Ambrosia trifida	Great ragweed	NE	48.4d	30.3	—	—	3.4	—	32.9	—	—	[23]
Ambrosia trifida	Great ragweed	TX	48.9d	30.9	—	—	2.3	—	32.1	—	v	[22]

Anaphalis margaritacea	Pearly everlasting	VA	44.3d	27.6	—	—	9.7	—	38.1	—	v	[24]
Arctium nemorosum	Petiole hollow	MD	41.7d	27.0	—	—	11.7	—	43.5	v	—	[24]
Artemisia vulgaris	Common mugwort	KS	42.2d	26.1	—	v	9.9	—	40.7	—	—	[24]
Artemisia sp.	Wormwood	VA	48.3d	31.1	—	—	10.7	—	34.7	—	—	[24]
Aster sp.	Starwort	DE	47.0d	30.2	—	—	5.6	—	37.3	—	—	[24]
Athemis colula	Mayweed	MD	44.6d	27.6	—	—	13.4	—	41.1	—	v	[[22]
Baccharis glutinosa	Seep willow	TX	40.7d	25.9	—	—	8.3	—	30.9	v	—	[22]
Baccharis sarothroides	—	AZ	39.2d	22.9	—	—	12.8	—	43.7	—	v	[22]
Isamorhiza annus seed nuts	Sunflower	—	32.0cc	30.0	27.0	27.0	—	v	—	—	—	[7]
Bidens frondosa	Beggar-ticks	CT	40.6d	26.3	—	—	18.9	—	44.9	—	—	[24]
Centaurea cyanus	Blue bottle	DE	44.2d	27.7	—	v	7.4	—	42.3	—	—	[22]
Centaurea cyanus	—	MD	46.4d	27.7	—	—	8.6	—	35.2	—	—	[24]
Cichorium intybus	Common chicory	MD	45.0d	29.6	—	—	6.3	—	40.7	—	—	[22]
Cirsium discolor	—	MD	42.1d	27.2	—	—	7.5	v	44.7	v	v	[23]
Chrysanthemum leucanthemum	Oxeye daisy	CT	40.3d	25.0	—	v	12.1	—	45.0	—	—	[24]
Erigeron annuus	White-top	MD	47.4d	29.6	—	v	8.2	—	38.3	—	—	[24]
Erigeron canadensis	Horse weed	VA	31.3d	31.9	—	—	5.5	—	32.0	—	—	[24]
Erigeron tweedya	—	Uruguay	44.0d	28.4	—	—	4.4	—	43.6	—	v	[22]
Eupatorum dubium or puroureum	Joe-Pye weed	MD	49.2d	33.0	—	v	7.8	—	31.9	—	—	[24]
Eupatorium hyssopifolium	—	ND	48.2d	31.8	—	—	6.2	—	28.0	—	v	[24]
Eupatorium laevigatum	—	Uruguay	53.4d	36.1	—	—	2.3	—	23.8	—	—	[22]
Eupatorium perfoliatum	Thoroughwort	Uruguay	50.6d	33.6	—	—	6.9	—	31.2	v	—	[22]
Eupatorium perfoliatum	Thoroughwort	MD	50.0d	32.0	—	v	4.2	v	31.4	v	v	[24]
Eupatorium serotinum	—	MD	45.8d	28.8	—	—	9.1	—	42.2	—	v	[24]
Eupatorium urticasfolium	—	CT	46.5d	28.8	—	—	10.6	—	42.6	—	—	[24]
Grindelia stricta ssp. Venulosa	Gumweed	CA	33.7d	20.8	—	v	19.3	—	54.6	—	—	[22]
Gutierrezia sarothrae	—	TX	24.5d	14.4	—	—	27.3	—	61.3	—	—	[22]
Heliopsis Laevis	—	MD	45.9d	29.3	—	—	9.0	—	38.8	—	v	[24]
Heterotheca subaxillaris	Camphorweed	DE	35.1d	22.0	—	—	22.7	—	55.1	—	v	[24]
Haplopappus ciliatus	—	KS	50.5d	32.4	—	—	6.4	v	30.1	v	v	[22]

Table 5.4 Chemical Properties of Agro-based Fibers

Fibers		Variety/or	Cross				Solubility					
		Place of	& Bevan	α-		Pento-	Alcohol	Hot	1%			
Botanical name	Common name	growth	Cellulose	Cellulose	Lignin	sans	Benzene	Water	NaOH	Ash	Others	Ref.
Helenium tenuifolium	Sneezweed	VA	40.1d	22.8	—	—	8.3	—	37.1	—	—	[22]
Helianthus annuus	Common sunflower	NE	46.9d	3n.s	—	—	7.2	—	36.9	—	—	[23]
Helianthus grosserratus	Sunflower	KS	46.2d	29.4	—	—	6.5	—	39.2	—	—	[23]
Helianthus maximiliani	Sunflower	KS	42.4d	26.8	—	—	10.1	—	40.9	—	—	[23]
Helianthus rigidis	Sunflower	KS	37.3d	23.3	—	—	13.9	—	51.3	—	—	[23]
Helianthus salicifolius	Sunflower	KS	37.2	29.9	—	—	17.4	—	49.3	—	—	[23]
Helianthus scaberrimus	Sunflower	NE	44.4d	28.1	—	—	11.0	—	40.7	—	—	[22]
Helianthus tuberosus	Sunflower	NE	48.5d	31.1	—	—	6.4	—	38.7	—	v	[23]
Helianthus tuberosus	Sunflower	IA	34.0d	22.1	—	—	21.6	—	55.1	—	—	[22]
Hulsea heterochroma	—	CA	32.6d	19.7	—	—	23.8	—	57.1	—	v	[22]
Hymenopappus sp.	Old-plainsman	KS	40.0d	23.9	—	—	9.2	—	44.2	—	—	[22]
Iva zanthifolia	—	NE	47.3d	30.0	—	—	5.5	—	36.9	—	—	[22]
Lactuca scaroila	—	MD	45.8d	30.0	—	—	11.8	—	41.4	—	—	[24]
Liatris punctata	—	KS	41.6d	24.1	—	—	6.9	—	43.0	—	—	[24]
Lactuca canadensis	Milkweed	MS	42.3d	29.4	—	—	11.3	—	47.1	—	—	[23]
Pluchea foetida	Stinking Fleabane	DE	47.2d	29.6	—	—	6.1	—	35.9	—	—	[22]
Ratibida columnifera	—	TX	48.0d	31.5	—	—	4.2	—	29.7	—	—	[22]
Rudbeckia serotina	—	MDS	39.1d	23.9	—	—	10.2	—	39.2	—	—	[22]
Senecio braisliensis	—	Uruguay	47.3d	30.6	—	—	3.2	—	36.3	—	—	[22]
Silphium intgrifolium	—	KS	43.3d	27.0	—	—	8.3	—	39.3	—	—	[23]
Silphium laciniatum	Rosinweed	KS	39.7d	25.1	—	—	10.1	—	45.6	—	—	[23]
Solidago gigantea	Goldenrod	CT	48.7d	31.5	—	—	7.9	—	34.3	—	—	[23]
Solidago graminifolia	Goldenrod	CT	38.2d	20.6	—	—	7.0	—	34.0	—	—	[22]
Solidago rugosa	Goldenrod	CT	53.7d	34.2	—	—	4.2	—	30.0	—	—	[22]
Solidago sempervirens	Goldenrod	MD	44.3d	27.1	—	—	7.9	—	39.3	—	—	[23]
Sonchus oleraceus	Sow-thistle	MD	38.9d	26.0	—	—	13.6	—	51.7	—	—	[22]

Tagetes patula	Marigold	MD	40.8d	26.0	—	—	7.3	—	42.4	—	—	[22]
Tragopogon pratensis	Goat's-beard	MD	39.8d	24.3	—	—	9.9	—	45.6	—	—	[22]
Verbesina occidenalis	Crown-beard	MD	42.2d	27.0	—	—	10.7	—	42.4	—	—	[22]
Veronia baldwini	Ironweed	NE	42.0d	26.3	—	—	11.7	—	37.8	—	—	[22]
Vernonia novelboracensis	Ironweed	MD	44.2d	27.7	—	—	8.7	—	36.7	—	—	[22]
Wyethia angustifolea	—	CA	42.0d	25.8	—	—	9.5	—	46.4	—	—	[22]
Wyethia helenioides	—	CA	34.4d	21.2	—	—	13.2	—	55.4	—	—	[22]
Xanthium pensylvanicum	—	MD	42.2d	26.9	—	—	5.7	—	39.6	—	—	[22]
Zinnia elegans	Zinnia	MD	43.8d	24.8	—	—	3.3	—	33.0	—	—	[22]
Cruciferae												
Berteroa incana	—	CT	46.0d	27.1	—	—	5.5	—	38.8	—	—	[24]
Brassica nigra	—	MD	37.7d	22.9	—	—	10.4	—	50.7	—	—	[24]
Brassica cf. nigra	—	CA	49.9d	30.3	—	—	4.5	—	33.2	—	—	[24]
Brassica rapa	—	MD	50.8d	31.8	—	—	3.5	—	30.6	—	—	[24]
Cakile edentula	—	MD	28.5d	17.5	—	—	15.4	—	57.9	—	—	[24]
Erysimum officinale	—	MD	40.3d	24.7	—	—	6.9	—	46.6	—	—	[24]
Lepidium virginiacum	—	MD	47.1d	28.0	—	—	3.8	—	32.8	—	—	[24]
Raphaus raphanistrum	—	MD	41.4d	25.9	—	—	7.6	—	46.0	—	—	[24]
Sisyembrium irio	—	TX	36.2d	22.2	—	—	7.9	—	52.7	—	—	[22]
Dipsacaceae												
Dipsacus sylvestri	—	MD	43.7d	28.5	—	—	8.8	—	41.1	—	—	[24]
Euphorbiaceae												
Acalypha viginica	—	MD	44.0d	27.8	—	—	15.4	—	47.4	—	—	[23]
Croton glandulosus	—	ND	47.0d	30.3	—	—	7.2	—	38.4	—	—	[24]
Croton texensis	—	KS	42.6d	27.8	—	—	10.6	—	43.2	—	—	[24]
Euphorbia maculanta	—	MD	42.0d	25.2	—	—	13.1	—	40.9	—	—	[24]
Euphorbia maginata	—	NE	49.9[d]	32. 5	—	—	4.6	—	36.2	—	—	[24]
Ricinus communis	—	NE	49.7[d]	34.0	—	—	2.5	—	32.7	—	—	[23]
Guttiferae												
Ascyrum hypericodes	—	VA	44.0[d]	26.7	—	—	9.7	—	36.6	—	—	[24]
Hypericum perforatum	—	KS	44.9[d]	27.1	—	—	9.2	—	35.0	—	—	[24]
Hypericum punctatum	—	MD	44.1[d]	27.2	—	—	10.3	—	39.7	—	—	[24]
Hydrophyllaceae												
Phacelia californica	—	CA	38.5[d]	23.7	—	—	8.7	—	50.1	—	—	[22]

Table 5.4 Chemical Properties of Agro-based Fibers

Fibers		Variety/or	Cross				Solubility					
Botanical name	Common name	Place of growth	& Bevan Cellulose	α-Cellulose	Lignin	Pentosans	Alcohol Benzene	Hot Water	1% NaOH	Ash	Others	Ref.
Labiatae												
Monarda citriodora	—	TX	50.1[d]	32.0	—	—	2.2	—	30.8	—	—	[22]
Monarda fistutulosa	—	MD	41.9[d]	26.8	—	—	11.3	—	43.0	—	—	[22]
Nepeta cataria	—	DE	45.0[d]	25.7	—	—	9.6	—	39.8	—	—	[22]
Salvia azure	—	KS	45.8[d]	28.0	—	—	6.4	—	38.6	—	—	[22]
Trichostema dichotomumt	—	MD	47.0[d]	28.5	—	—	10.9	—	39.0	—	—	[22]
Leguminosae												
Aeschynomene scabra	—	Mexico	48.2[d]	33.5	—	—	5.7	—	29.7	—	—	[22]
Alysicarpus rugosus	—	GA	40.5[d]	27.1	—	—	7.3	—	42.9	—	—	[24]
Alysicarpus vaginalis	—	GA	37.8[d]	25.3	—	—	9.8	—	44.7	—	—	[24]
Arachis hypogaea, hulls	Peanut	—	46.0[cc]	36.0	33.0	19.0	—	—	—	—	—	[7]
Astragalus cicer	—	KS	41.6[d]	28.2	—	—	11.8	—	47.9	—	—	[24]
Astragalus sp.	—	KS	38.8[d]	24.9	—	—	10.3	—	46.7	—	—	[24]
Astragalus sp.	—	NE	48.7[d]	32.4	—	—	7.0	—	36.0	—	—	[22]
Baptisia leucophaea	—	KS	44.0[d]	28.9	—	—	10.2	—	41.3	—	—	[22]
Baptisia minor	—	KS	49.7[d]	31.7	—	—	5.3	—	31.9	—	—	[24]
Baptisia tinctoria	—	MD	46.1[d]	29.6	—	—	13.1	—	38.8	—	—	[24]
Cassia fasciculata	—	IA	40.9[d]	20.5	—	—	12.1	—	44.3	—	—	[24]
Cassia marilandica	—	KS	49.5[d]	31.3	—	—	4.7	—	31.8	—	—	[22]
Cassia tora	—	NC	45.0[d]	28.0	—	—	14.5	—	44.5	—	—	[24]
Crotalaria eriocarpa	—	Mexico	52.9[d]	36.1	—	—	6.0	—	28.1	—	—	[24]
Crotalaria intermedia	—	GA	54.8[d]	38.1	—	—	3.5	—	32.0	—	—	[24]
Crotalaria incana	Sunn hemp	FL	51.0[d]	34.8	—	—	4.6	—	31.0	—	—	[22]
Crotalaria juncea	Sunn hemp	IL	54.1[d]	37.9	—	—	3.3	—	31.2	—	—	[23]
Crotalaria juncea (Brazilian)	Sunn hemp	IL	55.4[d]	39.2	—	—	3.3	—	28.7	—	—	[23]
Crotalaria juncea (ridged stem)	Sunn hemp	TX	54 .6[d]	36.3	—	—	5.1	—	29.3	—	—	[23]

Crotalaria juncea (smooth stem)	Sunn hemp	TX	53.7[d]	35.6	—	—	4.3	—	30.5	—	—	[24]
Crotalaria juncea	Sunn hemp	FL	—	78.3	—	—	—	—	—	0.3	1.5[a]	[25]
Crotalaria juncea	Sunn hemp	GA	54.9[cc]	38.7	—	3.6	4.5	—	29.5	—	—	[10]
Crotalaria juncea	Sunn hemp	IL	51.9[cc]	36.3	4.0	—	6.8	—	33.5	—	—	[10]
Crotalaria juncea	Sunn hemp	IN	53.3[cc]	36.9	—	—	3.3	—	31.4	—	—	[10]
Crotalaria juncea	Sunn hemp	LA	54.9[cc]	36.3	—	—	5.2	—	28.7	—	—	[10]
Crotalaria juncea	Sunn hemp	MI	60.5[cc]	41.5	—	—	1.9	—	23.2	—	—	[10]
Crotalaria juncea	Sunn hemp	MO	55.1[cc]	38.9	—	—	3.8	—	30.1	—	—	[10]
Crotalaria juncea	Sunn hemp	TX	54.6[cc]	37.6	—	—	4.7	—	29.3	—	—	[10]
Crotalaria juncea	Sunn hemp	—	53.7[cc]	35.6	—	—	3.6	—	29.1	—	—	[10]
Crotalaria mucronata	—	FL	55.3[d]	38.0	—	—	3.2	—	29.8	—	—	[22]
Crotalaria spectabilis	—	FL	40.6[d]	29.0	—	—	8.5	—	48.8	—	—	[22]
Crotalaria sp.	—	GA	49.6[d]	34.1	—	—	8.4	—	36.4	—	—	[24]
Crotalarta striata	—	FL	48.9[d]	34.2	—	—	5.4	—	38.4	—	—	[22]
Cyamopsis tetragonoloba	—	NE	49.6[d]	34.8	—	—	4.5	—	40.8	—	—	[24]
Dalea alopecuroides	Foxtail dalea	IA	47.2[d]	32.5	—	—	9.0	—	37.2	—	—	[22]
Dalea enneandra	—	KS	52.2[d]	35.9	—	—	7.2	—	32.0	—	—	[24]
Dalea deffusa	—	Mexico	52.2[d]	35.2	—	—	4.0	—	31.6	—	—	[22]
Dalea leporina	—	Mexico	50.8[d]	34.9	—	—	5.2	—	25.6	—	—	[22]
Dalea mutabies	—	Mexico	48.3[d]	33.3	—	—	7.0	—	34.6	—	—	[22]
Dalea vernicia	—	Mexico	45.7[d]	30.7	—	—	11.8	—	33.1	—	—	[22]
Desmanthus sp.	—	KS	46.8[d]	31.0	—	—	5.6	—	34.5	—	—	[23]
Desmanthus illinoensis	—	NE	49.0[d]	31.7	—	—	5.6	—	33.6	—	—	[23]
Desmanthus interior	—	Mexico	48.6[d]	31.8	—	—	4.9	—	33.6	—	—	[22]
Desmodium distortum	—	GA	46.5[d]	31.9	—	—	6.9	—	37.2	—	—	[24]
Desmodium gyrans	—	KS	43.9[d]	29.7	—	—	7.6	—	40.5	—	—	[22]
Desmodium illinoensis	—	KS	45.9[d]	31.1	—	—	8.0	—	40.8	—	—	[22]
Desmodium nicaraguense	—	Mexico	55.7[d]	40.2	—	—	3.2	—	28.5	—	—	[22]
Desmodium virgatum	—	GA	46.8[d]	30.1	—	—	9.3	—	36.0	—	—	[24]
Glycine max	Soy bean	IL	53.3[d]	34.8	—	—	3.1	—	31.1	—	—	[24]
Glycyrrhiza lepidota	Wild licorice	NE	42.1[d]	28.3	—	—	6.6	—	42.2	—	—	[24]
Indigofera hiruta	—	FL	52.8[d]	36.3	—	—	5.6	—	34.2	—	—	[24]

Table 5.4 Chemical Properties of Agro-based Fibers

Fibers		Variety/or	Cross				Solubility					
		Place of	& Bevan	α-		Pento-	Alcohol	Hot	1%			
Botanical name	Common name	growth	Cellulose	Cellulose	Lignin	sans	Benzene	Water	NaOH	Ash	Others	Ref.
Indigofera sp.	Indigo	GA	37.6[d]	24.3	—	—	6.2	—	41.6	—	—	[22]
Lespedeza capitata	—	NE	48.5[d]	31.5	—	—	4.4	—	29.3	—	—	[24]
Lespedeza hedysarioides	—	KS	42.8[d]	27.3	—	—	7.4	—	39.9	—	—	[24]
Lespedeza inschanica	—	KS	41.5[d]	26.9	—	—	8.4	—	42.9	—	—	[24]
Lespedeza sp.	Bush clover	GA	49.6[d]	34.6	—	—	6.4	—	32.3	—	—	[22]
Lotus scoparius	—	CA	34.8[d]	22.8	—	—	11.0	—	47.4	—	—	[22]
Lupinus formosus	—	CA	48.4[d]	32.6	—	—	9.6	—	35.6	—	—	[22]
Lupinus latifolius	—	CA	40.9[d]	25.9	—	—	8.7	—	47.0	—	—	[22]
Lupinus micranthus	—	CA	28.9[d]	18.3	—	—	12.0	—	59.8	—	—	[22]
Melilotus albus	—	IL	40.8[d]	26.4	—	—	12.2	—	44.7	—	—	[22]
Melilotus officinalis	Yellow melilot	MD	42.6[d]	27.0	—	—	10.1	—	42.8	—	—	[22]
Petalostemon mutiflorum	Prairie clover	KS	36.1[d]	23.3	—	—	9.9	—	47.9	—	—	[22]
Petalostemon purpureum	Prairie clover	KS	36.1[d]	23.3	—	—	9.9	—	47.9	—	—	[22]
Sesbania arabica	—	OK	51.8[d]	35.3	—	—	3.6	—	27.4	—	—	[22]
Sesbania cannabina	—	OK	50.7[d]	33.9	—	—	5.0	—	28.8	—	—	[22]
Sesbania cinerescens	—	GA	44.8[d]	30.3	—	—	6.4	—	38.8	—	—	[23]
Sesbania drummondii	—	TX	46.7[d]	31.6	—	—	5.3	—	31.8	—	—	[22]
Sesbania exaltata	Colorado River hemp	TX	51.3[d]	34.5	—	—	3.4	—	28.9	—	—	[22]
Sesbania sonorae	—	AZ	54.1[d]	36.3	—	—	3.3	—	27.8	—	—	[23]
Sesbania vesicaria	—	TX	50.2[d]	33.1	—	—	2.4	—	27.1	—	—	[22]
Sesbania sp.	—	TX	55.6[d]	40.2	—	—	2.1	—	23.9	—	—	[22]
Tephrosia virginiana	—	MD	45.1[d]	30.0	—	—	7.8	—	34.4	—	—	[22]
Tephrosia sp.	Horay pea	GA	42.8[d]	28.6	—	—	13.8	—	40.6	—	—	[22]
Swainsona salsula	—	CO	41.3[d]	27.3	—	—	8.4	—	36.4	—	—	[22]
Linaceae								—				
Linum usitatissimum	Flax	—	76-79	45-68	10-15	6-17	—	—	—	2-5	—	[17]
Linum usitatissimum	Flax	—	71.2[c]	—	2.2	—	—	—	—	—	2.0[k], 6.0[e]	[19]

Linum usitatissimum	Flax	IL	40.4[d]	27.2	—	—	8.9	—	39.6	—	—	[22]
Lythraceae												
Heimia salicifolia	—	Uruguay	46.4[d]	28.9	—	—	5.9	—	32.8	—	—	[22]
Malvaceae												
Abutilon americanum	—	Mexico	45.6[d]	28.8	—	—	4.6	—	31.6	—	—	[22]
Abutilon crispum	—	Mexico	48.0[d]	31.3	—	—	5.0	—	33.6	—	—	[22]
Abutilon theophrasti	—	MD	42.8[d]	27.6	—	—	9.4	—	43.6	—	—	[23]
Abutilon trisulcatum	—	Mexico	48.0[d]	32.1	—	—	4.8	—	30.0	—	—	[22]
Altaea rosea	—	MD	47.9[d]	30.5	—	—	6.5	—	36.8	—	—	[23]
Althaea cannabina	Hollyhock of gardens	Spain	44.3[d]	27.5	—	—	6.9	—	39.4	—	—	[22]
Althaea setosa	—	Israel	45.8[d]	28.8	—	—	4.3	—	38.5	—	—	[22]
Anoda pentaschista	—	Mexico	47.9[d]	30.7	—	—	5.2	—	36.8	—	—	[22]
Gossypium spp.	Cotton, staple	—	—	85-90	3-3.3	21-23	—	—	—	1-1.5	<1[s]	[17]
Gossypium spp.	Cotton, linters	Indonesia	—	80-85	3-3.5	—	—	—	—	1-2	<1[s]	[17]
Hibiscus cannabinus	Kenaf, hurds	IL	53.8[d]	34.7	15-18	—	3.7	—	30.9	—	—	[23]
Hibiscus cannabinus	Kenaf	FL	52.2[d]	34.0	10.5	21-23	3.4	—	29.4	—	—	[23]
Hibiscus cannabinus	Kenaf	—	47-57	31-39	12.1	18.3	—	—	—	2-5	—	[17]
Hibiscus cannabinus	Kenaf, stem	MD	53.1[cc]	36.5	13.2	22.7	4.3	11.2	33.0	—	—	[4]
Hibiscus cannabinus	Kenaf, stem	GA	58.0[cc]	40.2	7.7	19.7	3.3	7.4	28.4	—	—	[4]
Hibiscus cannabinus	Kenaf, whole	—	54.4[cc]	37.4	8.0	16.1	—	—	—	4.1	—	[28]
Hibiscus cannabinus	Kenaf, bast	—	57.2[cc]	42.2	17.4	16.0	—	—	—	5.5	—	[28]
Hibiscus cannabinus	Kenaf, core	—	51.2[cc]	33.7	13.4	19.0	—	—	—	2.9	—	[28]
Hibiscus cannabinus	Kenaf, bottom	—	53.9[d]	35.3	—	20.1	3.4	9.7	32.4	—	6[o]	[8]
Hibiscus cannabinus	Kenaf, top	—	46.4[d]	29.8	—	—	5.5	12.8	39.6	—	11.1[o]	[8]
Hibiscus cisplantinus	Kenaf	NC	46.1[d]	30.5	—	—	2.5	—	33.2	—	—	[22]
Hibiscus eelveldeanus	Kenaf	FL	51.9[d]	36.3	—	—	7.5	—	31.0	—	—	[22]
Hibiscus esculentus	Okra	NE	35.4[d]	23.3	—	—	5.0	—	45.6	—	—	[23]
Hibiscus esculentus (pods only	Okra	SC	52.8[d]	34.7	—	—	5.3	—	36.8	—	—	[23]
Hibiscus grandiflorus		FL	45.3[d]	31.6	—	—	5.2	—	29.1	—	—	[22]
Hibiscus militaris	Halberd-leaved rose mallow	Mexico	43.9[d]	28.9	—	—	8.3	—	38.0	—	—	[22]

Table 5.4 Chemical Properties of Agro-based Fibers

Fibers		Variety/or	Cross				Solubility					
Botanical name	Common name	Place of growth	& Bevan Cellulose	α-Cellulose	Lignin	Pento-sans	Alcohol Benzene	Hot Water	1% NaOH	Ash	Others	Ref.
Hibiscus lasiocarpus	—	Mexico	50.1[d]	33.2	—	—	2.8	—	30.1	—	—	[22]
Hibiscus rosa—inensis	Chinese rose	FL	44.2[d]	29.4	—	—	9.2	—	36.9	—	—	[22]
Hibiscus sabdariffa	Roselle	CA	48.6[d]	32.3	—	—	9.1	—	37.8	—	—	[23]
Hibiscus syriacus	Rose of Sharon	MD	36.8[d]	22.9	—	—	5.7	—	41.0	—	—	[22]
Hibiscus trionum	Flower of an Hour	MD	38.1[d]	23.1	—	—	6.8	—	39.8	—	—	[22]
Hibiscus sp.	Rose mallow	Uruguay	46.5[d]	32.2	—	—	4.6	—	38.8	—	—	[22]
Horsfordia newberryi	—	Mexico	50.3[d]	34.0	—	—	5.9	—	27.6	—	—	[22]
Kosteletzkya althacifolia	—	FL	46.7[d]	31.4	—	—	4.4	—	31.8	—	—	[22]
Kosteletzkya sagittata	—	Mexico	42.8[d]	28.5	—	—	7.0	—	42.6	—	—	[22]
Lavatera arborea	Tree mallow	CA	48.1[d]	28.9	—	—	4.3	—	38.2	—	—	[22]
Lavatera punctata	—	Israel	38.7[d]	25.0	—	—	9.4	—	48.8	—	—	[22]
Lavatera rotundata	—	Spain	42.3[d]	25.5	—	—	7.9	—	41.5	—	—	[22]
Malachra alceaefolia	—	Mexico	45.9[d]	30.4	—	—	3.8	—	35.3	—	—	[22]
Malva rotundifolia	—	MD	42.7[d]	27.2	—	—	9.0	—	55.4	—	—	[22]
Malva sylvestris	—	Israel	29.6[d]	18.4	—	—	22.7	—	59.6	—	—	[22]
Malva tournefortiana	—	Spain	44.2[d]	27.5	—	—	7.5	—	43.7	—	—	[22]
Malvastrum sp.	—	S. Africa	49.8[d]	33.0	—	—	4.9	—	29.0	—	—	[22]
Pavonia xanthogloca	—	Urguay	45.0[d]	29.0	—	—	5.1	—	41.6	—	—	[22]
Sida acuta	—	Mexico	44.8[d]	28.6	—	—	5.0	—	33.6	—	—	[22]
Sida carpinifolia	—	GA	40.5[d]	24.5	—	—	7.8	—	40.4	—	—	[22]
Sida inflexa	—	NC	38.2[d]	23.3	—	—	5.1	—	40.2	—	—	[22]
Sida rhombifolia	—	Mexico	50.6[d]	32.6	—	—	2.4	—	25.3	—	—	[22]
Sida sp.	—	FL	51.2[d]	34.1	—	—	4.0	—	28.3	—	—	[22]
Sphaeralcea angustifolia	—	Mexico	48.6[d]	30.3	—	—	3.0	—	34.2	—	—	[22]
Sphaeralcea bonariensis	—	Uruguay	47.0[d]	29.9	—	—	4.3	—	30.9	—	—	[22]
Sphaeralcea coccinea	Prairie mallow	NM	29.0[d]	16.4	—	—	8.6	—	53.8	—	—	[22]
Sphaeralcea emoryi	—	—	42.6[d]	26.3	—	—	6.0	—	40.7	—	—	[22]

Sphaeralcea sp.	Flase mallow	CA	47.8[d]	30.2	—	—	7.4	—	32.0	—	—	[22]
Urena lobata	—	FL	52.4[d]	34.4	—	—	5.6	—	35.2	—	—	[22]
Wissadula amplissima	—	Mexico	49.5[d]	33.3	—	—	4.7	—	29.9	—	—	[22]
Wissadula cineta	—	Mexico	50.0[d]	34.5	—	—	5.1	—	25.0	—	—	[22]
Moraceae												
Cannabis sativa	Hemp	—	74.9[c]	—	3.7	—	—	—	—	—	.9[k], 3.1[e]	[19]
Cannabis sativa	Hemp	IL	56.2[d]	37.6	—	—	4.0	—	30.1	—	—	[23]
Onagraceae												
Gaura parviflora	—	NE	42.6[d]	26.9	—	—	12.8	—	46.1	—	—	[24]
Oenothera affinis	—	Uruguay	53.0[d]	38.7	—	—	4.1	—	31.3	—	—	[22]
Oenothera biennis	Evening primrose	CT	47.2[d]	31.4	—	—	3.0	—	41.6	—	—	[24]
Oenothera humifusa	—	MD	37.4[d]	23.5	—	—	10.0	—	46.4	—	—	[24]
Oenothera laciniata	—	MD	35.7[d]	23.2	—	—	8.0	—	44.8	—	—	[24]
Phytolaccaceae								—		—	—	
Phytolacca americana	Virginian Pokeweed	MD	42.4[d]	28.4	—	—	11.0	—	47.9	—	—	[22]
Polemonuaceae								—		—	—	
Phlox paniculata	—	MD	53.0[d]	35.4	—	—	2.9	—	28.6	—	—	[22]
Polygonaceae								—		—	—	
Eriogonum annuum	—	KS	43.2[d]	25.6	—	—	10.6	—	46.3	—	—	[22]
Eriogonum fasciculatum	California buckwheat	CA	51.8[d]	30.4	—	—	3.7	—	21.1	—	—	[22]
Polygonoceae								—		—	—	
Polygonum persicaria	—	CT	43.0[d]	29.0	—	—	7.7	—	34 .7	—	—	[23]
Polygonum orientale	Prince's feather	NE	46.6[d]	30.4	—	—	7.4	—	40.4	—	—	[22]
Rumex crispus	—	MD	47.1[d]	30.9	—	—	7.7	—	43.9	—	—	[22]
Ranunculaceae												
Thalictrum polycarpum	—	CA	49.5[d]	30.6	—	—	4.6	—	29.5	—	—	[22]
Rosaceae								—		—	—	
Potentilla fruiticosa	Golden hardhack	CT	43.6[d]	27.1	—	—	9.1	—	31.3	—	—	[24]
Petentilla norvegica	—	MD	39.1[d]	23.9	—	—	6.6	—	44.1	—	—	[22]

Table 5.4 Chemical Properties of Agro-based Fibers

Fibers		Variety/or	Cross				Solubility					
Botanical name	Common name	Place of growth	& Bevan Cellulose	α-Cellulose	Lignin	Pento-sans	Alcohol Benzene	Hot Water	1% NaOH	Ash	Others	Ref.
Rubiaceae												
Diodia teres	—	VA	38.6[d]	22.8	—	—	7.3	—	43.2	—	—	[24]
Scrophulariceae												
Gerardia flava	—	NJ	42.4[d]	27.0	—	—	4.9	—	40.1	—	—	[24]
Linaria canadensis	Toadflax	MD	41.0[d]	24.9	—	—	16.0	—	46.6	—	—	[24]
Mimulus guttatus	Common monkey flower	CA	32.5[d]	20.1	—	—	19.5	—	63.5	—	—	[24]
Mimulus ringens	Allegheny monkey flower	CT	30.4[d]	16.2	—	—	17.4	—	55.1	—	—	[24]
Penstemon digitalis	—	MD	41.2[d]	23.6	—	—	14.1	—	47.0	—	—	[24]
Penstemon palmeri	—	CA	38.8[d]	21.1	—	—	16.4	—	50.5	—	—	[24]
Scrophularia californica	—	CA	42.8[d]	26.4	—	—	11.2	—	45.8	—	—	[24]
Scrophularia marilandica	Carpenter's square	MD	47.5[d]	30.6	—	—	8.3	—	37.5	—	—	[24]
Veronica peregrinia	—	MD	33.5[d]	17.7	—	—	5.8	—	49.7	—	—	[24]
Verbascum blattaria	Moth mullein	MD	43.8[d]	26.5	—	—	13.6	—	43.3	—	—	[22]
Vervascum sinuatum	—	MD	49.0[d]	29.8	—	—	7.7	—	35.0	—	—	[22]
Solanaceae												
Datura stramonium	Stramonium	NE	49.9[d]	33.1	—	—	9.4	—	34.2	—	—	[24]
Thymelaeaceae												
Gnidia oppositifolia	—	S. Africa	52.5[d]	34.9	—	—	4.3	—	30.0	—	—	[22]
Tiliaceae												
Corchorus capsularis	Jute	FL	56.3[d]	39.1	—	—	3.5	—	28.6	—	—	[22]
Corchorus capsularis	Jute	—	57-58	—	21-26	18-21	—	—	—	.5-1	<1[s]	[17]

Corchorus capsularis	Jute	—	58.0	—	26.8	—	0.6	1.4	25.9	1.9	—	[20]
Corchorus capsularis	Jute	—	71.5[c]	—	8.1	21.6	—	—	—	—	.2[k], 1.8	[19]
Corchorus capsularis	Jute	India	—	60.7	12.5	—	0.9	—	—	0.8	3.5[a], 4.8[p]	[25]
Corchorus olitorius	Jute	India	—	61.0	13.2	15.6	—	—	—	0.5	2.9[a], 5.2[p]	[25]
Corchorus capsularis	Jute	India	—	58.9	13.5	15.9	—	—	—	0.5	2.8[a], 4.9[p]	[25]
Corchorus capsularis	Jute	India	57.6[c]	—	21.3	17.0	—	1.1-1.8	—	0.3-5	—	[27]
Corchorus capsularis	Jute	Bangladesh	21-24[m]	58-63	12-14	18.8	—	—	—	—	—	[21]
Umbelliferae												
Cicuta maculata	Water hemlock	MD	48.9[d]	31.6	—	—	4.3	—	34.4	—	—	[23]
Daucus carata	Wild carrot	MD	42.8[d]	28.0	—	—	9.7	—	45.9	—	—	[24]
Heracleum lanatum	—	MD	45.6[d]	30.4	—	—	5.9	—	36.2	—	—	[23]
Pastinaca sativa	Parsnip	MD	46.0[d]	29.8	—	—	7.0	—	40.2	—	—	[22]
Pituranthos tortuosa	—	Israel	42.6[d]	28.0	—	—	6.4	—	37.7	—	—	[22]
Urticaceae												
Boehmeria cylindrica	Bog hemp	NC	42.5[d]	28.8	—	—	9.0	—	42.0	—	—	[22]
Boehmeria nivea	Ramie	LA	53.5[d]	37.7	—	—	7.1	—	33.2	—	—	[22]
Boehmeria nivea	Ramie	India	—	86.9	0.5	—	—	—	—	1.1	0.6[a]	[25]
Boehmeria nivea	Ramie	—	76.2[c]	—	0.7	3.9	—	—	—	—	2.1[k], 6.4[e]	[19]
Laportea canadensis	—	NC	40.3[d]	28.6	—	—	12.9	—	47.0	—	—	[22]
Verbenceae												
Verbena hastata	—	CT	43.2[d]	25.5	—	—	4.6	—	40.1	—	—	[22]
Verbena urticifolia	White vervain	MD	47.3[d]	29.4	—	—	6.5	—	36.8	—	—	[22]
Other fibers												
Botanical name not given	Seed flax tow	India	47.0	34.0	23.0	25.0	—	—	—	2-5	—	[17]
Botanical name not given	Sabai	—	54.5	—	22.0	23.9	4.1	9.5	39.7	6.0	—	[11]
Botanical name not given	Sabai	—	54-57	—	17-22	18-24	—	—	—	5-7	3-4[s]	[17]
Botanical name not given	Mesta	India	—	60.0	10.1	14.8	—	—	—	0.7	4.8[a]	[25]
Botanical name not given	Roselle	India	—	59.7	9.9	15.0	—	—	—	0.5	4.8[a]	[25]
Botanical name not given	Dhaincha	India	—	63.6	16.3	9.8	—	—	—	0.7	1.2[a]	[25]
Botanical name not given	Bhindi	—	—	53.5	10.4	21.0	—	—	—	1.3	4.9[a]	[25]
Botanical name not given	Palmyrah	—	40-52[c]	—	42-43	—	—	—	—	—	—	[26]
Botanical name not given	Talipot	India	67-68[c]	—	28-29	—	—	—	—	—	—	[26]
Botanical name not given	Bhabar	India	53.8[c]	—	22.2	24.0	—	15-33	—	4.8-5.0	—	[27]

Table 5.4 Chemical Properties of Agro-based Fibers

Fibers		Variety/or Place of growth	Cross & Bevan Cellulose	α-Cellulose	Lignin	Pento-sans	Solubility			Ash	Others	Ref.
Botanical name	Common name						Alcohol Benzene	Hot Water	1% NaOH			
Botanical name not given	Groundnut husk	India	50.6[c]	—	30.6	11.1	—	5-11	—	2-5	—	[27]
Botanical name not given	Munj	Pakistan	58.2[c]	—	20.5	23.7	—	6-10	—	2-3	—	[27]
Botanical name not given	Bindi	—	81.0[c]	—	9.4	—	—	—	—	0.6	0.3[w]	[31]

[a] Acetyl
[c] Cellulose
[cc] Crude cellulose
[d] Monoethanoamine procedure
[e] Extractives
[h] Holocellulose
[i] Insoluble ash
[k] Pectins
[m] Hemicellulose
[o] Protein
[p] Polyuronide
[s] Silica
[t] Total ash
[u] Uronic acid
[w] Wax

Table 5.4 Chemical Properties of Agro-based Fibers

Botanical Name	Common Name	Holo-cellulose	Cross & Bevan Cellulose	α-cellulose	Pento-sans	Klason Lignin	Solubility 1% NaOH	Solubility Hot Water	Solubility EtOH/Benzene	Ether	Ash	Ref
Hardwoods												
Acer macrophyllum	Bigleaf maple	—	—	46.0	22.0	25.0	18.0	2.0	3.0	0.7	0.5	[33]
Acer negundo	Boxelder	—	—	45.0	20.0	30.0	10.0	—	—	0.4	—	[33]
Acer rubrum	Red maple	77.0	61.0	47.0	18.0	21.0	16.0	3.0	2.0	0.7	0.4	[33]
Acer saccharinum	Silver maple	—	56.0	42.0	19.0	21.0	21.0	4.0	3.0	0.6	—	[33]
Acer saccharum	Sugar maple	—	60.0	45.0	17.0	22.0	15.0	3.0	3.0	0.5	0.2	[33]
Alnus rubra	Red alder	74.0	—	44.0	20.0	24.0	16.0	3.0	2.0	0.5	0.3	[33]
Arbutus menziesii	Pacific madrone	—	—	44.0	23.0	21.0	23.0	5.0	7.0	0.4	0.7	[33]
Betula alleghaniensis	Yellow birch	73.0	64.0	47.0	23.0	21.0	16.0	2.0	2.0	1.2	0.7	[33]
Betula nigra	River birch	—	57.0	41.0	23.0	21.0	21.0	4.0	2.0	0.5	—	[33]
Betula papyrifera	Paper birch	78.0	63.0	45.0	23.0	18.0	17.0	2.0	3.0	1.4	0.3	[33]
Carya cordiformus	Bitternut hickory	—	56.0	44.0	19.0	25.0	16.0	5.0	4.0	0.5	—	[33]
Carya glaubra	Sweet Pignut hickory	71.0	—	49.0	17.0	24.0	17.0	5.0	4.0	0.4	0.8	[33]
Carya ovata	Shagbark hickory	71.0	—	48.0	18.0	21.0	18.0	5.0	3.0	0.4	0.6	[33]
Carya pallida	Sand hickory	69.0	—	50.0	17.0	23.0	18.0	7.0	4.0	0.4	1.0	[33]
Carya tomentosa	Mockernut hickory	71.0	—	48.0	18.0	21.0	17.0	5.0	4.0	0.4	0.6	[33]
Celtis laeoigata	Sugarberry	—	54.0	40.0	22.0	21.0	23.0	6.0	3.0	0.3	—	[33]
Eucalyptus gigantea	—	72.0	—	49.0	14.0	22.0	16.0	7.0	4.0	0.3	0.2	[33]
Fagus grandifolia	American beech	77.0	61.0	49.0	20.0	22.0	14.0	2.0	2.0	0.8	0.4	[33]
Fraxinus americana	White ash	—	51.0	41.0	15.0	26.0	16.0	7.0	5.0	0.5	—	[33]
Fraxinus pennsyloanica	Green ash	—	53.0	40.0	18.0	26.0	19.0	7.0	5.0	0.4	—	[33]
Gleditsia triacanthos	Honey locust	—	—	52.0	22.0	21.0	19.0	—	—	0.4	—	[33]
Laguncularia racemosa	White mangrove	—	52.0	40.0	19.0	23.0	29.0	15.0	6.0	2.1	—	[33]
Liquidambar styraciflua	Sweetgum	—	60.0	46.0	20.0	21.0	15.0	3.0	2.0	0.7	0.3	[33]
Liriodendron tulipifera	Yellow poplar	—	62.0	45.0	19.0	20.0	17.0	2.0	1.0	0.2	1.0	[33]

Table 5.4 Chemical Properties of Agro-based Fibers

Botanical Name	Common Name	Holo-cellulose	Cross & Bevan Cellulose	α-cellulose	Pento-sans	Klason Lignin	Solubility 1% NaOH	Solubility Hot Water	Solubility EtOH/Benzene	Ether	Ash	Ref
Lithocarpus densiflorus	Tanoak	71.0	—	46.0	20.0	19.0	20.0	5.0	3.0	0.4	0.7	[33]
Milalenca quinqueneroi	Cajeput	—	56.0	43.0	19.0	27.0	21.0	4.0	2.0	0.5	—	[33]
Nyssa aquatica	Water tupelo	59.0	45.0	—	16.0	24.0	16.0	4.0	3.0	0.6	0.6	[33]
Nyssa syloatica	Black tupelo	72.0	57.0	45.0	17.0	27.0	15.0	3.0	2.0	0.4	0.5	[33]
Populus alba	White poplar	67.0	52.0	—	23.0	16.0	20.0	4.0	5.0	0.9	—	[33]
Populus deletoides	Eastern cottonwood	—	64.0	47.0	18.0	23.0	15.0	2.0	2.0	0.8	0.4	[33]
Populus tremoides	Quaking aspen	78.0	65.0	49.0	19.0	19.0	18.0	3.0	3.0	1.2	0.4	[33]
Populus trichocarpa	Black cottonwood	—	—	49.0	19.0	21.0	18.0	3.0	3.0	0.7	0.5	[33]
Prunus serotina	Black cherry	85.0	60.0	45.0	20.0	21.0	18.0	4.0	5.0	0.9	0.1	[33]
Quercus alba	White oak	67.0	—	47.0	20.0	27.0	19.0	6.0	3.0	0.5	0.4	[33]
Quercus coccinea	Scarlet oak	63.0	—	46.0	18.0	28.0	20.0	6.0	3.0	0.4	—	[33]
Quercus douglasii	Blue oak	59.0	—	40.0	22.1	27.0	23.0	11.0	5.0	1.4	1.4	[33]
Quercus falcata	Southern red oak	69.0	—	42.0	20.0	25.0	17.0	6.0	4.0	0.3	0.4	[33]
Quercus kelloggii	California black oak	60.0	—	37.0	23.0	26.0	26.0	10.0	5.0	1.5	0.4	[33]
Quercus lobata	Valley oak	70.0	—	43.0	19.0	19.0	23.0	5.0	7.0	1.0	0.9	[33]
Quercus lyrata	Overcup oak	—	—	40.0	18.0	28.0	24.0	9.0	5.0	1.2	0.3	[33]
Quercus marylandica	Blackjack oak	—	57.0	44.0	20.0	26.0	15.0	5.0	4.0	0.6	—	[33]
Quercus prinus	Chestnut oak	76.0	—	47.0	19.0	24.0	21.0	7.0	5.0	0.6	0.4	[33]
Quercus rubra	Northern red oak	69.0	—	46.0	22.0	24.0	22.0	6.0	5.0	1.2	0.4	[33]
Quercus stellata	Post oak	—	55.0	41.0	18.0	24.0	21.0	8.0	4.0	0.5	1.2	[33]
Quercus velutina	Black oak	71.0	—	48.0	20.0	24.0	18.0	6.0	5.0	0.2	0.2	[33]
Salix nigra	Black willow	—	61.0	46.0	19.0	21.0	19.0	4.0	2.0	0.6	—	[33]
Tilia heterophylla	Basswood	77.0	65.0	48.0	17.0	20.0	20.0	2.0	4.0	2.1	0.7	[33]
Ulmus americana	American elm	73.0	61.0	50.0	17.0	22.0	16.0	3.0	2.0	0.5	0.4	[33]
Ulmus crassifolia	Cedar elm	—	—	50.0	19.0	27.0	14.0	—	—	0.3	—	[33]

Softwoods

Abies amabilis	Forbes/Pacific silver fir	—	61.0	44.0	10.0	29.0	11.0	3.0	3.0	0.7	0.4	[33]
Abies balsamea	Balsam fir	—	58.0	42.0	11.0	29.0	11.0	4.0	3.0	1.0	0.4	[33]
Abies concolor	White fir	66.0	—	49.0	6.0	28.0	13.0	5.0	2.0	0.3	0.4	[33]
Abies lasiocarpa	Subalpine fir	67.0	—	46.0	9.0	29.0	12.0	3.0	3.0	0.6	0.5	[33]
Abies procera	Noble fir	61.0	—	43.0	9.0	29.0	10.0	2.0	3.0	0.6	0.4	[33]
Chamaecyparis thyoides	Atlantic white cedar	—	53.0	41.0	9.0	33.0	16.0	3.0	6.0	2.4	—	[33]
Juniperus deppeana	Alligator juniper	57.0	—	40.0	5.0	34.0	16.0	3.0	7.0	2.4	0.3	[33]
Larix larcina	Tamarack	64.0	—	44.0	8.0	26.0	14.0	7.0	3.0	0.9	0.3	[33]
Larix occidentalis	Western larch	65.0	56.0	48.0	9.0	27.0	16.0	6.0	2.0	0.8	0.4	[33]
Libocedrus decurrens	Incense cedar	56.0	—	37.0	12.0	34.0	9.0	3.0	3.0	0.8	0.3	[33]
Picea enlgellmanni	Engelman spruce	69.0	60.0	45.0	10.0	28.0	11.0	2.0	2.0	1.1	0.2	[33]
Picea glauca	White spruce	—	61.0	43.0	13.0	29.0	12.0	3.0	2.0	1.1	0.3	[33]

REFERENCES

Amin, M. B., Maadhah, A. G., and Usmani, A. M., Natural vegetable fibers: A Status Report, *Renewable-Resource Materials New Polymer Sources*, Carraher, C. E. Jr. and Sperling, L.H., Eds., Plenum Press, New York, 29, 1985.

Atchison, J. E., Data on non-wood plant fibers, *Pulp and Paper Manufacture, Volume 1, Properties of Fibrous Raw Materials and their Preparation for Pulping*, M. J. Kocurek, Series Editor, TAPPI Press, Atlanta, GA, XVII, 1983.

Atchison, J. E., IV, Bagasse, *Pulp and Paper Manufacture, Volume 3, Secondary Fibers and Non-wood Pulping*, M. J. Kocurek, Series Editor, TAPPI Press, Atlanta, GA, Chap. IV, 1987.

Bagby, M. O., Adamson, W. C., Clark, T. F., and White, G. A., Kenaf stem yield and composition: Influence of maturity and field storage, Report No. 58, Non-Wood Plant Fiber Pulping Progress, TAPPI CA, Report No. 6, 1975.

Bhargava, R. L. V, Bamboo, *Pulp and Paper Manufacture, Volume 3, Secondary Fibers and Non-wood Pulping*, M. J. Kocurek, Series Editor, The Joint Textbook Committee of the Paper Industry, 71, 1987.

Chawla, J. S., Poppy straw, A new source for fibre boards, *Indian Pulp & Paper*, December, 1977–January, 1978.

Chow, P., *Dry Formed Composite Board From Selected Agricultural Fiber Residues*, Food and Agriculture Organization of the United Nations, FO/WCWP/75 Doc. No. 110, 14, 1974.

Clark, T. F. and Wolff, I. A., A search for new fiber crops, XI, Compositional characteristics of Illinois kenaf at several population densities and maturities, *TAPPI*, 52(11), 1969.

Clark, T. F., Nelson, G. H., Cunningham, R. L., Kwolek, W. F., and Wolff, I. A., A search for new fiber crops; Potential of sorghums for pulp and paper, *TAPPI*, 56, 3, 1973.

Cunningham, R. L., Carr, M. E., and Bagby, M. O., Hemicellulose isolation from annual plants, *Biotechnology and Bioengineering Symp.*, No. 17, 159, 1987.

Cunningham, R. L., Clark, T. F., and Bagby, M. O., Crotalria juncea–annual source of papermaking fiber, *TAPPI*, 61, 2, 1978.

Duckart, L., Byers, E., and Thompson, N. S., The structure of a "xylan" from kenaf, *Cellulose Chem. Technol.*, 22, 29, 1988.

Eddi, M. and Shinagawa, S., Chemical compositions of coconut coir and its pulp properties, *Berita Selulosa*, XVII, TAPPI Press, 4, 1982.

Eroglu, H. and Usta, M., Oxygen bleaching of soda oxygen wheat straw pulp, *Nonwood Plant Fiber Pulping*, Progress Report No. 19, 1988.

Francia, P. C., Escolano, E. U., and Semana, J. A., Proximate chemical composition of the various parts of the coconut palm, *The Philippine Lumber*, 19(7), 1973.

Goyal, S. K. and Ray, A. K., Economic comparison of soda and soda–AQ pulping processes for cereal straw, *Non-wood Plant Fiber Pulping*, Progress Report No. 19, TAPPI Press, 1988.

Gupta, P. C., Day, A., and Mazumdar, A. K., Chemical characterization of banana (*Musa sapientum* Linn) fibre, *Research and Industries*, 17, 1972.

Hamid, S. H., Maadhah, A. G., and Usmani, A. M., Bagasse-based building materials, *Polym.-Plast. Technol. Eng.*, 21(2), 173, 1983.

Hurter, A. M., Utilization of annual plants and agricultural residues for the production of pulp and paper, *Non-wood Plant Fiber Pulping*, Progress Report No. 19, TAPPI Press, 1988.

Iiyama, K., Lam, T. B. T., Meikle, P. J., Ng, K., Rhodies, D. I., and Stone, B. A., Cell wall biosynthesis and its regulation, *Forage Cell Wall Structure and Digestibility*, Jung, H. G., Buxton, D. R., Hatfield, R. D., and Ralph, J. Eds., American Soc of Agronomy, Inc., Madison, WI, 621, 1993.

Jones, R. W., Krull, J. H., Blessin, C. W., and Inglett, G. E., Neutral sugars of hemicellulose fractions of pith from stalks of selected plants, *Cereal Chem.*, 56(5), 441, 1979.

Lewin, M. and Pearce, E. M., Volume IV: Fiber Chemistry, *Handbook of Fiber Science and Technology*, Marcel Dekker, Inc., New York, 1985.

Misra, D. K., VI, Part VI. Cereal straw, *Pulp and Paper Manufacture, Volume 3, Secondary Fibers and Non-wood Pulping*, M. J. Kocurek, Series Editor, The Joint Textbook Committee of the Paper Industry, 82, 1987.

McGovern, J. N., Fibers, Vegetable, in *Encyclopedia of Polymer Science and Engineering*, Vol. 7, John Wiley & Sons, Inc., New York, 1986.

McGovern, J. N., Coffelt, D. E. , Hurter, A. M., Ahuja, N.K., and Wiedermann, A., IX Other Fibers, in *Pulp and Paper Manufacture, Volume 3, Secondary Fibers and Non-wood Pulping*, M. J. Kocurek, Series Editor, The Joint Textbook Committee of the Paper Industry, 1987.

Nelson, G. H., Clark, T., Wolff, F., I. A., and Jones, Q., A search for new fiber crops: Analytical Evaluations, *TAPPI*, 49(1), 1966.

Nieschlag, H. J. , Nelson, G. H., Wolff, I. A., and Perdue, Jr., R.E., A Search For New Fiber Crops, *TAPPI*, 43(3), 1960. a&b

Nieschlag, H. J., Earle, F. R. , Nelson, G. H., and Perdue, Jr., R. E., A search for new fiber crops II. Analytical Evaluations, Continued, *TAPPI*, 43(12), 1960.

Obst, J., Analysis of plant cell walls, *Forage Cell Wall Structure and Digestibility*, Jung, H. G., Buxton, D. R., Hatfield, R. D., and Ralph, J. Eds., American Soc of Agronomy, Inc., Madison, WI, Chapter 7, 167, 1993.

Pandey, S. N., Anantha Krishnan, S. R., Fifty years of research 1939–89, Jute Technological Research Laboratories publication, 1990. Calculta, India

Pettersen, R. C., The chemical composition of wood, *The Chemistry of Solid Wood*, Rowell, R. M., Ed., Advances in Chemistry Series, 20, American Chemical Society, Washington, D.C., (ACS), 1984.

Ralph, J., Hatfield, R. D., Lu, F., Grabber, J. H., Jung, H. G., Han, J. S., and Ralph, S. A., Kenaf's amazing lignin, *Proceedings of the 8th International Symposium on Wood and Pulping Chemistry*, Helsinki, Finland, Vol. 2, 125, 1995.

Sarkanen, K. V. and Ludwig, C. H. (Eds.), *Lignins: Occurrence, Formation, Structure and Reactions*, Wiley-Interscience, New York, 1971.

Satyanarayana, K. G., Pai, B. C., Sukumaran, K., and Pillai, S. G. K., Fabrication and properties of lignoncellulosic fiber-incorporated polyester composites, *Handbook of Ceramic and Composites*, Marcel Dekker, New York, 1989.

Singh, S. M., Physico-chemical properties of agricultural residues and strength of portland cement-bound wood products, *Research and Industry*, 24, 1979.

Sjöström, E., Wood polysaccharides, *Wood Chemistry, Fundamentals and Applications*, Academic Press, New York, 51, 1981.

Stamm, A.J., *Wood and Cellulose Science*, The Ronald Press Co., New York, 1964.

Touzinsky, G. F., VIII, Kenaf, *Pulp and Paper Manufacture, Volume 3, Secondary Fibers and Non-wood Pulping*, M. J. Kocurek, Series Editor, The Joint Textbook Committee of the Paper Industry, 1987, p. 106.

Upadhyaya, J. S. and Singh, S. P., Studies on neutral sulphite pulping of *Cajanus cajan*, *Nonwood Plant Fiber Pulping*, Progress Report No. 19, TAPPI Press, 1988a.

Upadhyaya, J. S. and Singh, S. P., Studies on soda and soda-anthraquinone pulping of *Sesbania aculeata*–The over-all reaction pattern, *Nonwood Plant Fiber Pulping*, Progress Report No. 19, TAPPI Press, 1988b.
Wakil, A. A., Jamil, N., and Younis, M. T., The characteristics of *Hibiscus esculentus* okra (Bindi) fibers, *Pakistan Journal of Science & Industrial Research*, 16(3/4), 1973.
Wiedermann, A., VII, Reeds, *Pulp and Paper Manufacture, Volume 3, Secondary Fibers and Non-wood Pulping*, M. J. Kocurek, Series Editor, The Joint Textbook Committee of the Paper Industry, 1987, 94.

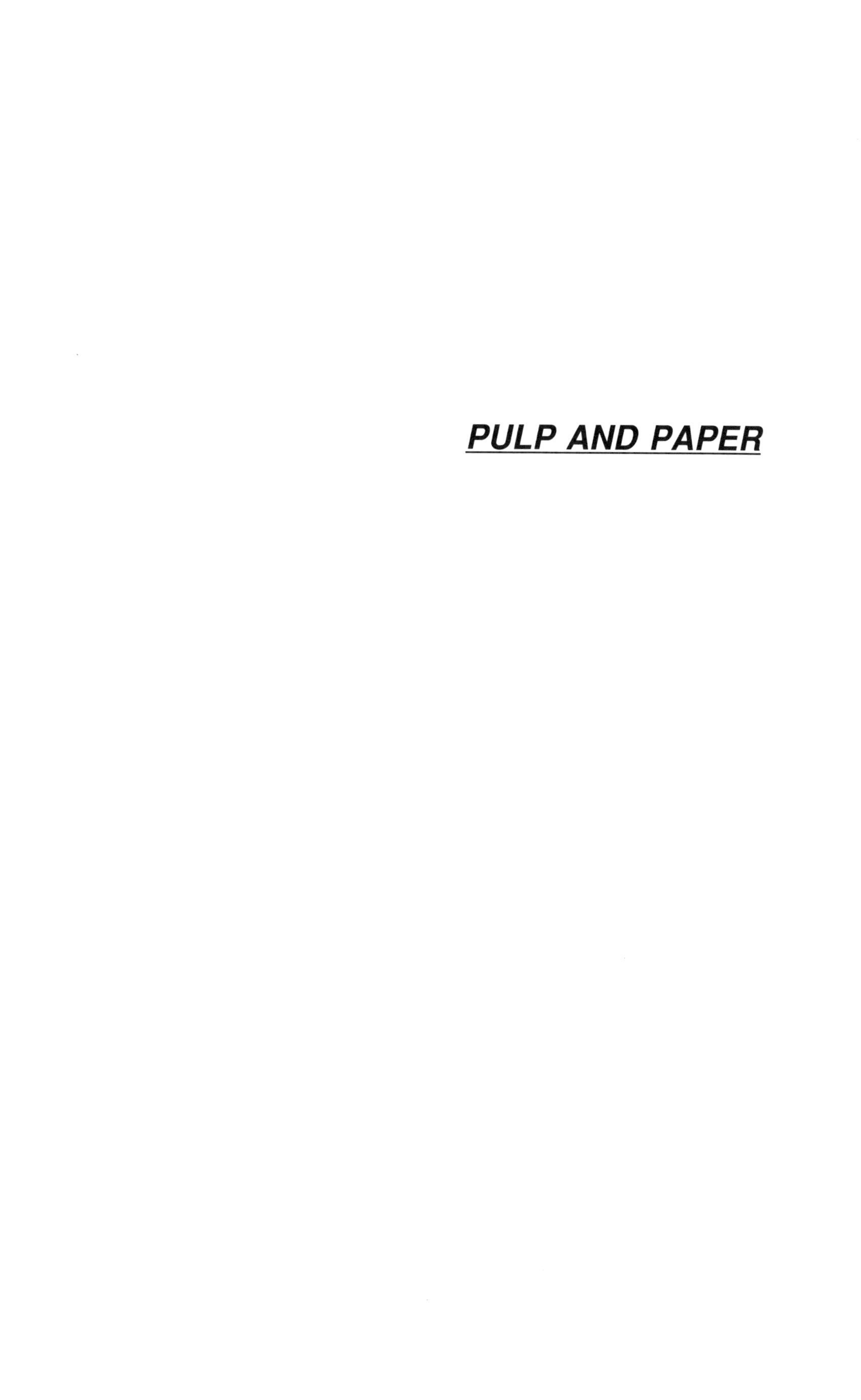

PULP AND PAPER

CHAPTER 6

Processing of Agro-Based Resources into Pulp and Paper

Raymond A. Young

CONTENTS

1. INTRODUCTION

The major sources of fibers for pulp and paper have historically been from plant sources other than wood. Although the early Chinese and Central American Indian civilizations made a paper-like substance from the inner bark of trees, Ts'ai Lun is credited with making the first paper from vegetable fibers in 105 A.D. in China. These papers were composed of macerated hemp and ramie (China grass)/(Atchison and McGovern, 1987; Hurter, 1991). The production of pulp and paper from wood is a relatively recent development. The use of wood was necessitated in the 19th century by an insufficient supply of the traditional raw materials such as cotton and linen rags. About 91% of the pulp and paper produced in the world today is now from wood, mostly in the developed countries. However, there is increasing pressure to use alternate raw material sources for pulp and paper in both the developed and developing countries, both from environmental and supply perspectives. The use of agro-based fibers is particularly relevant to many countries where demand for pulp and paper is increasing but there are only limited wood resources available (Figure 6.1).

Figure 6.1 The demand for agro-based materials has dramatically improved living conditions for farmers in regions of India. Shown here is an up-scale home of a sugar cane farmer in India whose material is sold to an integrated pulp, paper and sugar mill (Photo by R. A. Young).

In 1990 the global pulp production was more than 160 million metric tons, and Kaldor (1992) has estimated the demand will rise to about 238 million metric tons by the year 2000 and more than double by the year 2010–370 million metric tons. The total additional fiber which is known to be available is approximately 65 million metric tons per year (O.D. basis). Thus, the world will be running short of virgin fiber around the year 2000. Kaldor's analysis also indicated that the forest resources required by the year 2010 would need to be almost doubled, based on the present usage rate, which could not be achieved without a major plantation program. This, combined with environmental pressures to preserve forests, will necessitate the use of alternate fiber sources for pulp and paper.

The worldwide capacity for production of agro-based fiber pulps has increased substantially since 1975, rising from 6.9% to about 10.6% in 1993. The average annual increase in agro-based fiber pulps capacity is now more than triple the average annual increase in wood pulp capacity, at 6.0% vs. 2.0% for wood. The pulp capacity reached almost 21 million tons in 1993 and is estimated to be at least 23 million tons, or 11% of total capacity, by 1998 (Atchison, 1994). Straw, bagasse, and bamboo are the primary agro-based fibers used in pulp and paper, but many other agro-fibers are used in a variety of specialty papers. In many countries the pulp production is based entirely on agro-based fibers; over 25 countries depend on agro-based fibers for over 50% of their pulp production. As shown in Table 6.1, the leading countries for production of pulp and paper from agro-based fibers are China and India, with China having over 73% of the world's agro-based pulp capacity. The majority of the pulp produced in India is also from agro-based fibers (55.5%). China and India together account for about 80% of the total agro-based fiber pulping capacity (Atchison, 1994).

Table 6.1 Leading Countries in Total Agro-Based Plant Fiber Papermaking Pulp Production Capacity and Percentage of Total Capacity, Which Is Based on Agro-Based Fibers in 1993 with Estimates for 1998

	1993		1998 Estimate	
Country	Agro-Based Pulping Capacity	Percentage of Total Pulping Capacity Agro-Based Materials	Agro-Based Pulping Capacity 1000 mt	Percentage of Total Pulping Capacity from Agro-Based Materials
1. China	15,246	86.9	16,830	84.3
2. India	1,307	55.5	2,001	61.3
3. Pakistan	415	100	415	100
4. Mexico	321	29.2	324	29.3
5. Peru	296	95.2	296	95.2
6. Indonesia	267	22.1	267	10.1
7. Colombia	218	45.1	218	37.2
8. Thailand	209	100	509	100
9. Brazil	196	3.1	238	3.3
10. Venezuela	185	75.2	187	75.4
11. U.S.	179	0.3	204	0.3
12. Greece	150	85.7	160	84.2
13. Spain	140	7.9	141	7.7
14. Argentina	140	14.6	140	12.8
15. Egypt	127	100	127	100
16. Italy	120	13.3	120	13.3
17. Cuba	108	100	108	100
18. Turkey	103	16.5	103	16.5
19. Romania	102	10.9	102	10.9
20. Iraq	101	100	101	100
21. Taiwan	100	20.0	100	20.0
22. South Africa	99	6.4	99	6.4
23. Iran	90	25.0	90	25.0
24. Vietnam	86	60.1	100	40.0
SUBTOTAL FOR 24 COUNTRIES	= 20,305	20.6%	22,980	21.5%
TOTAL ALL COUNTRIES	20,736	10.6%	23,471	11.2%

Sources: Based on FAO Capacity Survey 1993–1998 and information obtained from individual countries. (From Atchison, J. E., Present Status and Future Prospects for Use of Non-Wood Plant Fibers for Paper Grade Pulps, paper presented at AF&PA Pulp and Fiber Fall Seminar, Tucson, AZ, November, 1994.)

For five of the 24 leading countries, agro-based fibers represent 100% of the pulping capacity and 11 of the countries depend on agro-based plants for more than 50% of their pulping capacity. When taking all countries into consideration, 14 depend on agro-based fibers for 100% of capacity and more than 20 depend on agro-based materials for over 50% of pulping capacity. In many other countries, agro-based fibers are only a small percentage of the pulp capacity, with the U.S. having only 0.3% of pulp capacity from these natural fibers. Clearly, there is considerable potential for much greater usage of agro-based fibers in the U.S. and many other countries (Atchison, 1994).

2. SUITABILITY OF AGRO-BASED FIBERS FOR PULP AND PAPER

Although there has been a considerable expenditure of time and money on the development of synthetic-type pulps for papermaking (Battista, 1964), for economic reasons the paper industry remains almost exclusively dependent on the vegetable kingdom for its raw material. There are more than 250,000 known species of higher plants; however, relatively few have been cultivated to any great extent and even fewer have been evaluated for use in papermaking.

The most extensive evaluation of alternate plant materials for pulp and paper was carried out by scientists at the U.S. Department of Agriculture (USDA), Northern Regional Research Center in Peoria, IL. They carried out an extensive screening of over 500 species of plants as potential sources of papermaking fibers. The species were rated for agronomic potential, chemical composition, fiber properties and physical characteristics. The ratings designated by the USDA scientists are given in Tables 6.2 and 6.3 and were assigned as follows (Nieschlag et al., 1960).

2.1 Botanical Evaluation

The species were classified in five groups according to their probable adaptability to growth in the United States and probable yield of vegetative matter. The most promising species were rated 1, and the least promising, 5.

2.2 Chemical Analysis

The results of chemical analyses were used to assign numerical ratings on the following basis:

Rating	Crude Cellulose, %	α-Cellulose, %	Soluble in 1% NaOH, %
1	>50	>34	<32
2	44–49	29–33	33–39
3	<44	<29	>39

Fiber Dimensions – Rating	Average Fiber Length, mm
1	>1.00
2	0.75–0.99
3	<0.75

The minimum fiber length necessary to produce acceptable paper strength properties is dependent on many factors, and fiber lengths are not unequivocally related to

Table 6.2 Potential Sources of Papermaking Fibers: Point Evaluation Ratings of Monocotyledoneae

	Rating by					
Sample	**Botanical evaluation**	**Chemical composition**	**Fiber dimension**	**Individual inspection**	**Maceration yield**	**Total point value**
Gramineae						
Gynerium sagittatum	1	1	1	1	1	5
Phyllostachys bambusoides	1	1	1	1	1	5
Zea mays[1]	1	1	1	1	1	5
Miscanthus sinensis	1	1	1	2	1	6
Sorghum almum	1	1	1	2	1	6
Stipa tenacissima	2	1	1	1	1	6
Andropogon gerardi	1	2	1	2	1	7
Arundo donax	1	2	1	2	1	7
Sorghum halepense	1	2	1	2	1	7
Bambusa vulgaris	4	1	1	1	1	8
Lygeum spartum	3	1	1	1	2	8
Pennisetum typhoides	1	3	1	1	2	8
Sorghum vulgare (broom corn)	1	2	2	2	1	8
Sorghum vulgare (grain)	1	2	2	2	1	8
Stipa viridula	2	2	1	1	2	8
Tripsacum dactyloides	1	3	1	1	2	8
Andropogon intermedius	1	2	2	2	2	9
Guadua amplexifolia	4	1	1	1	2	9
Guadua angustifolia	4	2	1	1	1	9
Stipa splendens	1	3	2	1	2	9
Andropogon hallii	1	2	2	3	2	10
Pennisetum spicatum	2	2	1	3	2	10
Stipa coronata	2	3	2	2	2	11
Stipa speciosa	3	3	2	1	2	11

[1] Fibrous residue from a variety of corn.

Source: Nieschlag, H. J. et al., *TAPPI*, 43(3), 193, 1960.

Table 6.3 Potential Sources of Papermaking Fibers: Point Evaluation Ratings of Dicotyledoneae

	Rating by					
Sample	Botanical evaluation	Chemical composition	Fiber dimension	Individual inspection	Maceration yield	Total Point value
Apocynaceae						
Apocynum cannabinum	2	2	1	2	1	8
Asclepiadaceae						
Asclepias syriaca	1	2	1	2	2	8
Caprifoliaceae						
Sambucus canadensis	1	3	3	2	1	10
Compositae						
Helianthus grossesserratus	1	2	2	2	1	8
Ambrosia trifida	1	2	3	2	1	9
Helianthus annuus	1	2	2	3	1	9
Helianthus tuberosus	1	2	3	2	2	9
Silphium laciniatum	1	3	1	2	2	9
Solidago gigantea v. leiophylla	1	2	3	3	1	9
Helianthus maximiliani	1	3	3	2	2	10
Silphium integrifolium	2	3	2	3	1	11
Achillea millefolium	2	3	3	3	2	12
Cirsium discolor	1	3	3	3	2	12
Solidago sempervirens	1	3	3	3	2	12
Ambrosia artemisiaefolia	1	3	3	3	3	13
Helianthus rigidis	1	3	3	3	3	13
Helianthus salicifolius	1	3	3	3	3	13
Lactuca canadensis	1	3	3	3	3	13
Euphorbiaceae						
Ricinus communis	1	1	1	2	1	6
Acalypha virginica	2	2	2	3	2	11
Leguminosae						
Crotalaria juncea (Brazilian)	1	1	1	1	1	5
Crotalaria juncea	1	1	2	1	1	6

Table 6.3 (continued) Potential Sources of Papermaking Fibers: Point Evaluation Ratings of Dicotyledoneae

	Rating by					
Sample	**Botanical evaluation**	**Chemical composition**	**Fiber dimension**	**Individual inspection**	**Maceration yield**	**Total Point value**
Crotalaria juncea (ridged stem)	1	1	1	1	2	6
Sesbania sonorae	1	1	1	1	2	6
Crotalaria juncea (smooth stem)	1	1	2	1	2	7
Desmanthus sp.	1	2	2	2	1	8
Desmanthus illinoensis	1	2	3	1	2	9
Sesbania cinerescens	5	2	2	2	2	13
Malvaceae						
Hibiscus cannabinus (average value)	1	1	1	1	1	5
Hibiscus esulentus (pods)	1	1	1	2	1	6
Hibiscus sabdariffa	1	2	1	1	1	6
Althaea rosea	1	2	1	2	1	7
Hibiscus cannabinus (hurds)	1	1	3	1	2	8
Hibiscus esculentus	1	3	1	1	2	8
Abutilon theophrasti	2	3	2	1	1	9
Moraceae						
Cannabis sativa	1	1	1	2	1	6
Polygonaceae						
Polygonum persicaria	2	2	2	3	2	11
Umbelliferae						
Cicuta maculata	1	2	3	2	3	11
Heracleum lanatum	1	2	3	2	3	11

Source: Nieschlag, H. J., et al., *TAPPI*, 43(3), 193, 1960.

paper strength properties. For example, certain bamboos are known to produce good pulps and were given a rating of 1.

Maceration Yield	Rating Yield, %[1]
1	>50%
2	40–50%
3	<40%

[1] Acetic acid–sodium chloride treatment.

2.3 Qualitative Inspection

This rating was based on a qualitative visual and physical estimation by five technical people of the amount of pith, strength of bast, proportion of bast, and texture of the woody portion (of the whole stalk of monocotyledons) considering its hardness and fibrous nature. The samples were rated as: 1–promising, 2–questionable, and 3–not promising.

2.4 Total Point Value

A cumulative value was obtained by adding the point evaluations for each category. Materials with a point value of 8 or less were considered as potential pulping materials meriting further study. Those rating 9–10 had some slight promise and might warrant additional study; ratings of 11 and higher were considered unsatisfactory.

After analysis of the 58 species representing 11 families, Nieschlag et al. (1960) concluded that representatives of the Gramineae, Leguminosae, and Malvaceae were the most promising. More rigorous appraisal of samples by a small-scale pulping procedure and handsheet evaluations were then carried out by Nelson et al. (1961) to permit further selection and narrowing in the choice of preferred raw materials and to serve as a guide to the validity of the analytical screening. Data obtained for 61 species pulped by the kraft process are shown in Table 6.4. A few lower quality samples were included to provide a wider range of properties, and several standard fibrous raw materials were given identical treatment for comparative purposes. The authors appropriately noted that this one pulping procedure may not be adequate for deciding whether the pulp yield and properties are characteristic of a raw material or merely indicate varying response to the chosen experimental conditions. This limitation is especially true since beater evaluations to assure maximum strength development were not performed. However, compromises are necessary to evaluate such a wide range of species in a reasonable time frame, and subsequent statistical evaluation indicated the validity of the data for comparative evaluations.

Based on their extensive analysis of pulp and papermaking properties of the 61 plant species, Nelson et al. (1961) concluded that mallow plants have favorable characteristics for use in pulp and paper. Under the conditions used, the pulp yields ranged up to 57%, burst factors up to 50, tear factors to 64, and Schopper folds to 1,000. Certain legumes and grasses also possessed favorable characteristics.

Table 6.4 Kraft Pulping and Paper Properties of Selected Species of Agro-Based Plants[1]

Material	Common name	Type of plant[2]	Screened yield %	Permanganate number	Freeness[5]	Burst[3] factor	Tear[4] factor	Fold, Schopper
Picea sp.	Spruce	—	57.8	74.5	695	70	77	609
Acer sp.	Maple	D	56.4	39.3	645	61	65	699
Stipa lenacissima	Esparto	M	53.6	5.2	665	56	136	902
Glycine max.	Soybean	D	52.9	59.1	615	15	70	4
Triticum aestivum (kawvalo)	Wheat	M	47.3	14.6	575	44	60	175
Crotalaria juncea (Brazilian)	Sunn hemp	D	60.2	55.0	655	33	72	200
Sesbania sonorae	Seaban	D	58.8	60.3	625	37	49	239
Hibiscus cannabinus variety 52–41	Kenaf	D	57.3	52.0	610	38	64	348
Asclepias incarnata	Swamp milkweed	D	57.2	62.1	540	22	78	93
Abutilon theophrasti	Velvet weed	D	57.1	60.8	595	42	56	349
Crotalaria juncea (smooth stem)	Sunn hemp	D	56.3	57.4	640	45	51	501
Crotalaria juncea	Sunn hemp	D	55.9	54.8	625	37	81	277
Sorghum vulgare PQX65584	Sorghum	M	55.7	50.9	595	36	55	178
Ambrosia trifida	Giant ragweed	D	55.5	67.9	590	24	40	34
Crotalaria juncea (ridged stem)	Sunn hemp	D	55.5	55.6	610	46	62	643
Hibiscus esculentus (Louisiana market)	Okra	D	55.2	61.0	610	38	42	617
Hibiscus esculentus (Louisiana green velvet)	Okra	D	54.8	46.4	625	50	46	1,008
Hibiscus cannabinus variety 52–41	Kenaf	D	54.6	53.8	570	44	59	417
Desmanthus sp.	Bundle flower	D	54.3	63.3	575	27	49	64
Lupinus formosus	Lunaria lupine	D	54.2	61.1	640	20	40	21
Helianthus grosseserratus	Sawtooth sunflower	D	53.5	60.9	595	26	46	32
Hibiscus sabdariffa	Roselle	D	53.3	64.3	585	48	59	537
Crotalaria intermedia	Kotschy	D	53.1	55.9	630	30	64	140
Iva zanthifolia	Rag sumpweed	D	52.6	65.9	585	24	43	27
Dalea alopecuroides	Foxtail dalea	D	52.5	62.3	635	32	65	102

Table 6.4 (continued) Kraft Pulping and Paper Properties of Selected Species of Agro-Based Plants[1]

Material	Common name	Type of plant[2]	Screened yield %	Permanganate number	Freeness[5]	Burst[3] factor	Tear[4] factor	Fold, Schopper
Ricinus communis	Castor	D	52.1	67.4	560	37	54	455
Dalea ennandra		D	50.8	56.6	655	28	54	41
Sorghum alum		M	50.7	21.7	610	42	52	217
Gynerium sagitallum	Uva grass	M	50.3	35.9	685	43	92	289
Sorghum vulgare PI177549	Sorghum	M	49.8	21.5	515	37	34	144
Phyllostachys bambusoides	Bamboo	M	49.7	48.5	700	46	81	284
Pennisetum spicatum	Pearl millet	M	49.3	39.1	625	40	44	194
Althaea rosea	Hollyhock	D	49.0	66.0	535	30	46	146
Sorghum vulgare (kaoliang PI71310)	Sorghum	M	48.9	29.4	640	28	41	36
Pennisetum typhoides		M	48.8	24.9	595	42	40	173
Sorghum vulgare PI62610	Sorghum	M	48.8	23.8	620	38	36	153
Helianthus maximiliani	Maximilian sunflower	D	48.7	67.8	575	20	39	8
Arundo donax	Giant reed	M	48.6	44.9	585	32	44	82
Sorghum vulgare (kaoliang PI88000)	Sorghum	M	48.6	21.9	585	39	36	174
Asclepias incarnata	Swamp milkweed	D	48.3	63.3	555	16	46	11
Hibiscus cannabinus (variety 52–41)	Kenaf	D	48.3	51.3	575	38	54	465
Sorghum vulgare (kaoliang thick rind CI792)	Sorghum	M	48.2	28.7	600	35	39	88
Helianthus scaberrimus	Blackhead sunflower	D	47.9	63.0	605	26	46	34
Andropogon gerardi	Blue stem	M	47.6	15.1	555	38	43	120
Miscanthus sinensis	Chinese silvergrass	M	47.6	22.7	595	36	42	113
Sorghum vulgare (kaoliang PI71309)	Sorghum	M	47.4	19.5	625	36	44	148
Sorghum halepense	Johnson grass	M	47.1	19.3	600	32	55	97
Asclepias syriaca	Common milkweed	D	46.9	51.5	465	18	62	19
Apocynum cannabinum	Hemp dogbane	D	45.6	55.1	480	21	48	19

Table 6.4 (continued) Kraft Pulping and Paper Properties of Selected Species of Agro-Based Plants[1]

Material	Common name	Type of plant[2]	Screened yield %	Permanganate number	Freeness[5]	Burst[3] factor	Tear[4] factor	Fold, Schopper
Andropogon intermedius	Australian bluestem	M	45.5	28.9	630	26	34	17
Sorghum halepense	Johnson grass	M	45.5	22.5	600	35	34	96
Andropogon hallii	Sand bluestem	M	45.3	25.7	580	36	42	88
Lygeum spartum	Albardine grass	M	45.2	13.9	615	40	114	218
Sorghastrum nutans	Yellow Indian grass	M	43.8	16.7	675	31	41	16
Cenchrus sp.	Sandbur	M	42.7	38.2	655	30	38	30
Sorghum vulgare (Sudanense 34641FC)	Sudan grass	M	42.1	25.4	590	33	34	78
Tripsacum dactyloides	Eastern gama grass	M	40.9	27.8	620	40	40	86
Hibiscus esculentus	Okra	D	40.6	50.5	580	40	46	442
Andropogon intermedius	Australian bluestem	M	40.4	15.7	615	32	46	37
Arundinaria tecta	Switch cane	M	38.9	48.0	625	17	38	4
Stipa viridula	Green needle grass	M	38.8	16.6	595	50	46	222

[1] 15% Kraft (active alkali as Na_2O, 11.7%; Sulfidity, 33.9%).
[2] M = monocotyledon, D = dicotyledon.
[3] Burst in $g/cm^2/(g/m^2)$.
[4] Tear in g/gsm.
[5] CSF.

Source: Nelson, G. H., et al., *TAPPI*, 44(5), 319, 1961.

Nelson et al. (1961) also noted large qualitative differences between the grasses (monocotyledons) and the dicotyledonous plants in response to pulping chemicals and properties. The ease of pulping of the raw materials, judged by extent of delignification, was significantly greater for grasses than for the dicotyledons. However, burst and fold strengths for papers from dicotyledons were significantly better than those from the grasses. No significant differences in tear strength were observed between the two groups of plant materials. Relationships between fiber dimensions and pulp properties were highly significant for dicotyledons but not for grasses. Nelson and coworkers also found that the α-cellulose content of the raw material was a excellent predictor of both pulp yield and quality of a given species.

Later work by Cunningham et al. (1970) expanded the search for new fiber crops to a total of 78 species and included laboratory scale kraft pulping trials of selected species. The results of their chemical and physical characterization of the beater-refined kraft pulps of selected agro-based species are given in Table 6.5. The new species produced screened pulps varying in yields from 38–59%, with permanganate numbers ranging from 12–73. Strength indices were determined for all 78 species and an esparto grass (*Stipa tenacissima*) had the highest index (2826) in the series. Grass species, including three bamboos plus *Gynerium sagittatum*, *Lygeum spartum*, and *Sinarundinaria murielae*, occupied the next six positions. Two of the bamboos (*Guadua amplexifolia* and *Arundinaria alpina*) gave good pulp yields, were easy to bleach, and had good strength and cold tolerance. Hemp (*Cannabis sativa*) also gave good strength pulps. Cunningham and coworkers suggested that *Sorghum almum* might also be a good choice as a raw material even though the strength index rating was lower (21st). This is because this plant is adaptable to a variety of soil types, has high productivity in the field, and has excellent drought-resistance. They also felt that the *Hibiscus* genus with representatives in strength index positions of 10–16 was worthy of further attention, particularly kenaf because of its good productivity and other positive agronomic characteristics.

It should be clear from the prior discussion that practically any plant material can be digested to yield a pulp for papermaking. The choice of the plant material will depend on a great many factors in addition to yield and strength such as: suitability to climate; growth rate and productivity; cost of harvesting and collection; extent of contamination, etc. These factors have not only limited the numbers of species utilized but have also limited the extent to which agro-based fibers in general have been utilized for pulp and paper production. The species of plants now utilized for commercial production of pulp and paper are listed in Table 6.6 (Hurter, 1991). Also shown in this table is the pulping process most commonly employed for pulp production from each species, the bleached and unbleached yields, and the common end-use of the paper product. A more detailed description of the pulping characteristics and paper properties for these plant materials is given in a later section of this chapter. However, there are a number of common characteristics for most of the agro-based fiber crops that deserve consideration.

Table 6.5 Kraft Pulping Data and Pulp Properties for Selected Agro-Based Plants

Material	Common name	Pulp Yield Screened, %	Permanganate number	Freeness, SR, mL	Burst factor $g/cm^2/(g/m^2)$	Tear factor $g/(g/m^2)$	Fold, Schopper
		Monocotyledoneae					
Gramineae							
Acena barbata	Slender oat	40.9	12.2	560	45	31	434
Bambusa vulgaris	Common bamboo	41.8	42.8	640	51	114	482
Bothriochloa intermedia	Australian bluestem	38.3	17.6	600	44	35	244
Guadua amplexifolia	Bamboo	49.4	30.1	690	53	118	690
Phyllostachys dulcis	Sweetshoot bamboo	44.2	54.8	720	30	58	44
Phyllostachys makinoi	Makino bamboo	44.4	37.8	700	34	85	161
Phyllostachys viridigllaucescens	Bamboo	45.6	45.4	690	43	79	211
Sinarundinaria murielae	Muriel chinacane	43.3	32.1	650	46	68	510
Stipa coronata	Crested needlegrass	38.4	31.2	595	26	37	14
		Dicotyledoneae					
Capparaceae							
Cleome serrulata	Bee spiderflower	54.4	58.4	590	26	44	56
Compositae							
Anthemis cotula	Mayweed camomile	43.0	62.6	500	34	37	233
Erigeron canadensis	Horseweed fleabane	53.9	57.0	570	19	38	12
Eupatorium perfoliatum	Boneset	52.6	65.1	565	22	37	13
Euphorbiaceae							
Croton texensis	Texas croton	38.0	63.6	555	38	45	407
Euphorbia marginata	Snow-on-the-mountain	52.3	48.7	550	34	55	518
Hydrophyllaceae							
Phacelia californica	California phacelia	41.3	72.7	510	17	26	3
Leguminosae							
Crotalaria incana	Shackshack crotalaria	54.0	55.5	625	30	43	184

Table 6.5 (continued) Kraft Pulping Data and Pulp Properties for Selected Agro-Based Plants

Material	Common name	Pulp Yield Screened, %	Permanganate number	Freeness, SR, mL	Burst factor $g/cm^2/(g/m^2)$	Tear factor $g/(g/m^2)$	Fold, Schopper
Crotalaria mucronata	Stiped crotalaria	58.2	44.1	670	46	49	665
Crotalaria spectabilis	Showy crotalaria	44.0	57.0	530	22	49	55
Lespedeza capitata	Roundhead lespedeza	51.8	58.8	575	24	39	24
Sesbania drummondii	Drummond sesbania	50.7	55.4	680	35	53	116
Sesbania exaltata	Hemp sesbania	53.0	49.1	550	31	48	211
Sesbania vesicaria	Bagpod sesbania	54.7	62.4	660	34	43	87
Malvaceae							
Hibiscus eetveldeanus		55.3	62.8	640	42	56	430
Sida sp.		58.8	59.0	660	43	56	378
Onagraceae							
Oenothera biennis	Common evening primrose	56.2	62.4	630	34	50	276
Polemoniaceae							
Phlox paniculata	Summer phlox	54.2	54.2	630	24	46	21
Polygonaceae							
Polygonum persicaria	Spotted lady's thumb	41.8	51.0	600	29	44	154

Source: Cunningham, R. L. et al., *TAPPI*, 53(9), 167, 1970.

Table 6.6 Pulping Processes for Agro-Based Fibrous Materials

Raw materials	Pulping process	Use of pulp	Pulp Yield (%)	
			Unbleached	Bleached
Mixed cereal straw	Lime	Coarse paper	55–65	
	Lime	Strawboard	70–82	
	Soda or kraft	Paper	44–46	40–42
	Soda or kraft	Corrugating	65–68	
Rice straw	Soda	Paper	40–43	34–39
Esparto	Soda	Paper	45–50	42–46
Sabai grass	Soda	Paper	45–50	42–47
Phragmites	NSSC	Paper	50–53	48–50
communis reeds	Soda or kraft	Paper	46–51	42–48
Papyrus	Soda	Paper	35–38	27
Sugarcane bagasse	Soda or kraft	Industrial paper	60	
(depithed)		Corrugating	70	
		Linerboard	63	
		Bleached paper	50–52	45–48
Bamboo	Soda	Paper	44–45	40–41
	Kraft	Paper	46–47	42–43
Seed flax tow	Soda	Cigarette paper	42–45	35–40
Textile flax tow	Soda	Paper	60–67	56–62
Jute	Lime	Industrial paper	62	58
	Soda	Paper	55	50
Kenaf (bast)	Soda or kraft	Paper	46–51	41–46
Abaca	Monosulphite	Thin paper	60–63	56–62
	Soda or kraft	Paper	45–54	43–50
Sisal (agava)	Soda	Paper	69	60
Cotton linters	Soda or kraft	Paper		70–80
		Dissolving pulp		65–75
Cotton staple	Soda or kraft	Paper		75–83
Cotton rags	Lime soda	Paper		70–85
	Soda	Paper		70–85

Source: Hurter, A. M., Utilization of annual plants and agricultural residue for the production of pulp and paper, *Non-Wood Plant Fiber Pulping*, Prog. Rept. No. 19, TAPPI Press, Atlanta, 49, 1991.

3. UNIQUE FEATURES OF AGRO-BASED RESOURCES FOR PULP AND PAPER

3.1 Preparation, Storage, and Handling

Similar to forests, other plant material can be either harvested from the wild or cultivated for use in pulp and paper. Esparto, sabai grass, reeds and bamboo are often collected in the wild, while most of the other plant materials are cultivated. However, harvesting, collection, and storage of agro-based materials can be a much greater problem than with wood. The handling, storage, and preparation of agro-based materials have not received nearly the amount of attention that wood materials have and this has caused serious problems in the past. The situation is further complicated because only a few of the agro-based raw materials, such as reeds, bamboos, and some grasses, are harvested specifically for pulp and paper production. Often the material is obtained as agricultural residues and wastes from other industries; consequently, the handling system must be coordinated with the primary use of the material. Also, the costs are often too prohibitive to transport the material for long distances due to its bulky nature (Hurter, 1991; Jeyasingam, 1991a; Assumpcao, 1991; Misra, 1980; Clark, 1969; Rydholm, 1965).

Material handling arrangements for agro-based resources differ considerably from one raw material to another. Large cut bamboos are handled and stored in much the same way as wood, while smaller bamboos and reeds are handled and stored in bundles. Straws and grasses are stored in the form of bales and are transported in trucks or other motorized vehicles. Bagasse is somewhat unique in that the sugarcane raw material has usually already been collected by the sugar mills, where the residue bagasse is then handled in bulk or baled for handling and transportation (Figure 6.2). Of course, baling adds additional expense to the process.

Figure 6.2 Sugarcane field in India. Bagasse is the byproduct fiber for pulp and paper (Photo by R. A. Young).

Many agro-based materials contain a pith or woody core that usually must be removed to improve the quality of the resulting pulp. For the bast fibers, this traditionally involves retting of the stems in streams or ponds, whereby the bast fibers are loosened from the woody core through the biological action of bacteria in the aqueous medium. Retting is a primary source of contaminants such as sand, silt, and dirt, particularly in situations where water supplies are limited, e.g., when there is not enough rain in the monsoon season in tropical climates and the retting ponds do not have sufficient water to keep the plants from dragging the bottom of the pond. The stalks are then dried and passed through fluted rolls to break or reduce the woody portion into small particles, which are then separated by scutching. The scutching is done by beating with blunt wooden or metal blades by hand or mechanically. The bundles are hackled or combed to separate the short and long fibers. This is done by drawing the fibers through sets of pins, each finer than the previous one. In this way the fibers are further cleaned and aligned (Young, 1994a).

Frequently the agro-based fibers must be cut prior to processing. For straw materials the cutting is often done as part of the digester filling operation. Hemp, reeds, esparto, small bamboos, and agro-based rags are usually cut with strong drum cutters to avoid damage to the equipment. A dusting and sorting operation may be necessary before cutting of collected rags. Special dusters are used for esparto to remove the waxy epidermal material on this species.

With some plants, such as kenaf, the entire stem is sometimes chipped and pulped (Figure 6.3) (Young, 1987, see also Chapter 1, Figure 1.1). The pulp then contains a mixture of long fibers from the bast and very short fibers from the woody core. For plants such as abaca and sisal, the whole plant is also sometimes cut and shredded, but the highest quality pulp from these two species is produced by mechanical decortication, whereby the leaves, which contain the high quality, long fiber stock, are crushed, scraped, and washed for further processing to pulp. Further details of preparation, storage, and handling of specific agro-based plants will be given in the sections designated to each of the species.

Figure 6.3 Harvesting of kenaf in Rio Grande Valley, USA (USDA Photo).

3.2 Chemical Composition

Agro-based plant materials generally have lower lignin, often one-half that of wood. The lower lignin content results in a more easily-pulped material; typically, agro-based resources can be pulped in one-fourth to one-third the time of that for wood. Most agro-plants also have a more accessible cell wall structure such that penetration of the pulping chemicals proceeds much more readily, which also enhances the rate of pulping (Hurter, 1991; Misra, 1980; Clark, 1969).

The high content of hemicelluloses in stalk fibers (monocotyledons) influences the interaction with water, and it has been found that stalk fiber pulps hydrate rapidly. This results in reduced power consumption in pulp refining (beating) but requires that machine speeds be reduced due to the slower drainage rates. However, newer machine configurations have allowed substantial improvements in paper production with the stalk fibers as shown in Table 6.7 (Hurter, 1991).

Table 6.7 Typical Paper Machine Speeds for Straw and Bagasse (meters per minute)[1]

Furnish[2]	Older Machines (open draws)	New Machines[3] (no draw press section)
Rice Straw	100–180	200–400
Other Straws	150–250	300–500
Bagasse	200–350	400–650
Wood	250–500	500–800

[1] Without using drainage aids.
[2] Bleached softwood kraft pulp at 15% of furnish for agro-based pulps.
[3] Established technology, 1985.

Source: Hurter, A. M., Utilization of annual plants and agricultural residue for the production of pulp and paper, *Non-Wood Plant Fiber Pulping*, Prog. Rept. No. 19, TAPPI Press, Atlanta, 49, 1991.

A major disadvantage of agro-based pulps is the presence of high contents of ash and, particularly, silica. The silica is present to the extent of 12–18% in rice straw, 4–6% in wheat straw, 6–7% in esparto, 4–5% in reeds, 2–3% in bamboo, 1.5–2% in depithed bagasse, and 2–3% in cleaned bast and leaf fibers. With alkaline pulping processes, a high portion of the silica is dissolved in the spent liquor, where it creates severe operating problems in a number of the stages of the recovery operation. Heat transfer is inhibited by silica deposits on the inner walls of evaporator tubes where it forms a hard scale. This scale must be removed frequently to maintain the evaporation rate and minimum steam consumption; this is a laborious task involving both chemical and mechanical treatments. The tube surfaces of the superheater banks and economizers are also often occluded with silica, which retards the passage of flue gases. Silica also retards the settling rate in the clarifiers in the recausticizing operation (Misra, 1980, 1987). Handling the silica problem is of such importance in alkaline pulping of agro-based fibers that a separate section is devoted to this topic.

3.3 Silica Dissolution and Desilication

As already mentioned, the agro-based materials contain a substantial amount of silica, which is accumulated in the epithelial cells of the plant, derived from the soil during growth. The amount of silica in the plant is extremely variable depending on the site, species, and soil conditions. The silica content of rice straw alone can vary between 6–22%. However, there is also a considerable amount of silica adhering to the plant from the harvesting, collection, and other processing steps. This latter source of silica can amount to as much as 30% of the total silica dissolved in the spent black liquor from the pulping operation. Thus, where possible, the raw material should be washed before entrance into the digester. The washing operation is relatively straightforward for woody type materials such as bamboo. However, cleaning of straw-type feedstocks can be more difficult and many small mills do not have the necessary dusting and cleaning equipment to avoid additional extraneous silica from entering the system. Straws carry a significant proportion of dust, mud lumps, and sand granules. Inadequate cleaning of the raw material not only results in increased silica in the black liquors but also increased alkali consumption and dead load in the digesters (Misra, 1980; Bleier, 1991).

Amorphous silica is very soluble in water, and its solubility in alkaline black liquor is about the same, in the range of 0.3–0.4 g/L, with the upper end of the range at a higher pH (pH = 10). The dissolution of silica is also dependent on the cooking process and the pulping conditions. Although about 90% of the silica is dissolved when pulping at high alkali concentration (60–80 g/L), only 50% of the silica in straw, for example, is dissolved when pulping under milder conditions, that is, lower alkali concentration (20 g/L) and lower cooking temperature (150°C). It seems clear that a higher concentration of alkali in the cooking liquors enhances the dissolution of silica. If calcium salts are also present, there is a tendency to also form calcium silcates. Calcium silicates with calcium sulphates form scales in the liquor heat exchangers of indirectly heated digesters. This results in an economic loss in terms of steam consumption for the digester system. The location of these deposits is shown in Table 6.8 (Tandon et al., 1991).

Table 6.8 Location of Silica Deposits in a Bamboo Kraft Pulp Mill

Particulars	Silica, %
Furnace walls	22–30
Boiler and superheater tubes	22–26
Air heater and economizer tubes	8–9
Evaporator tubes	43–45

Source: Tandon, R. et al., Properties of black liquors from pulping of non-woody raw materials, in *Proc. Intl. Seminar & Workshop on Desilication*, UNIDO, Vienna, 23, 1991.

Because of the definite economic advantage realized when silica is removed from waste liquors, a considerable effort has been expended to develop methods for removal of this contaminant. Panda (1991) recently reviewed the various desilication methods. The removal of the dissolved, or occluded silica can be from the lime sludge, the green liquor, or the black liquor. Desilication of the lime sludge is energy intensive and does not solve the problems in the evaporators, recovery boilers, and in the recausticization stage. Therefore, this is not really a viable alternative.

Desilication of both green and black liquors can be done with lime treatment or by pH adjustment. The lime precipitates the silica as calcium silicates according to the following reaction:

$$Na_2SiO_3 + CaO + H_2O \rightarrow CaSiO_3 + 2NaOH$$

The calcium silicate precipitate can then be filtered out of the solution. However with black liquor, a part of the lime also reacts with sodium carbonate and sodium sulphate such that considerable excess lime must be added beyond the stochiometric requirement.

Silica can also be precipitated from pulping liquors by reduction of the pH. This is done by passing carbon dioxide, or more conveniently flue gases, through the liquor, which precipitates silicon dioxide. The silicon dioxide is also then removed by filtration.

It has been found that selective precipitation of silica from the black liquor can be achieved even if, in theory, some lignin should precipitate first due to weaker acidic functional groups. The acidity of the phenol groups in lignin is in the range $pK_a = 9.4–10.8$, while the silicic acid acidity is between these values at $pK_a = 9.8$. However, preferential precipitation of silica from black liquors is possible due to the unique condensation or polymerization of silicates to form large molecules bound by siloxane linkages (Bleier, 1991). Lowering the pH of alkali silicate solutions results first in formation of monosilicic acid. This monomer undergoes ionic polymerization in proportion to the hydroxyl concentration, above pH = 2. The mechanism of polymerization and the growth of stable aggregated particles are shown in Figure 6.4 (Kulkarni et al., 1991). Since the threshold pH for precipitation of silica and lignin is very close, it is important to have a thorough understanding of the chemical composition of the black liquor to optimize pH, temperature, and concentration of the black liquor (Tandon et al., 1991).

In early studies, difficulties were encountered with avoiding co-precipitation of lignin and silica. There were also problems with uncontrolled foaming of acidified black liquor in the absorption towers. Therefore, pH reduction of the green liquor was evaluated. However, with green liquor pH reduction, the operational problems in evaporators and the recovery furnace were not solved, and the carbonation led to losses of sulfur as hydrogen sulfide and sodium sulfide in the green liquor. Black liquor acidification with flue gases is consequently considered to be the method of choice for desilication (Panda, 1991).

It has been reported that the most easily filterable and washable precipitates are produced by slowly carbonating the kraft black liquor. Elevated temperature helps the growth of the silica particles by enhancing the transfer of silica to the growing particle. Reverse bubble reactors, rather than packed columns, have proven advantageous for the gentle carbonation of the black liquor. Foam formation is also avoided with the bubble type reactors (Bleier, 1991).

A larger problem is the removal of the silicon dioxide precipitate from the acidified black liquor. However the use of endless band filters has been quite successful and filtration has not posed a serious problem (Bleier, 1991).

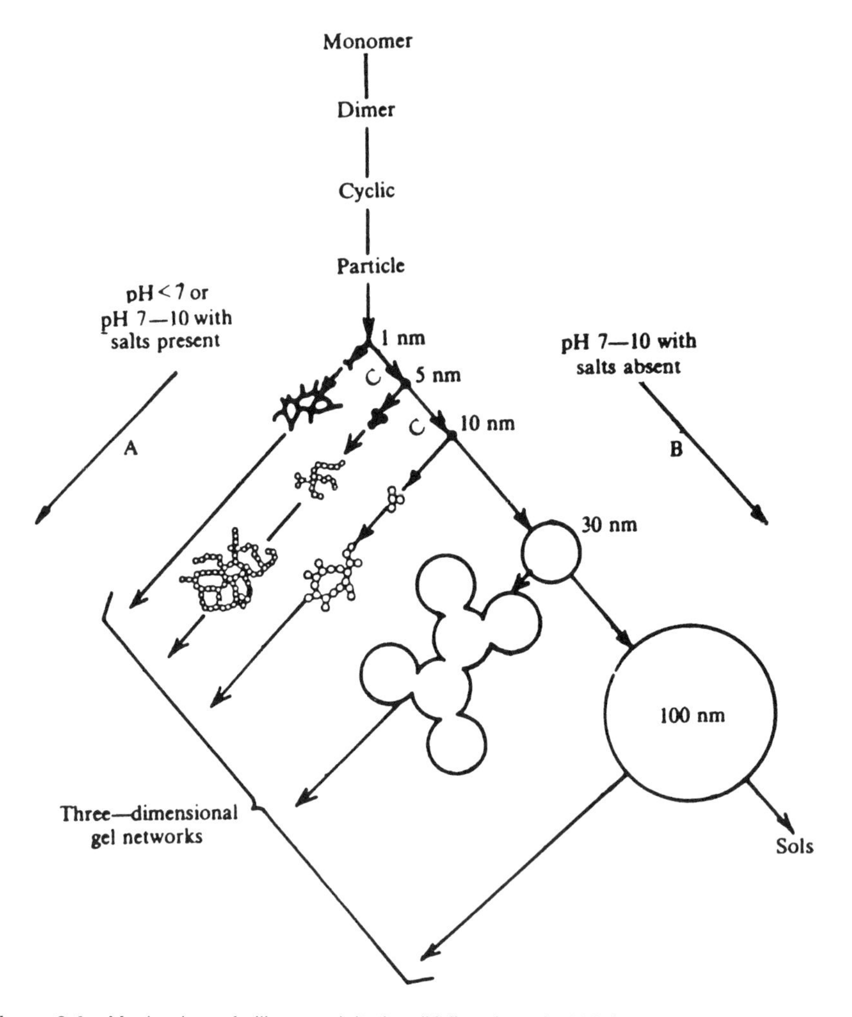

Figure 6.4 Mechanism of silica precipitation (Kulkarni et al., 1991).

Realkalization of the desilicated black liquor is important to economize on the alkali consumption. Vacuum stripping of carbon dioxide at elevated temperatures decomposes the majority of the bicarbonates formed by carbonation. However, this can be a rather delicate process since, in desilication, the pH of the black liquor is always taken to near the lignin precipitation point, which leaves the lignin sol in a very unstable state. Addition of alkali should take place after passage of the desilicated black liquor through the first evaporation stage, since almost complete stripping of carbon dioxide occurs after this stage. Following this procedure, a 20% reduction in the realkalization demand can be realized (Bleier, 1991).

3.4 Pulping Processes

The pulping processes used for agro-based materials are fundamentally the same as those used for pulping of wood, although some are more desirable than others for a variety of reasons which will be covered in this and the following sections. There are a considerable number of descriptions available in the literature of the principles of the various pulping processes (Rydholm, 1965; Casey, 1980; Biermann, 1993; Britt, 1970). The intention here is to point out features favoring different processes and special modifications which have been developed to handle agro-based fibers. Further descriptions of pulping processes are also given in the following sections designated to each of the various specific agro-based fibers.

Agro-based materials have been successfully converted to pulps by both mechanical and chemical processes and combinations of the two. Although it is usually not possible to produce good groundwood pulps by the conventional stone groundwood process from agro-based fibers, due to the structure of the agro-based material, refiner mechanical pulps have been produced from bagasse, reeds, kenaf and other agro-plants. More promising approaches for pulp from agro-based materials are the "enhanced" refiner mechanical processes that also employ steam for thermomechanical pulp (TMP), chemicals for chemimechanical pulps (CMP) and combinations for chemithermomechanical pulps (CTMP). Such mechanical pulps are currently being produced commercially from bagasse for newsprint applications; this will be discussed further in the section devoted to bagasse. Very recently it was demonstrated by Sabharwal et al. (1994a, 1994b, 1995) that excellent biomechanical pulps (BMP) can be produced from kenaf and jute. This process involves fungal treatment of the agro-material prior to refiner disintegration. The fungus biologically degrades the lignin, loosens the plant structure, and reduces energy consumption in the refining stage. The higher yields, lower costs, and reduced pollution hazards of the "enhanced" mechanical processes bode a promising future for these approaches. The kenaf and jute sections contain further details.

The primary chemimechanical pulping process utilized for agro-based materials is the cold soda process. The agro-material is steeped in a caustic soda solution (about 30 g/L) for 15–90 min and then fiberized in a disk refiner. Both bamboo and straw have been successfully pulped by the cold soda process.

Scientists at the USDA Northern Regional Research Laboratory in Peoria, IL developed a mechanochemical process for pulping straws and bagasse. The agro-material is charged to a large hydrapulper with an oversized rotor containing a solution of caustic soda or a caustic soda/sodium sulfite mixture at atmospheric pressure. A high liquor to material ratio (10–15:1) is employed and direct steam is used to raise the temperature to 95–100°C. The strong mechanical action of the rotor on the plant material in the presence of the alkali results in exposure of a significantly greater surface area resulting in very rapid pulping of the agro-material. The yields from mechanochemical pulping are typically 6–8% higher than conventional kraft or soda cooking, and the method offers high production capacity, simplicity of operation, and modest capital investments (Misra, 1980; Clark, 1969).

Chemical pulping of agro-based materials has been carried out by alkaline (kraft, soda, soda–anthraquinone), acidic (sulfite, nitric acid), neutral (neutral sulfite), organosolv (Alcell–ethanol, ester), and several specially developed processes. The latter include the Celdecor–Pomilio (soda–chlorine), the two-stage Cusi, and the NACO processes. Lime and lime–soda processes have also been used in special cases for lower grade pulps. Lime cooking is carried out with 60–120 kg of CaO per ton of plant material or 120–180 kg when using dolomite (Rydholm, 1965). The acid sulfite process has not proven very suitable for agro-based fibers since the strength properties of the pulps are usually lower, and in some cases the yields are lower as well. However, sulfite pulping can produce good pulps from some bast and leaf fibers, although persistent problems of pollution and chemical recovery from sulfite pulping severely limit the utility of this process. Ammonium sulfite or nitric acid pulping could be used in small mill situations where the expense of a chemical recovery operation cannot be warranted. With these processes the effluent can be utilized as a fertilizer for crop production.

The alkaline processes have proven very suitable for pulping agro-plants, giving good strength properties and high yields. However, it is important to note that since many agro-based materials are pulped so rapidly, there may be no advantage to using sodium sulfide with the caustic soda (kraft pulping) for pulp production. Usually the soda process is sufficient to give good yields and good strength properties without the attendant air pollution problems associated with the kraft process. However, with the denser bamboo species, the kraft process is preferred to the soda process, since in this case it gives better strength and higher yields (Hurter, 1991). Also a number of mills in India utilize the kraft process to pulp jute and bagasse. The paper properties required by the end user ultimately dictate the choice of the process.

Organosolv pulping has only recently been intensively evaluated for use with agro-based materials. Organosolv processes offer the advantage of reduced pollution loads, improved by-product recovery and smaller scale operation. Modern, economically-sized kraft mills currently being built around the world are, for example, designed to handle 1,000 tons of pulp per day or larger and cost over $600 million. These mills are designed for use with a large volume wood feedstock and are not always suitable for agro-based materials that are low density and often widely dispersed, such that transportation costs for delivery of such large volumes of materials can be cost prohibitive. Thus, the advantage of a smaller scale operation can be very important, particularly in developing countries without an extensive transportation infrastructure. Organosolv systems can usually be installed at one-quarter or less than the minimum economic size of a kraft mill. Both aqueous alcohol and acetic acid-based organosolv pulping processes have been evaluated for use with agro-based materials (Aziz and Sarkanen, 1989; Young, 1992; Lora and Pye, 1992; Cai, 1988). Pye, Lora and associates have published extensively on the Alcell process, which is a multistage, ethanol/water system originally designed for delignification of hardwoods (Lora and Pye, 1992; Aziz and Sarkanen, 1989). A diagram of the process is shown in Figure 6.5. They have shown that this process produces good pulps from kenaf, bagasse, and wheat straw, with the added advantage of by-product production of furfural at 2–3% yield (based on dry weight of raw material).

Agro-materials can be pulped by either batch or continuous processes. Due to matting and compaction of agro-based materials in the digester, it is necessary to use spherical rotary or tumbling digesters and a high liquor-to-material ratio for

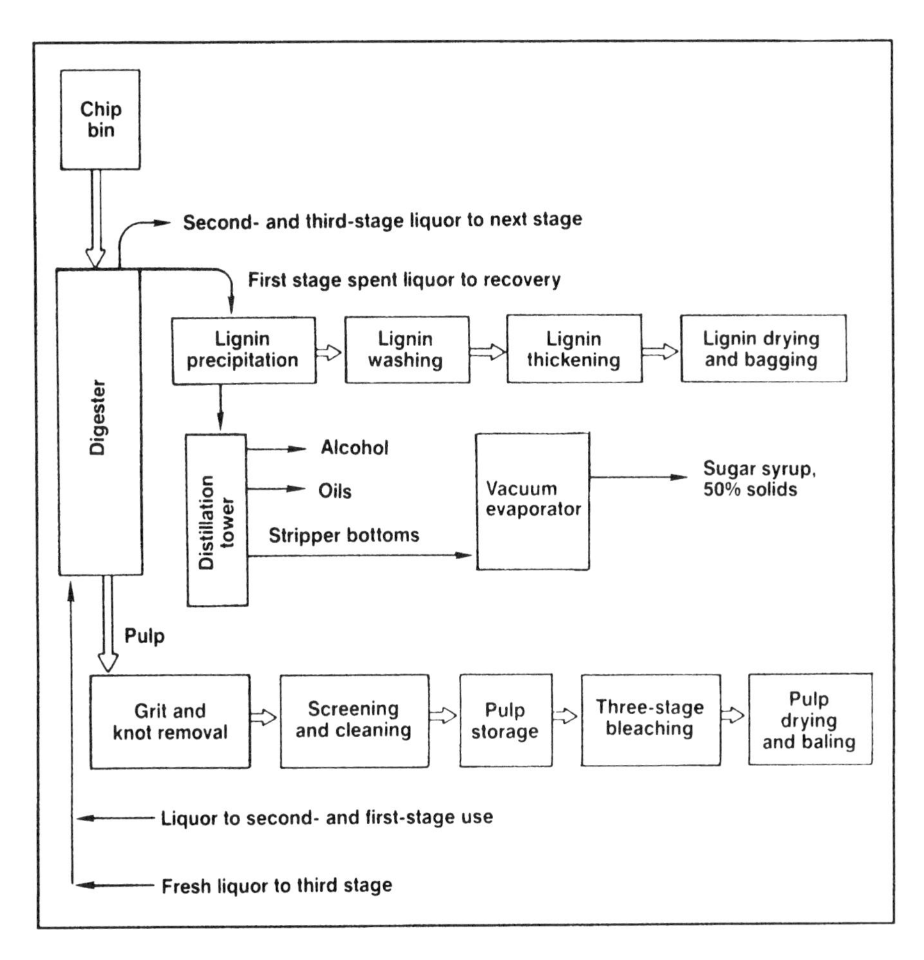

Figure 6.5 Flowsheet of Alcell™ pulping process and subsequent recovery of byproduct chemicals (Aziz and Sarkanen, 1989).

batch pulping (Figure 6.6). A considerable amount of air is also entrapped in the matted material during digester charging, which must be vented in the early stages of cooking. It is even more important to remove the trapped oxygen if cooking with an alkaline process, since this will lead to oxidative alkaline degradation of the polysaccharides and concomitant yield and strength losses for the pulp. These added steps can result in longer cooking cycles, from six to ten hours, for agro-based materials and, thus, lower output capacity and increased energy costs.

Continuous pulping is highly desirable for agro-based materials because of the rapid pulping properties. As previously mentioned, this is due to the lower lignin content and open structure of the agro-based material compared to wood. The pulping liquors rapidly penetrate and react with the lignin in the agro-plants. Jeyasingam (1988a) and Misra (1980) have listed a number of other advantages of continuous pulping for agro-materials: ease of handling of the raw material, whereby the bulk is reduced by mechanical compaction when the material is forced into the cooking zone; greater process flexibility; more uniform pulp quality; lower steam consump-

Figure 6.6 Spherical type digester utilized for pulping of agro-based materials (Photo by R. A. Young).

tion due to lower fiber-to-liquor ratio; higher black liquor concentration to the evaporators; and increased output of pulp. The disadvantages are a lack of flexibility, and the entire pulp mill is likely to be shut down for maintenance problems and breakdowns. High maintenance upkeep standards are necessary. Uniform feeding of the raw material is also critical and the availability of spare parts is crucial (Jeyasingam, 1988a).

A Celdecor–Pomilio continuous pulping process has been successfully applied to variety of agro-based materials including cereal straw, bagasse, and esparto. The process involves mild digestion of the agro-material with caustic soda, followed by gas phase chlorination, as shown in the flow diagram in Figure 6.7 (Clark, 1969). The alkali-treated material is fed to the top of the cylindrical reaction tower and is gravity fed downward. The towers are custom designed to different sizes to alter retention times and volume throughput. The dwelling time, temperature, and caustic soda concentration are varied, dependent on the raw material and degree of lignin modification desired. The partially-cooked pulp is then shredded and sent to the chlorination tower, washed, thickened, and sent to a high density caustic extraction tower. The pulp may also be bleached with hypochlorite (Misra, 1980; Clark, 1969).

Pandia and Kamyr continuous digesting systems have also been utilized with a wide variety of agro-based materials. These systems have been described in detail elsewhere (Rydholm, 1965; Biermann, 1993; Casey, 1980; Britt, 1970). A diagram of the Pandia system is shown in Figure 6.8 (Clark, 1969). It consists of a screw feeder, an impregnation chamber, conveyor screws, a rotary digester, and a cyclone-type collector. Chopped and cleaned raw material is conveyed to a mixer-impregnator where the cooking liquor is added, and then to the screw feeder where the density

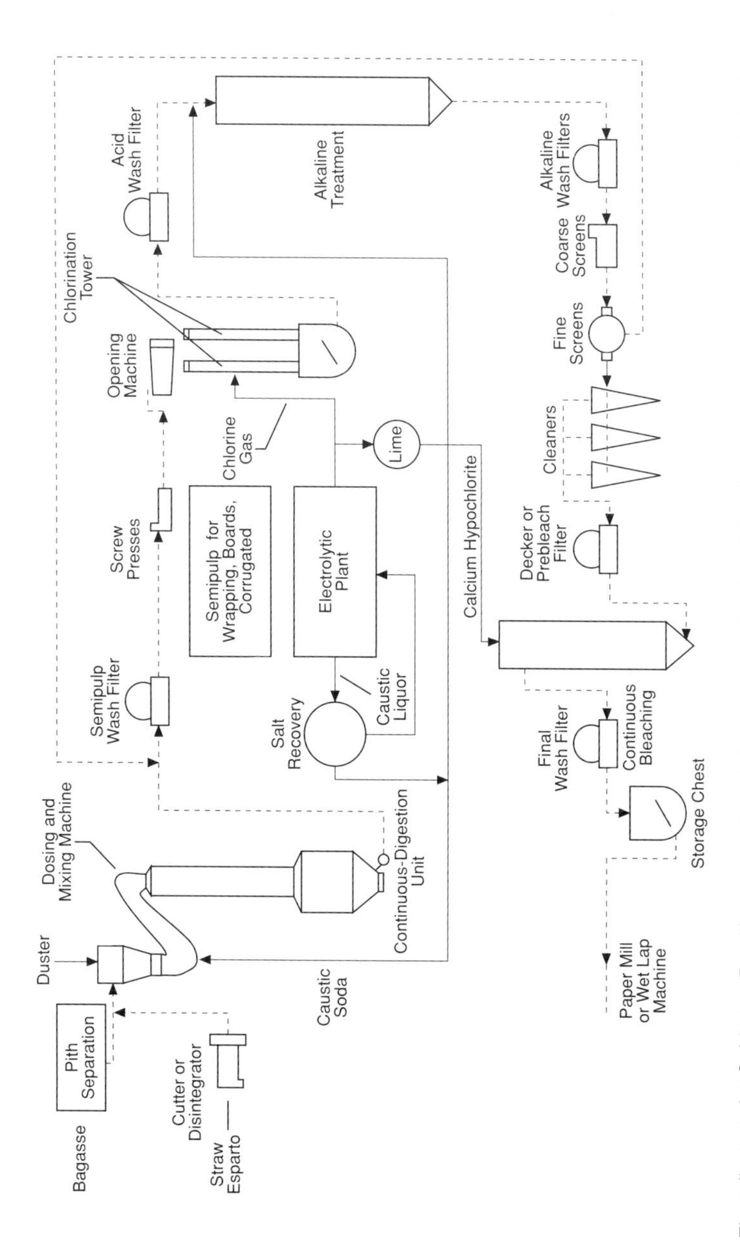

Figure 6.7 Flow diagram for Celdecor–Pomilio process for producing unbleached and bleached pulps from agro-plant materials such as bagasse, esparto, and straw. (Cellulose Development Corporation, Ltd.) (Clark, 1969).

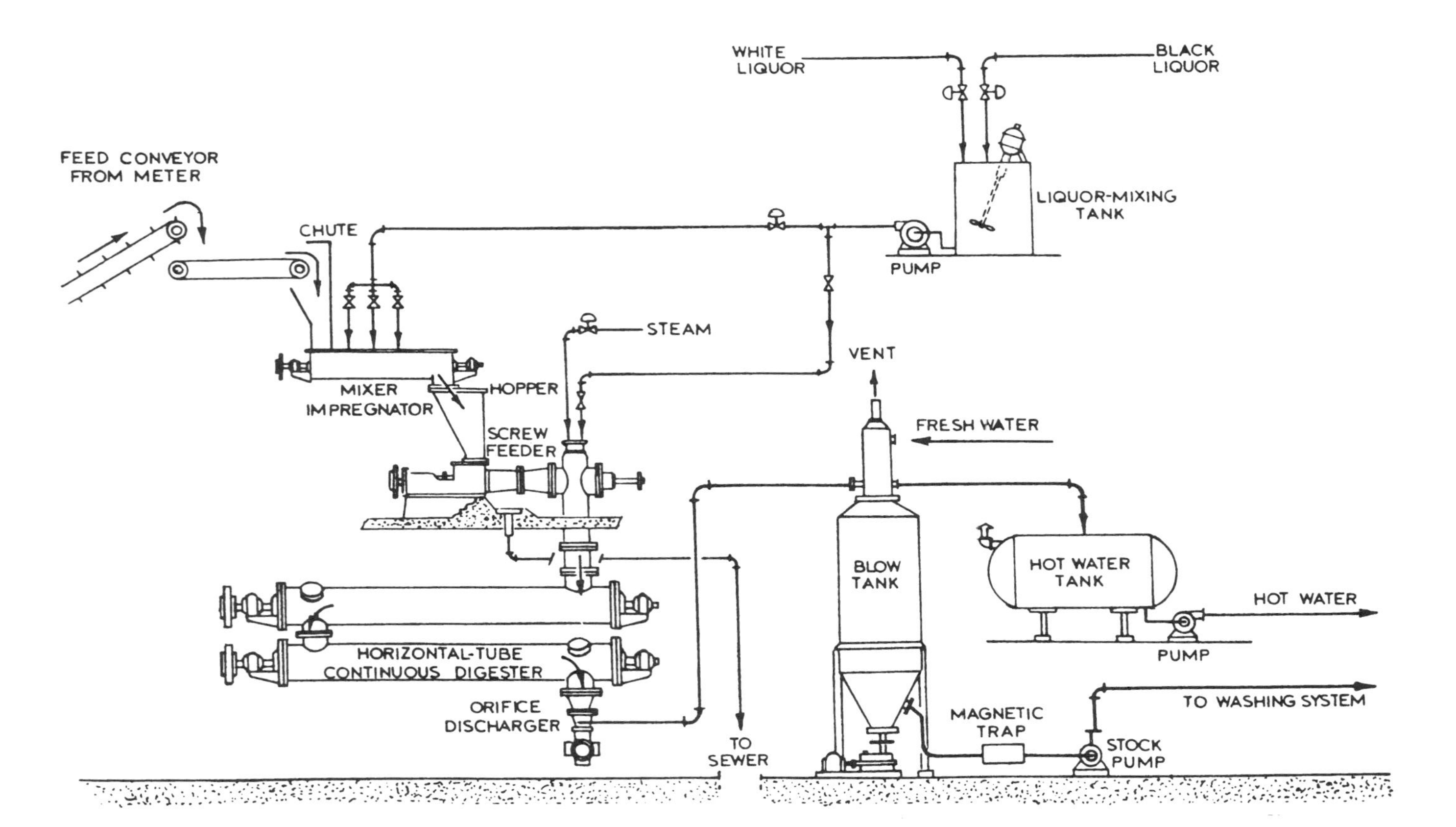

Figure 6.8 Pandia system for pulping of agro-plant materials (Parsons & Whittemore, Inc.) (Clark, 1969).

is increased. Additional liquor is added and the material is dropped into the first horizontal tube. Retention time in the tube is controlled by the speed of the internal conveyors. Due to the compacting action, the liquor-to-biomass ratio can be reduced to about 4.5:1 or less. This provides an advantage in terms of heat requirements and steam losses. The versatility of the system is demonstrated for a variety of materials as shown in Table 6.9 (Clark, 1969).

Table 6.9 Cooking Conditions and Results for Short-Period Continuous Pulping of Several Agro-Based Plants

Raw material	Pulping process, chemical	Chemical applied, as Na_2O, %	Dwell time, min	Steam pressure, psi	Pulp yield, %	Permanganate No.
Wheat straw[1]	Soda	4.6	8	75	67	—
	Soda	10.0	8	80	50	—
Rice straw[1]	Soda	9.8	5.5	100	39	4.9
Sugarcane bagasse[2]	Kraft	12.0	10	130	52	7.5
Reeds[1]	Soda	13.8	20	130	48	15.0
Esparto	Soda	12.4	20	120	52	6.0
Napier grass[1]	Soda	15.0	30	150	44	15.0
Bamboo (*Bambusa arundinacea*)	Kraft	18.2	30	130	45	10.0
Reeds[3]	Neutral-sulfite (Na_2SO_3)	19.1	25	150	53	12.3
	Neutral-sulfite (Na_2SO_3)	10.7	10	150	62	24.4

[1] Uncleaned raw fiber.
[2] Depithed or cleaned.
[3] Cleaned reeds; chemical applied as Na_2SO_3.

Source: J. E. Atchison, Indian Pulp Paper 17(12): 681–689, 694 (June, 1963), 18(2): 159–161, 163–165, 167–168, 171 (August, 1963), Clark (1969).

A modified Kaymr continuous digestion system, shown in Figure 6.9, is utilized for the annual crops (Misra, 1980). Direct steam, rather than forced circulation preheated liquors, is used in the modified digester. The pre-impregnator and the low pressure steaming vessel provide thorough mixing of cooking chemicals and uniform preheating to provide optimum conditions for the pulping reactions. Misra (1980) has suggested that this system offers such advantages as efficient utilization of steam, minimum damage to fibers, prevention of blowback problems due to the presence of the rotary valve and LP and HP feeders, and good control of the cooking time.

The Cusi pulping process and modifications thereof were designed specifically to handle bagasse but could be applied to a variety of other agro-based materials. The process involves first depithing in a specially designed mechanical scaping depither. This is followed by a mild alkaline cook (kraft or soda) of the fibrous fraction. The unique feature of the process is the separation of the cooked pulp into two fractions by screening, the vascular fibers rich in pentosan that need little or no refining, and the rind-vascular bundles, which are subjected to further pulping and refining. The pentosan-rich fiber fraction provides good bonding properties while the vascular-rind fraction has longer fibers that impart good tear strength. When the

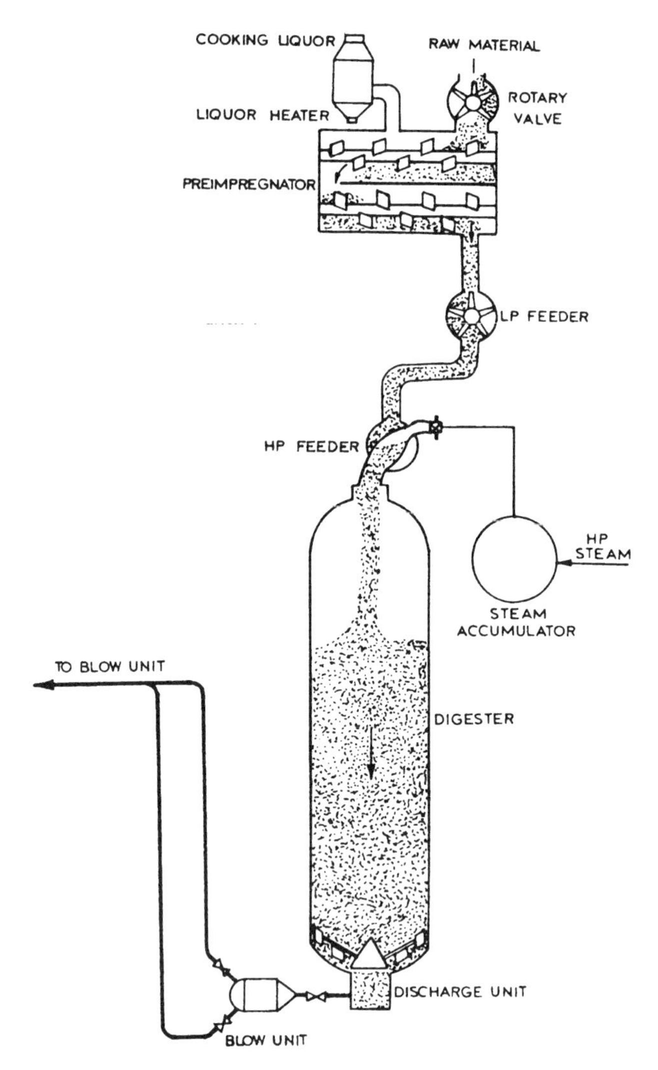

Figure 6.9 Celdecor–Kamyr continuous pressure digester for agro-plant materials (Courtesy of Cellulose Development Corporation) (Misra, 1980).

two fractions are then recombined, they give a pulp with good properties for a variety of paper end-uses (Misra, 1980; Clark, 1969; Jeyasingam, 1988a).

Steam explosion pulping has also been evaluated for application to agro-based materials. Marchessault (1991) has characterized the process as a "flash hydrolysis." Finely divided plant material is brought rapidly to a temperature of 230°C and a pressure of 500 psi and then is explosively ejected from the reactor. The organic acids in the plant material serve for pH control, and acetic acid is always found in the effluent gas. The products recovered include cellulose-rich fiber, a xylose-rich stream and a lignin-rich stream. Process options after steam explosion include: no fractionation, water extraction, aqueous alkali extraction or solvent extraction with solids recovery by several techniques.

The steam explosion process can also be either batch or continuous. The batch process is similar to the Masonite gun technology developed many years ago. The continuous processes are based on a high pressure feeding system, followed by a digester equipped with a transfer screw and a blow-down valve that acts as an orifice. The residence time of the fiber material through the digester is controlled by the characteristics of the rotating transfer screw and the opening sequence of the feeder choke cone and the discharge blow valve. A commercially available steam explosion reactor is marketed by Stake Technology of Canada.

A variety of agro-based materials have been evaluated by steam explosion technology, including peanut shells, bagasse, hemp, flax, and straw. Marmers et al. (1981) have developed a "Siropulper" for agro-based materials based on steam explosion pulping. A distinct advantage of the steam explosion process is that the whole plant material can be utilized without going through the usual retting process to separate the bast and core materials. The possibility of marketing a wider range of products from the various extraction streams is also possible and could be particularly important for developing countries who also have a need for liquid fuels. Ethanol could be produced as a fermentation product from the sugar streams of the discharged exploded biomass (Chornet and Overend, 1991).

3.5 Bleaching of Agro-Based Pulps

Agro-based chemical and semi-chemical pulps are relatively easy to bleach and the design of the bleachery is very similar to that for wood. A three-stage continuous system would typically involve chlorination, caustic extraction, and hypochlorite (CEH) stages. Sulfur dioxide is sometimes utilized as an antichlor in the final washer. Chlorine dioxide is used in only three mills, two of which produce dissolving pulps from reeds and bamboo.

The chlorination towers are usually operated upflow at low density and at ambient temperature, while the extraction and hypochlorite towers are usually operated downflow at high density and elevated temperatures. Some mills, however, operate at either all upflow or all downflow (Misra, 1980).

Some mills also split the hypochlorite bleaching into two stages. This allows the desired amount of retention time without the necessity of an excessively large single hypochlorite tower. In some cases the hypochlorite stage is followed by a treatment with an alkaline buffered hypochlorite solution (Misra, 1980; Clark, 1969; McGovern, 1967).

Thorough screening and cleaning are required for a number of bast and leaf materials prior to bleaching to remove shive and epidermal material. Thorough washing of the pulp can reduce excessive consumption of the bleach chemicals. The chemical requirements for pulping and bleaching a number of plant fibers are given in Table 6.10 (McGovern, 1967). The higher chlorine consumption for the rice straw relative to the κ number is because the pulp was bleached to a higher brightness. The Indian bamboo *Dendrocalamus strictus* is apparently more resistant to pulping and bleaching compared with other bamboos. Different bamboos from both Pakistan and the Philippines were much easier to bleach (McGovern, 1967).

Table 6.10 Chemical Requirements for Pulping and Bleaching of Agro-Based and Wood Materials

Fibrous material		Pulping		Bleaching %	
Kind	Lignin	Active alkali,[1]%	K no.	Total chlorine,[2]	Brightness
		Non-wood[3]			
Fibers					
Rice straw	12–14	7–8	6–8	6–7	88
Esparto	17–19	10–11	7–9	4–5	80
Bagasse	19–21	11–12	9–12	5–8	80–83
Wheat straw	17–19	14–16	8–10	7–8	80–85[4]
Bamboo (India)	22–30	15–17	16–18	11–13	75–80
(Pakistan)		13–14		9	80
		Wood Fibers[5]			
Hardwoods	18–25	15–16	14–16	6–8	85–90
Softwoods	26–30	20–22	16–20	7–9	88–90

[1] Based on raw material.
[2] Based on unbleached pulp.
[3] CEH or CEHH systems.
[4] Mill with CEHD system.
[5] CEDED or CEHDED systems.

Source: McGovern, J. N., *TAPPI*, 50(11), 63A, 1967.

4. PULP AND PAPER PROPERTIES OF AGRO-BASED MATERIALS

In this section the pulping characteristics, pulping processes, and paper properties for specific plants will be reviewed. Greater emphasis is placed on those materials that have been more extensively studied or have been of greater commercial significance for the pulp and paper industry. There is usually a regional emphasis dependent on the availability of the resource, lack of availability of wood, transportation infrastructure, and economic structure of the country. Thus, bagasse is relatively more important in Mexico, straw in China, and bamboo in India. A summary of the various pulping processes used for agro-based materials and the yields is shown in Table 6.6.

In general, abaca pulps show the greatest tear strength, as well as having high burst and tensile strengths. Sisal, kenaf bast, sunn hemp, and bamboo also produce long fiber pulps that have high tear strength. These pulps can be utilized as substitutes for softwood pulps (Atchison, 1994).

Pulps from reeds and some grasses have medium length fibers (about 2 mm) and are excellent for blending with other pulps for many grades of paper. Cereal and rice straw and bagasse pulps, however, have short fibers and lower tearing strengths. These pulps are more suited as substitutes for hardwood pulps. In some countries, quality writing and printing papers are produced with 90% bagasse or straw in the furnish; and corrugating medium and linerboard with 75% and 50% of these pulps in the furnish, respectively (Atchison, 1994). Typical properties of agro-based pulps are shown in Table 6.11 (Hurter, 1991).

Table 6.11 Typical Physical Properties of Agro-Based Pulps

Raw material	Process	Type of pulp	Freeness	Breaking Length (m)	Burst g/cm² /gsm	Tear g/cm² /gsm	Fold	Bulk cc/g
Wheat straw	Soda	Unbleached	50 SR	7200	40	40	100	1.50
		Bleached	50 SR	6800	42	56	70	1.48
Rice straw	Soda	Unbleached	50 SR	5420	38	54	110	1.45
		Bleached	50 SR	5300	41	50	90	1.43
Reeds	Kraft	Bleached	50 SR	6700	30	45	250	1.65
Bagasse	Soda	Unbleached	20 SR	7700	40	72	300	1.58
		Bleached	25 SR	6800	42	70	210	1.54
Bamboo	Kraft	Bleached	40 SR	6500	40	155	450	1.40
Esparto	Soda	Bleached	40 SR	4000	40	48	100	2.00
Kenaf (bast)	Kraft	Bleached	40 SR	10300	56	72	570	1.60
Jute	Lime	Unbleached	35 SR	5800	45	135	700	1.75
Flax	Soda	Bleached	28 SR	5000	35	200	200	1.90
Hemp	Soda	Bleached	28 SR	7000	53	180	460	1.85
Sisal	Soda	Bleached	28 SR	9000	79	310	1470	2.10
Softwood (spruce)	Kraft	Bleached	28 SR	11000	90	110	1900	1.70
Hardwood (beech)	Kraft	Bleached	45 SR	4000	22	45	20	1.54
(birch)	Kraft	Bleached	45 SR	7600	50	82	350	1.40

Source: Hurter, A. M., Utilization of annual plants and agricultural residues for the production of pulp and paper, *Non-Wood Plant Fiber Pulping*, Prog. Rept. No. 19, TAPPI Press, Atlanta, 49, 1991.

4.1 Bagasse

Bagasse is readily available and easily accessible in many tropical and subtropical countries in the world and is often very abundant in countries without large supplies of wood resources. Sugarcane, *Saccharum officinarum*, is of course grown for its high sucrose content, and bagasse is the fibrous residue left over after the crushing and extraction process. Although the fibrous bagasse residue is usually burned in mill boilers for production of steam and energy, the bagasse has much greater economic value as a raw material for pulp and paper production. An important advantage with bagasse is that there is usually very little problem associated with collection of the fibrous resource; the costs of collection, processing, and washing are borne by the sugar mill. However, the bagasse must be properly depithed and stored to produce high quality pulps (Figure 6.10).

Many of the early attempts to utilize bagasse were for production of newsprint. Most of these attempts were based on the use of 100% full chemical pulps for the newsprint and all were unsuccessful. It is usually necessary to use both mechanical and chemical pulps in the furnish to produce a suitable quality newsprint, and this simple axiom was apparently ignored in many of the early attempts to use bagasse pulps. However great success has been achieved in recent years for utilization of

Figure 6.10 Bagasse is stored in carefully stacked bales in India to avoid degradation (Photo by R. A. Young).

bagasse in practically all grades of paper (Atchison, 1992; Atchison, 1987a; Atchison, 1987b; Orgill, 1988; Abril et al., 1988; Venkataraman, 1988; Sawhney, 1988).

Bagasse pulping in 1992 was estimated to be up to almost 2.3 million metric tons, from the meager capacity of only 120,000 metric tons in 1950. The grades of paper in which bagasse pulps are now utilized include: bag, wrapping, printing, writing, toilet tissue, toweling, glassine, bleached boards, and coating base-stock (Atchison, 1992). Saavedra et al. (1988) have also shown that bagasse pulps are suitable for fluff pulp and absorbent disposable products. It took much longer to develop a suitable newsprint grade of paper from bagasse. This was accomplished in the 1980s by use of newer technology based on "enhanced" mechanical pulping where chemical pretreatments are utilized prior to disc refining, combined with appropriate combinations with chemical pulps to give a suitable newsprint furnish.

4.1.1 Depithing of Bagasse

It is now well established that efficient depithing of bagasse is necessary to obtain a high quality pulp from bagasse both for mechanical and chemical processing. If the pith is retained with the pulp, it reduces the strength, opacity, and brightness of the paper; and, the general appearance of the paper is diminished because the pith retains considerable amounts of dirt and extraneous materials accumulated from processing in the sugar mill. The pith also causes excessive consumption of chemicals in the pulping process, and the cooked pith becomes gelatinous which necessitates reduction in paper machine speeds (Atchison, 1993, 1987a; Clark, 1969; Misra, 1980).

Several methods have been developed for depithing of bagasse and are classified as dry, humid or moist, and wet depithing. Combinations of the methods have proven to be more effective for removal of the pith. Dry depithing involves mechanical disintegration in a bale breaker and screening to remove the pith, dirt, and extraneous matter. About two-thirds of the pith is removed with this process but an appreciable amount of bast fiber is also lost (Clark, 1969; Misra, 1980).

Moist depithing is generally done directly at the sugar mill with the plant containing about 50% moisture. The pith material is then used mainly for fuel in the mill boilers. There are a variety of machines designed for the moist depithing such as the Horkel, Rietz, Peadco, Gunkel, etc. Basically all the machines work on the same principle; that is, they are designed to break open the fiber bundles and dislodge the pith by rubbing and crushing actions. The Horkel machine, for example, employs a modified swing hammer surrounded by a perforated plate through which the fragmented pith is discharged. The moist methods generally remove about two-thirds of the pith fraction (Atchison, 1987a; Clark, 1969; Misra, 1980).

Wet depithing methods are usually carried out at the pulp mill and yield a high quality bast fiber. This approach generally involves disintegration of the whole plant material in a hydrapulper (Figure 6.11). The disintegrated material is dewatered and then sent to a depithing machine composed of swing hammers or a disk mill. The milled material is further dewatered and screened to remove the pith material (Clark, 1969).

Figure 6.11 Wet depithing of bagasse provides a high quality pulp for papermaking (Photo by R. A. Young).

Atchinson (1993) recommends a combination of moist and wet depithing to obtain the highest quality pulp, especially for mechanical and chemimechanical pulps from bagasse. He recommends that a minimum of 40% of the dry weight be removed in the moist depithing stage and that an additional 10–12% of the residual weight of the moist depithed bagasse be removed in the final washing and centrifugal wet depithing stage. Atchison recommends the following equipment to properly carry out the depithing of bagasse:

1. A metering device such as a twin-pin wheel feeder or metering bin.
2. A belt conveyor to receive the bagassse from the metering system and convey it to the pulper.
3. A hydrapulper to slush the moist depithed bagasse at 2.5–3% consistency. The equipment should include a Weir box to control the level in the hydrapulper and a junk box to remove extraneous materials.
4. A centrifugal pump to discharge the pulper through low gravity drop centrifugal cleaners.
5. A distribution box to distribute the bagasse slurry to drainers and the wet depither.
6. A drainer to dewater the bagasse to 10–12% consistency for wet depithing. This device can be perforated drums, vibrating screens, double screws with perforated bottoms, etc.
7. Centrifugal wet depithers.
8. Belt conveyors for feeding and conveying.
9. An equalizing screw or metering system to properly feed the disc refiners.
10. A dewatering and disposal system for the wet pith for fuel.

4.1.2 Mechanical Pulping

Attempts to produce satisfactory groundwood pulps from bagasse have been unsuccessful. The strength properties are excessively low due to extensive disintegration of the bagasse fiber in the grinding process. In 1961 however, a U.S. patent was granted to the Hawaiian Sugar Planters Association and the former Crown-Zellerbach Corporation for production of mechanical pulps from depithed bagasse with disc refiners to give a refiner mechanical pulp (RMP). The pulp was refined at 6–7% consistency and then thickened in a screw press to 30% consistency. The pulp had characteristics similar to mechanical wood pulp and imparted good opacity, good ink absorbency, and good printability to a newsprint sheet (Lathrop, 1957; Wethern and Captein, 1960; Henderson and Knapp, 1961; Knapp and Wethern, 1957). The physical properties of an RMP-type pulp from bagasse are shown in Table 6.12 (Luna and Torres, 1981). Bleaching of bagasse RMP with sodium hydrosulfite gives a pulp with a brightness of 60–63, which was suitable for blending with semibleached softwood pulp in a ratio of 70:30 to give good newsprint paper. It was also possible to produce high opacity newsprint from a furnish of 60% bleached bagasse mechanical pulp, 30% bleached bagasse chemical pulp, and 10% bleached or semibleached softwood kraft pulp (Luna and Torres, 1981).

Thermomechanical pulps (TMPs) have also been produced from bagasse by a process developed by Asplund-Defibrator of Sweden, by preheating to 130°C prior to disc refining. This resulted in improved strength properties of the pulp. Fuentes (1981) evaluated thermomechanical pulping of bagasse at the C. E. Bauer pilot plant in Springfield, Ohio. The steamed bagasse (5 min at 20 psig) was discharged into

Table 6.12 Comparison of Characteristics of Bagasse Mechanical (RMP) and Chemimechanical Pulps

		Chemimechanical	
Characterictics	**Mechanical RMP**	**Modified Soda**	**Alkaline Sulphite**
Freeness, °SR	65	65	65
Density, g/cm³	0.309	0.337	0.374
Tear Factor, g/gsm	16	40	44
Burst Factor, (g/cm²/gsm	8.2	11.4	13.7
Breaking Length, m	1250	2200	2466
Elongation, %	—	1.3	1.6
Brightness, %	45.0	38.0	44.6
Opacity, %	99.0	99.4	99.7
Specific Scattering Coefficient, cm² g	450	440	412

Source: Luna, G. V. and Torres, C. A., High yield pulp from bagasse, *Proc. Pulping Conf.*, TAPPI Press, Atlanta, 69, 1981.

a 36-inch double disc refiner and the pulp cleaned on pressurized screens. Fiber classification showed that 63.6% of the fibers were retained on a 16 mesh screen. The properties of the pulps, however, were not good with a tear index of 2.5 and a breaking length of only 1026 at 275 CSF. Energy consumption was also high.

The Hawaiian Sugar Planters Association, in cooperation with the former Crown-Zellerbach Corporation, also evaluated a number of chemimechanical approaches such as treatment with 1–7% sodium hydroxide, or 0.1–4% sodium sulfite, or mixtures of the two chemicals prior to disc refining (Lathrop, 1957; Wethern and Captein, 1960; Henderson and Knapp, 1961; Knapp and Wethern, 1957). Later work as part of the "Cuba-9" project brought the chemimechanical process to commercial fruition (Luna and Torres, 1981; Bambanaste et al., 1988). Luna and Torres (1981) evaluated chemimechanical pulping of bagasse and a comparison of the properties of RMP and two chemimechanical pulps (CMPs) from their work are shown in Table 6.12. For production of the RMP, the bagasse was pretreated in hot water for 60 min; for the CMP, the bagasse was treated either with caustic soda (6–10 g/L) at 80–100°C for 10–20 min or a mixture of sodium hydroxide and sodium sulfite (alkaline sulfite) to give 6–10 g/L sulfite concentration, for 20–40 min at 80–100°C. The yield of the washed unbleached pulps was about 85%. As shown in Table 6.12, the chemical pretreatment increased the strength properties about 50% compared to the RMP. The addition of the sodium sulfite also improved the unbleached pulp brightness. The improvement in the strength properties can be partially attributed to the retention of a greater amount of long fibers in the CMP material as shown in Table 6.13 (Luna and Torres, 1981).

Zanuttini and Christensen (1991) performed a comprehensive study of the effect of alkali charge on chemimechanical pulping of bagasse. The tensile and tear strengths clearly increased and the light scattering coefficient decreased with increasing alkali charge for handsheets prepared for the chemimechanical pulps (Figure 6.12). Tear and tensile strengths doubled when 5% caustic soda was charged in comparison to RMP. Quirarte (1991) found that the paper properties could be

Table 6.13 Fiber Classification and Physicomechanical Properties of Fractions and Whole Bagasse Pulp

Type of Pulp	Alkaline Sulphite		Modified Soda		Mechanical Pulp
Classification	%	mm	%	mm	%
R–14	9.5	1.92	18.0	1.86	1.5
R–28	17.9	1.56	12.0	1.70	3.9
R–48	18.7	1.15	13.0	1.18	11.6
R–100	18.7	0.78	17.0	0.85	12.6
R–200	12.3	0.46	13.0	0.56	22.6
R–200	22.6	—	27.0	—	47.6

Source: Luna, G. V. and Torres, C. A., High yield pulp from bagasse, *Proc. Pulping Conf.*, TAPPI Press, Atlanta, 69, 1981.

improved by increasing levels of addition of peroxide to such caustic pretreatment solutions, but economic factors limit the use of peroxide to no more than 3%.

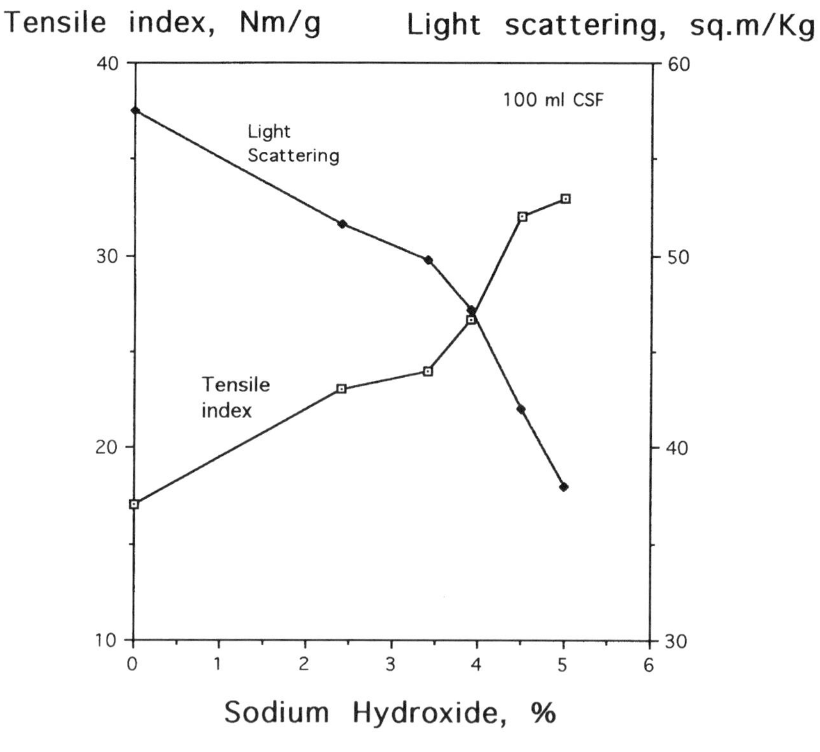

Figure 6.12 Effect of alkali charge on tensile strength and light scattering at 100 ml CSF for chemimechanical pulp from bagasse (Zanuttini and Christensen, 1991).

It was also shown by Zanuttini and Christensen (1991) that the increase in color that takes place in a cold soda pretreatment of bagasse is notably dependent on the temperature and caustic soda charge. At temperatures near 90°C there was a very detrimental effect; high alkali charge treatments at 80°C also considerably darkened the bagasse. Therefore, the use of lower temperatures (near 60°C) was recommended.

However, at the lower temperatures, a longer retention time is required. The significant drop in the light scattering coefficient, loss of porosity, and increase in fine shive content when the alkali charge was increased beyond 4% suggest that a 4% alkali charge is the highest practical limit. The addition of 2% sodium sulfite to the pretreatment liquor reduced the formation of colored chromophores, even to a greater degree than chelating agents. Stored bagasse darkened less than fresh bagasse and was more responsive to the sodium sulfite treatment.

Lai et al. (1988) evaluated various pretreatments for chemimechanical pulping of bagasse and found that treatment with neutral sulfite solutions gave superior properties compared to alkali or percarbonate treatments. Under the neutral sulfite conditions, a higher extent of sulfonation was achieved for bagasse compared to wood species, which was attributed to a higher free phenolic hydroxyl content in the bagasse lignin. Compared at the same pulp yield, bagasse and aspen CMP had similar strength properties.

A variety of chemimechanical processes have been developed specifically for bagasse over the years. In Cuba, de la Rosa produced CMP from bagasse in a two-stage digestion sequence; the first stage was a prehydrolysis with water at 150°C, and the second stage was treatment with kraft liquor at 12% active alkali. The final yield of unbleached pulp was 65% (Misra, 1980). The Beloit–SPB process takes into account the heterogeneous nature of the bagasse fiber similar to the Cusi process previously described (Ranqamannar et al., 1991; Venkataraman and Torza, 1987). Different types of treatments are specified for the different hard and soft fractions of the bagasse. The Beloit–SPB process involves thermomechanical pulping of the bagasse followed by a fractionation step to separate the fines fraction from the hard fiber fraction. The hard fiber fraction is then further treated by chemimechanical pulping. The suggested advantages of this process are: (a) high opacity fines (TMP) are generated from bagasse with low refining energy; (b) high quality, high strength, chemimechanical pulp (CMP) is produced from the long flexible hard fiber; and, (c) the yield, strength, and opacity of the final blended pulp are high due to the selective treatments (Ranqamannar et al., 1991; Venkataraman and Torza, 1987).

4.1.3 Chemical Pulping

The acid and bisulfite processes have proven to be unsuitable for pulping of bagasse. The process yields a brittle pulp that has low strength compared to pulps produced by alkaline processes. The traditional soda, kraft, and neutral sodium sulfite semichemical (NSSC) processes have all proven very suitable for production of excellent chemical pulps from bagasse. Generally, the soda-additive processes have been favored over the NSSC process.

As shown in Figure 6.13, depithed bagasse is readily delignified by the soda process (Granfeldt et al., 1991). At a cooking temperature of 165°C the κ number can be reduced to 20 in about 20 min. The alkali charge in these trials was 18%, but a further reduction in κ number requires a much higher alkali charge. Granfeldt et al. (1991) suggested the subsequent use of soda-oxygen treatments to further reduce the κ number to about 10 without causing attendant environmental hazards of chlorine treatments.

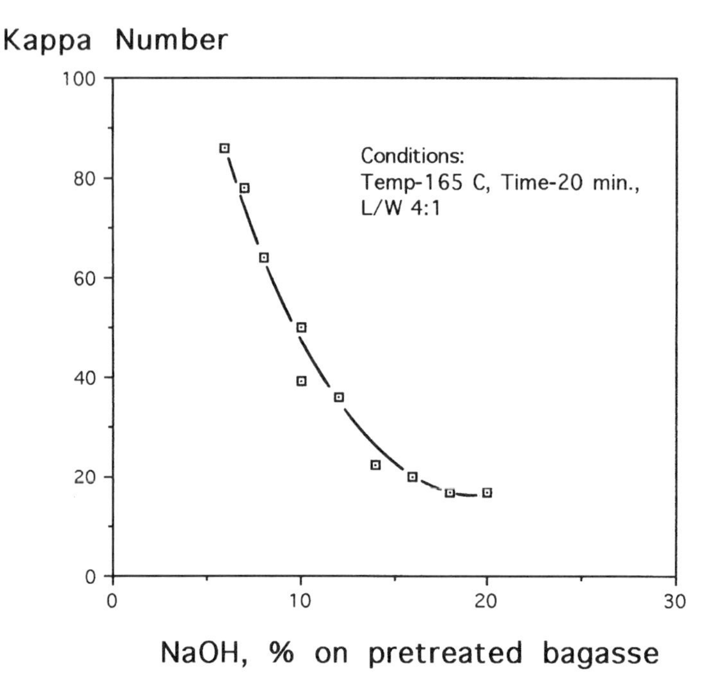

Figure 6.13 Kappa number versus sodium hydroxide concentration for soda pulping of bagasse (Granfeldt et al., 1991).

Knapp and Wethern (1957) carried out a comprehensive study of kraft pulping of bagasse. They found that depithed bagasse could be successfully pulped in times as short as 2 min at 340°F to give a 55–60% yield and a permanganate number of 7–9. Moderate strengths were obtained at 200 CSF as follows: burst factor–55, tear factor–71, breaking length–9,000 and fold–1,000. As shown in Figure 6.14, the burst and tensile strengths tended to decrease with longer pulping times. The maximum possible strength was achieved at very short times with relatively high chemical charges. In general, kraft pulping closely resembled soda pulping except the brightness was slightly higher and the strength properties slightly lower with the soda process.

Krishnamachari et al. (1981) carried out continuous pulping of bagasse via the soda process in a Pandia continuous digester in their mill in India. The cook was done at a pressure of 80–100 psi, using 12–14% caustic solution for 10–12 min. The solids-to-liquor ratio was about 1:3. The unbleached yield was 42–44% with a permanganate number of 9–11. This rapid cook produced pulps of acceptable quality as shown in Table 6.14. The pulps were readily bleachable (CEHH sequence) and the properties of the bleached pulps are also shown in Table 6.14. A wide variety of paper grades were produced from the bagasse pulp, in combination with wood pulps, and the properties of some of these papers are shown in Table 6.15 (Krishnamachari et al., 1981).

Mannar (1988) evaluated a number of variables for kraft pulping of bagasse in a Sunds continuous digester. He concluded that a certain minimum volume and

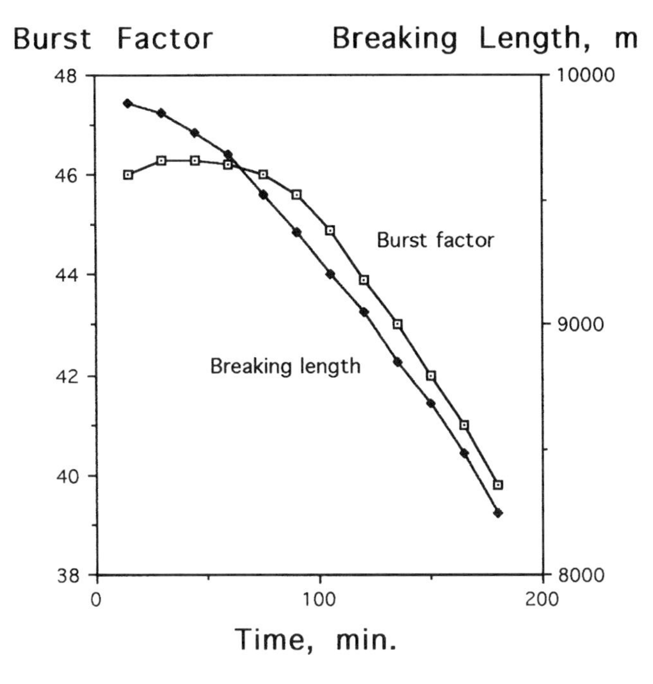

Figure 6.14 Strength versus cooking time at 171°C and 11% chemical charge for kraft pulping of bagasse (Knapp and Wethern, 1957).

Table 6.14 Properties of Unbleached and Bleached Bagasse Soda Pulp from a Pandia Continuous System

Freeness, °SR	Burst Factor ($g/cm^2/gsm$)	Tear Factor (g/gsm)	Breaking Length (m)	Double Folds
		Unbleached bagasse pulp		
45.0	43.0	48.3	5,757	195
45.4	45.6	48.3	6,583	225
42.0	46.1	49.1	6,114	282
		Bleached bagasse pulp		
39.0	33.6	35.4	6,012	72
45.5	40.8	42.9	6,115	50
48.2	35.4	33.9	5,787	47

Source: Krishnamachari, K. S., et al., Experiences of bagasse pulping with rapid continuous digester, Proc. Pulping Conference, TAPPI Press, Atlanta, 61, 1981.

concentration of cooking liquor are necessary to achieve uniform liquor distribution in the digester. The chemical uptake of the bagasse fiber is increased by cooking at higher temperatures and pressures at the optimal cooking liquor concentration of 75 g/L (as sodium oxide). This proved to be the best solution for keeping the shive level within reasonable limits for bleachable grades.

Table 6.15 Physical Properties of Paper Grades from Bleached Bagasse Soda Pulp Combined with Softwood Sulfite Pulp

Quality	Basic Weight g/m²	Burst Factor (g/cm²/gsm)	Tear Factor (g/gsm)	Breaking length (m)	Double Folds	Furnish
Creamwove	60	16	42	3000	8	70% bagasse: 30% softwood sulfite pulp
Duplicating	75	14	57	3150	7	80% bagasse: 20% softwood sulfite pulp
Duplicating	75	17	48	3100	9	100% bagasse pulp
Manifold	33	22	57	4000	7	70% bagasse: 30% softwood sulfite pulp
White Printing	60	18	60	3707	10	90% bagasse: 10% softwood sulfite pulp
White Printing	60	17	50	3061	8	100% bagasse pulp
White Printing	67	17	50	3400	8	100% bagasse pulp
White Printing	43	11	47	5332	5	—
Badhami Paper	55	10	44	2672	16	—

Source: Krishnamachari, K. S. et al., Experiences of bagasse pulping with rapid continuous digester, Proc. Pulping Conference, TAPPI Press, Atlanta, 61, 1981.

Several investigators (Upadhyaya et al., 1991; Tapanes et al., 1985; Sadawarte et al., 1981) have evaluated soda–anthraquinone (AQ) pulping of bagasse. Compared to a straight soda cook, Sadawarte et al. (1981) found that the yield was significantly increased, up to 4%, when 0.05–0.10% AQ was present in the cooking liquor. However the pulp yield was similar for kraft and soda–AQ cooking of bagasse. All three bagasse pulps (soda, soda–AQ, and kraft) had similar strength properties. Tapanes et al. (1985) also obtained higher yields and similar strength properties with soda and soda–AQ pulping of bagasse. They reported that the soda–AQ pulps were much easier to bleach. Upadhyaya et al. (1991) carried out pulping studies of bagasse with caustic soda, soda–oxygen, and soda–oxygen-AQ. They found that the addition of AQ to the soda–oxygen process resulted in increased yields. The strength properties of these pulps are shown in Table 6.16 (Upadhyaya et al., 1991).

Neutral sulfite and bisulfite processes (soda and ammonia bases) for pulping of bagasse were also evaluated by Captein et al. (1957). They found that the NSSC process gave pulps with slightly lower strengths but higher brightness and yield compared to the alkaline processes. However, the NSSC process required longer cooking times and more chemicals than either the kraft and soda processes. Bisulfite pulping of bagasse gave pulps with low strengths, low brightness, and high lignin contents. Thus, Captein et al. (1957) recommended the alkaline processes for chemical pulping of bagasse.

Table 6.16. Comparison of Soda, Soda–Oxygen and Soda–AQ Bagasse Pulps

Oxygen pressure Kg/cm²	AQ dose %	Pulp yield %	Screened yield %	Kappa number	Burst factor	Tear factor	Breaking length m	Brightness %
0	0	60.5	55.0	37.0	16.7	26.0	5233	30.2
5	0	64.6	62.4	35.2	18.2	27.3	5400	34.2
5	0.05	66.1	64.0	34.5	18.9	36.0	5422	34.3
5	0.1	68.1	66.1	33.0	22.8	41.0	5450	35.5
5	0.2	67.5	65.7	32.0	22.0	32.5	5475	35.0
5	0.5	67.2	65.5	32.5	23.8	35.6	5410	35.3

Cooking Parameters: Alkali charge – 10% (as Na_2O), Temperature – 140°C, Time to Temperature – 90 min., Time at Temperature – 30 min., Bath Ratio – 1:7, Freeness – 32 ± 2°SR.

Source: Upadhyaya, J. S. et al., Studies on Alkali–Oxygen–AQ delignification of bagasse, *Non-Wood Fiber Plant Pulping*, Prog. Rept. No. 20, TAPPI Press, Atlanta, 55, 1991.

A number of mills around the world (i.e., in India, Brazil, Philippines, Spain, and South Africa) have adopted the Celdecor–Pomilio process for pulping of bagasse. This process was developed in England based on the Italian Pomilio process. The process is continuous and can be viewed as having four stages starting with a mild soda cook followed by alternate bleaching and extraction stages (Figure 6.7). The primary advantage of these processes is that the main chemical required is just salt (sodium chloride), since the process uses sodium hydroxide and chlorine in roughly equal proportions that correspond to those obtained from an electrolytic cell. However, there is usually some excess of chlorine from these mills. Pulp strengths are slightly lower and yields higher from the Celdecor–Pomilio process compared to the kraft process. In countries where chemicals are difficult to obtain, the advantages of the readily obtained sodium chloride, and slightly higher yields from the Celdecor–Pomilio process may outweigh the disadvantage of the loss in strength (Hinrichs et al., 1957).

The Peadco process is another pulping process developed for bagasse that includes both the depithing and pulping operations (Misra, 1980). Fiber and pith are separated by the rubbing action when the bagasse is passed through a depither of swinging blades attached to a rotor. The depithed fiber is cleaned on screens and fed to a horizontal high pressure digester. Vapor phase pulping is carried out at 860 kPa for about 20 minutes. The pulp is blown from the digester at 145°C and then mechanically disintegrated in a disc refiner. Several mills in Central and South America and Taiwan utilize this process.

Hydrotropic pulping has also been evaluated for pulping of bagasse (Hinrich et al., 1957). This process involves cooking of the raw material in a concentrated solution of sodium xylene sulfonate, which was found to effectively delignify the bagasse. However, the hydrotropic pulps had low strength properties and low brightness values. Also, it was difficult to avoid precipitation of the lignin when the residual hydrotropic solution was diluted.

A number of investigators have evaluated organosolv pulping of bagasse (Winner et al., 1991; Lin et al., 1988; Duenas et al., 1993; Valladares et al., 1985; Carvelo et al., 1990). Winner et al. (1991) found that pulps comparable to market hardwood pulps could be produced from organosolv pulping of bagasse by the Alcell process previously described. High yield pulps (60–68%) were produced in the range of kappa numbers of 40–75. The strength properties of unbleached and bleached (EoDED) Alcell organosolv pulps are shown in Table 6.17 (Winner et al., 1991). Earlier reports that ethanol–water organosolv pulping was not suitable for bagasse were erroneous since proper pulping conditions were not utilized. A number of previous investigators have suggested that organosolv systems are not suitable for pulping of lignocellulosic materials due to lack of knowledge of the special operating conditions that must be utilized with this newer type of approach to pulping (Aziz and McDonough, 1987). Organosolv pulping is viable with practically all lignocellulosic materials and offers a number of advantages, as previously outlined, for pulping of agro-based materials, especially in developing countries.

Table 6.17 Bagasse and Kenaf Organosolv Pulp Properties

Pulp	Burst Index kPa/m²/g	Breaking Length km	Tear Index mN/m²/g	Bulk cm³/g
Kenaf				
Unbleached	4.4	7.6	9.9	1.8
Bleached DEDP	3.8	6.2	12.7	1.8
Diffusion Bagasse				
Unbleached	3.0	6.0	6.0	1.7
Bleached (E_0DED)	3.8	6.0	6.1	1.4
Rind Chip (sugarcane)				
Unbleached	3.4	6.1	5.0	1.6
Bleached (E_0DED)	3.5	4.6	5.4	1.7

Source: Winner, S. R. et al., Pulping of agricultural residues by the Alcell process, *Non-Wood Plant Fiber Pulping Conf.,* Prog. Rept. No. 20, TAPPI Press, Atlanta, 171, 1991.

Refining or beating of pulp is a necessary step prior to any papermaking process. The refining step results in both internal and external fibrillation of the pulp fibers that imparts good bonding characteristics to the fibers and, thus, good strength properties to the final sheet of paper (Young, 1986). Some amount of fiber shortening also occurs in the refining process, which means the refining must be well understood and carefully controlled. Different agro-based fibers have been shown to exhibit distinctly different responses to refining (Iyengar, 1981). For example, the stretch of bleached bagasse CTMP was found to be considerably lower than that of hardwood CTMP at the same energy input level. Rao et al. (1988) felt that this was due to the greater coarseness of the bagasse fiber that makes it less flexible. He found that bleaching rendered the pulps more flexible.

Bagasse soda pulps were shown to be especially delicate, and multipass refining was found to give the highest strength pulps, but at the expense of extra energy requirements. All strength parameters increased 20–25% with refining, with only a minor drop in tear strength. Bleached bagasse pulps were found to be even more delicate than unbleached, and little or no refining was recommended for maximum strength pulps (Iyengar, 1981; Liu, 1988). Similarly, it has been found with organosolv pulps that gentle refining yields superior pulp properties, and pretreatments have been found to give enhanced strength properties (Young, 1994c). It is probable that similar advantages could be realized with agro-based pulps since the agro-fibers also tend to be more brittle. Liu (1988) found that the brittleness of bagasse pulps was related to the average fiber length and the amount of non-fibrous cells in the pulp. He suggested that proper handling of the sugarcane to avoid over-crushing and adequate depithing helped to overcome the existence of brittleness in the pulp. Addition of long fibers and fillers to the bagasse pulp furnish also helped to alleviate the problem. The density of the finished sheet was found to be inversely proportional to the brittleness and shrinkage of the bagasse paper.

High quality dissolving pulps have also been produced from bagasse by prehydrolysis and kraft cooking (Misra, 1980; Mittal, 1991). Mittal (1991) optimized the parameters for production of the dissolving pulps and suggested the following conditions: acid prehydrolysis with 0.25% sulfuric acid at 170°C for 90 minutes with a 1:5 bath ratio which removed 49.4% of the pentosans; this was followed by cooking with 15–16% active alkali (as sodium oxide) and multistage bleaching with a CEHH(SO_2) sequence. The bleached pulp yield was 27.1% and the brightness and viscosity were 93% and 10.7 cps, respectively. The characteristics of water and acid-prehydrolyzed kraft cooked and bleached dissolving pulps from bagasse are shown in Table 6.18 (Mittal, 1991).

Table 6.18 Bleached Bagasse Dissolving Pulp Characteristics

	Prehydrolysis[1]	
Characteristic	**Water**	**Acid (0.25% H_2SO_4)**
Bleaching sequence	CEHH	CEHH
Bleached pulp yield on oven dry weight of original raw material, %	30.5	27.1
Alpha cellulose in (ash free) bleached pulp, %	90.5	94.4
Pentosans content, %	7.6	3.5
Viscosity of pulp in (Cuprammonium), cps	14.5	10.7
Brightness, %	91.0	93.0
Ash content, %	0.45	0.22
Silica content, %	0.08	0.07
1.0% NaOH solubility, %	2.5	2.7
18% NaOH solubility, %	3.8	3.5

[1] Prehydrolysis followed by kraft pulping.

Source: Mittal, K. C., Optimization of prehydrolysis – Kraft delignification of bagasse, *Non-Wood Plant Fiber Pulping,* Prog. Rept. No. 19, TAPPI Press, Atlanta, 23, 1991.

Steam explosion pulping (SEP) of bagasse has shown great promise for production of high quality pulps. Kokta and Ahmed (1993) have shown that SEP pulps from bagasse are superior to both bagasse CMP and CTMP. As shown in Figures 6.15 and 6.16, the bagasse SEP has much better breaking length with an 8% sodium sulfite pretreatment and a lower specific refining energy when the pretreatment includes 1% sodium hydroxide in addition to the 8% sodium sulfite. SEP also offers the possibility for by-product recovery that would add to the economics of the process.

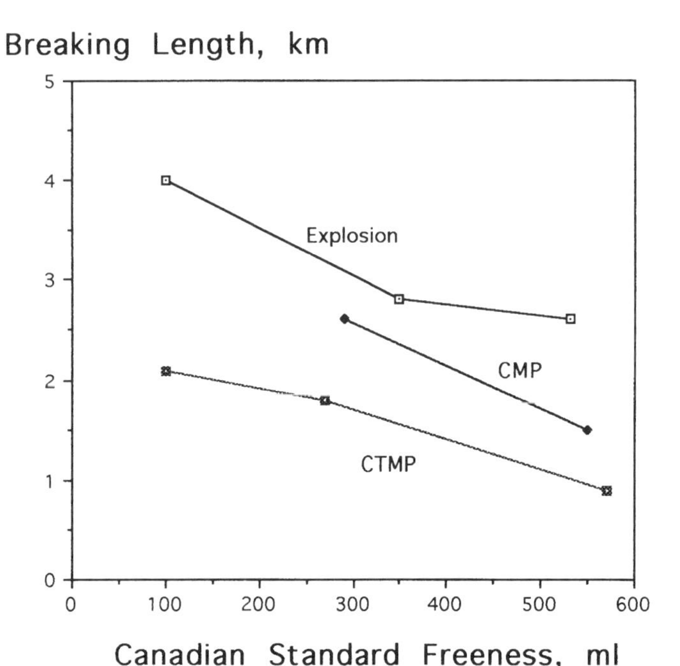

Figure 6.15 Comparison of the breaking length of bagasse explosion pulp, CMP and CTMP at various CSF levels (SEP pretreatment solution: 8% Na_2SO_3) (Kokta and Ahmed, 1993).

The conditions for a bagasse pulp bleachery are shown in Table 6.19 (McGovern, 1967). A brightness of 80–83% can be achieved with a low total chlorine consumption of about 4–5%, based on unbleached pulp. Fernandez et al. (1985) found that a considerable reduction in the retention time in bleaching could be realized in a C-E/H bleach sequence if higher temperatures were used during chlorination and there was a partial or total addition of hypochlorite to the extraction stage. Benitez and Bottan (1985) found that the use of an oxidative extraction stage (Eo), by applying oxygen during caustic extraction, considerably reduced chemical consumption in bagasse bleaching. The kappa number was reduced by 35% with a brightness increase of about 11 points. Use of the Eo stage also allowed reduction of applied hypochlorite with the beneficial effect of increased viscosity and yield, and lower shive content. Less polluting bagasse bleaching alternatives such as hydrogen peroxide and oxygen (O) have also been evaluated by several investigators with positive results (Chin et al., 1988; Dou, 1988).

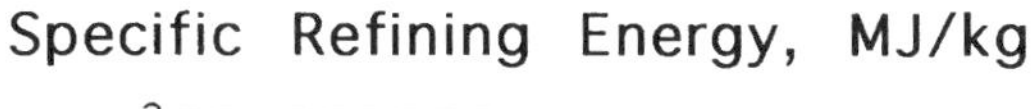

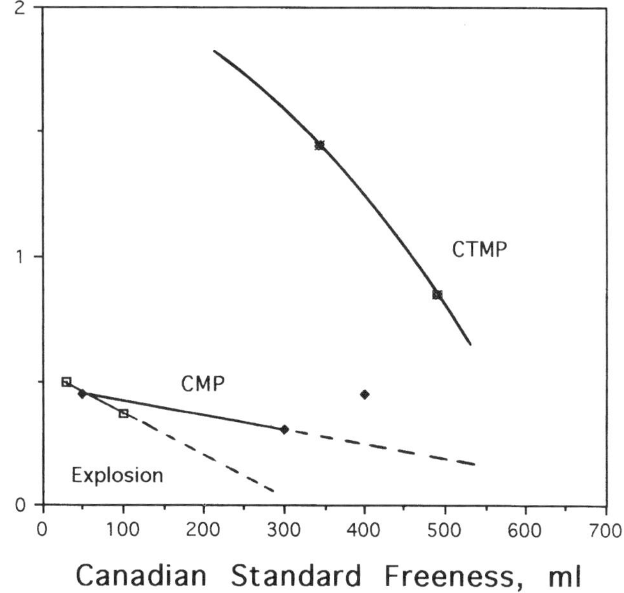

Figure 6.16 Comparison of the specific refining energy of bagasse explosion pulp, CMP and CTMP at various CSF levels (SEP pretreatment solution: 8% Na_2SO_3 + 1% NaOH) (Kokta and Ahmed, 1993).

Table 6.19 Conditions for Bleaching of Bagasse Pulps

Chemical		Amount %	Consistency %	Temperature °C	Time h
Chlorination	Cl_2	2.3–3.5	3.5	40	1
Caustic extraction	NaOH	1.4–1.9	11.5	60	2
Hypochlorite bleaching	$Ca(OCl)_2$	1.6[a]	11.5	30	3

[a] Calculated as available chlorine.

Source: McGovern, J. N., TAPPI, 50(11), 63A, 1967.

4.2 Straw

Straw has been utilized for centuries as a source of papermaking fibers, especially in China. It was a major source of fiber in Europe and North America until the replacement by wood in the early 1920s. The increased labor costs associated with the collection, storage, and handling of straw and changes in harvesting methods spelled the demise of straw as a raw material for papermaking in the U.S. and much of Europe. However, straw continues to be an important pulp and paper raw material in some Eastern European, Middle Eastern, South American, and Asian countries, particularly in China. One of the main reasons for using straw for pulp is that it is readily available as a residue from food crops, and straw pulping has dominated in developing countries where inexpensive manual labor is available to cut and handle the straw. The crop is cut close to the ground and large quantities are available after separation of the grain. Unbleached straw pulps are utilized for corrugating medium and boards, packaging, and bleached straw pulps for writing and printing papers (Clark, 1969; Misra, 1987).

It is possible that straw pulping could be reintroduced in the U.S. in the future with the adoption of modern technology for collection, baling, and handling of straw and with horizontal tube digesters where cooking times are 10–12 min. Substantial amounts of cereal straw are available in almost all 48 states of the U.S., with 78 million tons of wheat straw alone available (Atchison, 1994).

All types of straw (rice straw, and the cereal straws, wheat, barley, oats, etc.) can be utilized to produce papermaking pulps (Clark, 1969; Patel et al., 1985; Jeyasingam, 1991; Brink et al., 1988). Lower yields are typical from rice straw since it contains greater amounts of parenchyma and epithelial cells as well as other fine debris. The high silica content together with the fines present in the stem results in papers with good ink receptivity and high opacity but presents problems in equipment wear and chemical recovery as previously discussed (Brink et al., 1988).

Wheat and rye straw have generally been preferred for pulping. Rye straw has longer fibers than oat straw, and barley straw is difficult to pulp. Paper from wheat and rye straw is generally stiffer and stronger. The straw fibers have a relatively high length-to-diameter ratio, which results in good paper properties. However, straw fibers tend to be much more heterogeneous than wood pulp fibers (Misra, 1980; Misra, 1987). Jin et al. (1988) found differences in the properties of pulps from different types of straw from a soda CMP process. After separation of the fines, rice straw gave a sheet which had a remarkably greater tear index and specific scattering coefficient, whereas wheat straw produced paper with much greater folding endurance. The greater tear strength of the rice straw pulp was unexpected due to the shorter fiber length of this pulp.

For efficient processing of straw to paper, the moisture content of the straw should be kept under 15%, preferably between 10–12%. At higher moisture contents the straw is susceptible to microbiological attack and decay, and fire hazards could result from spontaneous combustion. Degraded straw consumes more chemicals in pulping, gives lower yields, lower brightness, and lower strength pulp (Jeyasingam, 1988a,b; Misra, 1987).

Straw is prepared for pulping by either a dry or wet method. The dry method is the most common and involves cutting of the straw in disc or rotary drum type cutters. The straw must then be thoroughly cleaned by conveying to screens and cyclone separators to remove the grain, sand, and dust mixed in with the fibrous stock. To utilize the dry method the straw must be kept below 15% moisture content, otherwise it is difficult to convey the straw with pneumatic systems. The advantage of the dry system is that it requires lower capital and energy costs (Jeyasingam, 1988a,b).

The wet preparation system involves treatment of the straw in a specially designed hydrapulper. The sand, grit, and leafy material are washed away and removed through a valve. About 25% of the total weight of rice straw and 10% of the total weight of wheat straw are removed by this cleaning method. An advantage of the wet system is that about 25% of the silica is removed from the straw as a result of the rubbing action in the hydrapulper. The addition of caustic soda to the hydrapulper in the cleaning process has also been utilized as part of the NACO process described below (Jeyasingam, 1988a,b).

4.2.1 Mechanical and Mechanochemical Pulping

There are limited reports on straight mechanical pulping of straw, but a number of mechanochemical processes have been developed for pulping directly in a hydrapulper. A high yield coarse mechanical pulp can be produced in a hydrapulper at about 100°°C without any chemicals. The pulp is produced in about 30–90 min at high consistency (10%) but it is not of high quality and is used for insulating boards. Lathrop and Aronovsky (1949) utilized caustic soda or a mixture of caustic soda (5–6% on O.D. wt.) and sulfite charged with the straw to a hydrapulper at about 95–100°C to produce a suitable pulp in about an hour. The yield was high, at about 75%; a bleachable grade pulp is produced by increasing the caustic charge to 12–13% and pulping for about one-half hour. A three-stage bleaching sequence gave a pulp with 80% brightness. Lime (10%) or lime plus caustic soda have also been used in the hydrapulper for pulp production from straw (Jeyasingam, 1988a; Misra, 1981).

The NACO process is a newer continuous process based on an Italian patent for production of a high yield, semi-chemical type pulp from straw. The process utilizes no chlorine or sulfur and involves three stages: pretreatment, delignification, and bleaching. The pretreatment is carried out in a hydrapulper in the presence of 1.5–2% caustic solution. After pressing, the delignification is performed in a turbopulper containing pressurized oxygen (6–8 kg/cm^2) with the alkali-fiber suspension. Intensive mechanical action is provided in the turbopulper, which aids the fiber separation process. The pulp is then continuously pumped into an oxygen reactor with a flow-down arrangement. After pressing, the pulp is bleached with hydrogen peroxide (2%) and hypochlorite (3%) or with ozone to a brightness of 80. The concentrated black liquor is burned in a recovery boiler to yield a smelt that is dissolved to produce sodium carbonate for direct use in the turbopulper. The advantages of the process include reduced pollution hazards and lower capital investment for the recovery system. However, there is poor heat recovery from the black liquor and additional capital investment may be required to produce oxygen. The NACO process is in operation in Foggia, Italy (Jeyasingam, 1988a, 1991).

High yield semi-chemical pulp can also be produced from straw by the Hojbygaard Fabrick (HF) process. This process was developed in Denmark for production of corrugating medium from wheat and rice straw. The straw is moved through a diffuser via a screw conveyor. The first stage involves impregnation with the chemicals (6–7% NaOH), followed by heating to about 100°C and finally washing. After about four hours followed by mechanical disintegration, a straw pulp at a yield of about 75% is obtained. The steam and water requirements are low in this simple operation. A similar process using caustic solutions was developed in Spain by the Sociedad Anonima De Industrias Celulos Icas Aragonesas. Cooking is carried out in an inclined horizontal continuous digester with the pulp conveyed by two parallel rotating screws. The resulting semi-chemical pulp is suitable for corrugating medium (Misra, 1980, 1987; Jeyasingam, 1988a). The Escher Wyss and MCP processes are further variations on the use of a tubular type digester with a screw feed. With these processes there is a pre-impregnation stage followed by a cook at 100°C in 6–8% caustic or lime with intense mechanical action. Three mills in France, Spain and the former Yugoslavia utilize this process (Jeyasingam, 1988a,b).

4.2.2 Chemical Pulping

Similar to bagasse, soda, kraft and neutral sulfite semi-chemical pulping are suitable for producing pulps from straw. The lime process is also still utilized for straw pulping. Acid sulfite pulping produces brittle, low strength pulps not suitable for papermaking. As with other agro-based fibers, spherical or tumbling digesters are utilized with batch systems to obtain uniform pulping.

The lime process consists of cooking with 7–10% calcium oxide at 130–140°C for 3–5 h. The pulp is used for corrugating medium and the effluent is sometimes used for irrigation of crops and soil modification. The soda process is the most widely utilized for straw pulping. A bleachable grade pulp is produced in spherical digesters with 10–12% caustic (based on O.D. straw). El-Taraboulsi and Abou–Salem (1967) found that very mild conditions could be utilized to pulp rice straw by the soda process with the advantages of lower ash content, higher yield, and improved physical properties, particularly fold and tear. Their mild cooks were performed at 50–90°C in 13% NaOH for 5–10 h.

The pulp strength properties realized from kraft pulping of straws are only marginally better than those realized from the soda process; therefore, the kraft process has not been widely utilized for straw pulping (Jeyasingam, 1988a,b). Combine this fact with the additional cost of salt cake chemical make-up and pollution problems associated with the sulfur from the kraft process, the decision to favor soda pulping is easy. However, many small soda mills do not have chemical recovery systems and effluents are discharged directly into aquatic systems compounding pollution problems in many developing countries. Typical conditions for pulping of straw by the various chemical processes are shown in Table 6.20 (Misra, 1980, 1987; Jeyasingam, 1988a).

Neutral sulfite pulping is an effective method for obtaining good pulps from straws; however it is somewhat more complicated than NSSC pulping of wood and some mills have not realized the expected yields of 70–75%, recovering only 55–58% of the straw as pulp. Jeyasingam (1987, 1988b) has identified a number of reasons for such low yields in mill operations. His recommendations for avoiding problems in NSSC pulping of straw were in four major areas—cooking, washing, screening, and refining—as follows:

Table 6.20 Conditions for Batch Pulping of Straw

	Lime	Lime–Soda	Soda	Kraft	Neutral Sulfite
Chemical (% on moisture-free straw)	CaO 6–12 or MgO 12–18	5–10, CaO + 3–4 Na_2CO_3	NaOH, 6–15	NaOH 10–12 + Na_2S 2–3	10 Na_2SO_3 alone or 4–10 Na_2SO_3 + 2–5 NaOH
Liquor-to-straw ratio after steaming	4:1	5:1	4:1–7:1	4:1–6:1	3:1–7:1
Cooking time, h	6–10	6–8	2.5–4	2.5–4	2–4
Cooking temperature, °C	125–140	120–140	150–170	150–170	160–170
Variety of pulp produced	Coarse	Coarse	High-yield semi-chemical and bleachable grade	High-yield semi-chemical and bleachable grade	High-yield semi-chemical and bleachable grade
Yield (crude yield) from					
Wheat + rye straws	70–85%	70–80%	48–70%	50–70%	53–65%
Rice straw	70–80%	65–70%	40–45%	42–47%	45–48%

Source: Misra, D. K., Pulping and bleaching of non-wood fibers, *Pulp and Paper, Chemistry and Chemical Technology*, Vol. 1, 3rd ed., Casey, J. P., Ed., Wiley-Interscience, New York, 504, 1980.

1. Cooking: (a) Batch vs. continuous is preferred because of the greater flexibility with batch operations. The flexibility in cooking is necessary because of the variable nature of the raw material, such as moisture, cleanliness, and morphological characteristics of the straw; (b) Digesters with mechanical screws in a horizontal tube should be avoided because of difficulty in obtaining uniform feeding of the bulky straw; and (c) Pressure cooking is preferred over atmospheric to provide better and more uniform liquor penetration thus reducing excessive fines formation.
2. Washing: (a) Pulpers must have suitably designed rotors to sufficiently fiberize the cooked material before washing; (b) Washing with belt thickeners is not possible if the yield is greater than 55%; and (c) Washing with simple decker type washers is preferred over belt, screw-type, and other washers.
3. Screening: (a) Flat screens are preferred for coarse screening; (b) After sufficient mechanical defibration, centrifugal screens can be utilized.
4. Refining: (a) Refining at high consistency and high shear is necessary to avoid excessive production of fines; (b) Refining below 3.5% consistency should be avoided.

By proper choice of equipment and processing techniques, Jeyasingam (1987, 1988b) felt that the major problems experienced by semi-chemical straw pulp mills could be avoided. Indeed, Gohre (1988), who described the successful operation of a semi-chemical straw pulp mill in Turkey, found that the flexibility of batch cooking was necessary to compensate for the variable nature of the straw raw material. He also outlined additional factors important to the operation of this mill producing fluting liner, wrapping paper, and linerboard as follows:

1. The use of choppers with rotative cutting knives, which give a more uniform length straw raw material. Although more expensive, this approach—combined with dust removal—gives higher yields, reduced chemical consumption, reduced wear on equipment due to abrasion, and improved quality of the paper product.
2. Impregnation of the chopped straw with hot cooking liquor to reduce cooking time and increase production.
3. Addition of calcium oxide with the caustic soda to give a buffered reaction that is sufficient to soften the wheat straw without excessive alkalinity. This also gives higher yields and a black liquor of very low toxicity.
4. Subdivision of the disintegration stage into a thermomechanical (first) stage and a refining (second) stage, which gives better control of the final pulp characteristics.

Annus and Szekeres (1988) have also emphasized the importance of gentle handling of the swollen straw pulp in their discussion of the operation of a kraft straw pulp mill in Hungary. They also strongly emphasized that wood pulping technology is not always appropriate for annual plants such as straw, and ignorance of this fact has caused considerable problems in straw pulp mill operations around the world.

Neutral ammonium sulfite pulping (NASP) has been adopted for many small pulp mills in China (Liu et al., 1988). The establishment of these mills has been coordinated with agricultural uses of the waste NASP liquors as fertilizer. The enhanced growth realized by incorporation of the NASP waste liquor in wheat cropping compared to other common fertilizers is shown in Table 6.21 (Liu et al., 1988). The NASP gives superior yield increases to the other commonly employed

Table 6.21 Comparison of Pulping Waste Liquor with Conventional Fertilizers for Wheat Cropping

Type of Fertilizers	Yield kg/hectare	Yield Increases %	Yield Increases kg/kg of nitrogen
NASP waste liquor[a]	5,310.0	59.2	16.5
Ammonium sulfate	4,849.5	45.4	12.6
Ammonium sulfite	4,492.5	34.7	9.6
Urea	4,488.0	34.6	9.6
Without fertilizers	3,334.5	—	—

[a] NASP = neutral ammonium sulfite pulping.

Source: Liu, M. G. et al., Neutral ammonium sulfite production of wheat straws and the utilization of its black liquor, *Int'l Non-Wood Fiber and Paper Making Conf.*, Beijing, P. R. China, July, 299, 1988.

fertilizers. Liu et al. (1988) described the operation of the Taiwan NASP Paper Mill in China, which produces pulp for printing and writing paper. They suggested the use of cooking liquors well above pH 7 to avoid corrosion and the use of non-corrosive linings in the digester wherever possible.

Both Kaymr and Pandia continuous digesters have also been utilized for pulping of straws. However, problems with clogging of the blow lines and piping are sometimes encountered with the straws. For alkaline processes, about 5–6% active alkali is used, while 4% sodium sulfite and 2% sodium hydroxide are used for neutral sulfite semi-chemical pulping (corrugating medium). The total retention time in the digester is only about 20 min at 160–170°C. Higher active alkali (12–14%) is required for bleachable grades. Some physical properties of unbleached and bleached (CEH) straw pulps are shown in Table 6.22 (Misra, 1980, 1987).

Table 6.22 Strength Properties of Unbleached Wheat Straw Pulp and Bleached Wheat and Rice Straw Pulps Cooked by the Soda Process

	Unbleached wheat straw pulp	Bleached wheat straw pulp	Bleached rice straw pulp
Permanganate number	12.8	—	—
Viscosity (cP)	35.8	15.0	—
Freeness (°S-R)	50.0	50.0	51.0
Basis weight (g/m²)	78.1	77.1	—
Breaking length (km)	7.6	7.3	5.3
Tear index (mN·m²/g)	4.23	3.74	6.27
Burst index (kPa·m²/g)	4.1	4.2	3.8
Folding endurance, double folds	65.0	40.0	149.0
Brightness	37.0	80+	78.0
Screened yield, %	45–46	40–41	—

Source Misra, D. K., Pulping and bleaching of non-wood fibers, *Pulp and Paper, Chemistry and Chemical Technology*, Vol. 1, 3rd edition, J. P. Casey, Ed., Wiley-Interscience, New York, 504, 1980.

The modified Celdecor–Pomilio process employing caustic soda and chlorine has also been utilized for pulping of straws (Figure 6.7). Although the process offers certain advantages as previously discussed, it has not proven viable because of corrosion problems. A few mills still operate with this system but no new mills have been established in the last 25 years for straw pulping (Jeyasingam, 1988a,b).

Soda–oxygen and soda–AQ delignification of straws have been evaluated by a number of investigators (Usta and Eroglu, 1988; Eroglu, 1988; Rao et al., 1991; Goyal, 1991; Qian and Tang, 1988). Rao et al. (1991) reported that the use of oxygen at 4 kg/cm^2 partial pressure in the soda cook improved the yield, strength, and optical properties of rice straw pulps. Goyal et al. (1991) reported that addition of small amounts of AQ to the caustic soda liquors enhanced that rate of delignification but resulted in only small improvements in the strength properties of the rice straw pulps.

Brink et al. (1988) evaluated a number of combinations of pretreatments and pulping processes for rice straw. The pulping methods utilized were soda, soda–AQ and kraft and the pretreatments were nitric oxide followed by oxygen (NOS), nitric acid (NAN), and sulfuric acid (NAS) followed by soda pulping. They found that virtually all the ash and silica in the rice straw could be removed in a three-stage pulping process: an acidic pretreatment, an extraction with sodium carbonate, and finally, a soda pulping. Nitric oxide impregnation followed by gaseous oxygen (NOS) proved to be the best pretreatment. Maintenance of a sufficient caustic soda level so that the pH remained about 11 throughout the soda cook was critical for silica removal. The NOS pulping method gave the highest yields and the best physical properties.

Organosolv pulping of straw has been evaluated by a number of investigators (Lora and Pye, 1994; Cai et al., 1988; Liang et al., 1988). Liang et al. (1988) found that the ethanol–water system gave superior delignification selectivity to conventional systems with the point of defibration 12% higher. Lora and Pye (1994) obtained pulps from wheat straw that would be competitive with bleached hardwood kraft pulps by organosolv pulping with the Alcell process. Figure 6.17 shows the influence of cooking time on the kappa number after aqueous ethanol pulping with 45% ethanol (195°C, 4:1 liquor-to-straw ratio) and also after a subsequent mild extraction. In the pulping trials the kappa number decreased and then leveled out at about 40, which is high for bleaching purposes; however, these pulps were found to be very receptive to both mild alkaline extraction and alkaline oxygen delignification. A significant reduction in kappa number could be achieved by extraction with 2% sodium hydroxide at 70°C for 30 minutes (κ no. = 9–20). The properties of these pulps are shown in Table 6.23 (Lora and Pye, 1994).

Cai et al. (1988) evaluated organosolv pulping of straw by the ester process developed by Young and coworkers (1989). This process employs acetic acid, ethyl acetate, and water in a homogenous system for pulping of a variety of lignocellulosic materials. Typical conditions utilized by Cai et al., with varying ratios of the three chemical components were: maximum temperature, 156°C; time, 5.7 h; pressure, 7.6 kg/cm^2, to give a yield of 50% with a permanganate number of 14.7. Two-stage bleaching (EH) gave a pulp of 67.2% brightness, which was suitable for printing and writing papers.

Another distinct advantage of organosolv pulping is that the silica does not accumulate in the system. The silica originally present in the straw is preferentially retained with the pulp; and, Lora and Pye (1994) found that the portion of silica leached into the spent pulping liquor could be purged out with either the lignin co-byproduct

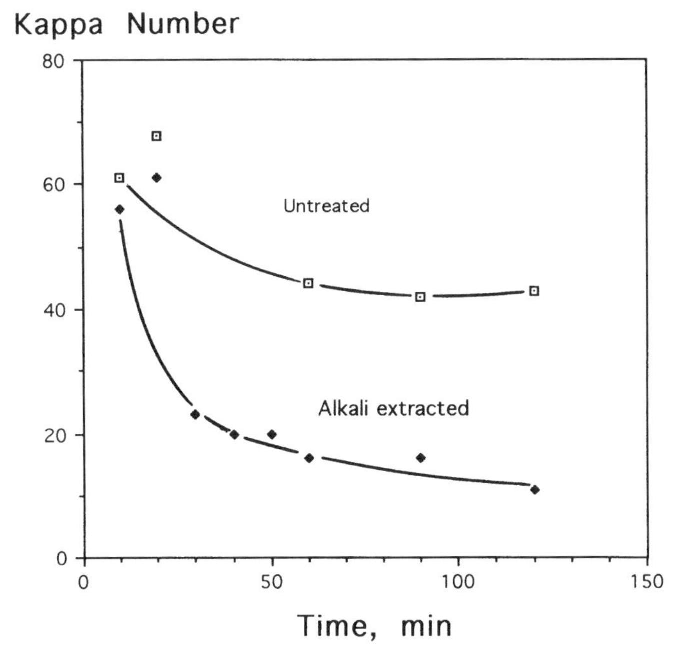

Figure 6.17 Effect of cooking time at 195°°C on the kappa number after Alcell pulping of wheat straw and after a subsequent mild alkali extraction (Lora and Pye, 1994).

Table 6.23 Characteristics of Organosolv Wheat Straw Pulps Produced by the Alcell Process

	After Alkaline Extraction	After EDED Bleaching
Kappa number, mL/g	11.8	—
Viscosity, cp	20.7	19.9
PFI, revolutions	1,000	1,000
CSF, mL	303	289
Bulk, cm^3/g	1.60	1.53
Tear index, $mN{\cdot}m^2/g$	6.1	6.6
Burst index, $kPa{\cdot}m^2/g$	3.1	3.1
Breaking length, km	4.9	4.9
Brightness, ISO	—	80
Opacity, % (TAPPI)	—	77

Source Lora, J. H. and Pye, E. K., The Alcell process, an environmentally sound appoach to annual fibers pulping, in *Proc. Non-Wood Fiber for Industry Conf.*, Vol. II, Pira/Silsoe Res. Inst., Pira, Leatherhead, Surrey, U.K., 1994.

or with the stillage obtained after alcohol recovery. The lignin recovered from the wheat straw cooks had properties that resembled those exhibited by conventional hardwood Alcell lignin.

Misra (1980) reported that wheat straw soda pulp was relatively more resistant to bleaching compared to bagasse and consumed 8–10% chlorine in a CEH sequence to achieve a brightness of 80%. Rao et al. (1987) found that a CEH bleaching sequence gave the highest strength pulps compared with CHH and HH sequences.

The HH bleaching sequence gave the weakest pulps. Bleaching conditions for a wheat straw kraft pulp are shown in Table 6.24 (McGovern, 1967). A pulp brightness of 81.5% was obtained with a total chlorine application of 8.1%. The total chlorine consumption for bleaching in an CEH bleaching sequence of a rice straw pulp was less, 7% to a brightness of 88% (McGovern, 1967). However, El-Taraboulsi and Abou–Salem (1967) found that the conventional CEH sequence gave only intermediate brightness of about 75% and lacked brightness stability. They suggested the use of a double chlorination stage to give stronger and higher yield straw pulps. Eroglu and Usta (1991) evaluated oxygen bleaching of soda–oxygen pulps as a much less polluting alternative for wheat straw pulps. The optimum bleaching conditions for the soda–oxygen pulps were: alkali charge, 4%; treatment time, 20 min; oxygen pressure, 10 kg/cm^2, at a consistency of 12.5%. The properties of some of these pulps are shown in Table 6.25 (Eroglu and Usta, 1991).

Table 6.24 Conditions for Wheat Straw Pulp Bleaching

	Chemical	Amount, %	Consistency, %	Temperature, °C	Time, h
Chlorination	Cl_2	0.8	3.0	30	1.5
Caustic extraction	NaOH	0.9	8.3	38	2.0
Hypochlorite bleaching	$Ca(OCl)_2$	Included in Cl_2	5.9	38	4.8
Neutralization	SO_2	0.15	—	—	—

Source: McGovern, J. N., TAPPI, 50(11), 63A, 1967.

With wheat straw as the short fiber component, a paper with a very smooth and dense surface for printing can be produced; however, incorporation of wheat straw pulp into the furnish can cause running problems on both the paper machine and the printing presses, particularly when printing with the offset process (Fineman et al., 1988; Kuang and Fineman, 1988; Youchang, 1988). Fineman et al. (1988) outlined the major problems as:

1. Poor runnability at the wet end of the paper machine due to poor drainage and high adhesion to the press rolls.
2. Linting in the offset printing due to a high content of epidermal and parenchyma cells in the pulp. Adhesion at the press rolls also contributes to linting problems in the printing.
3. Large dimensional changes in the offset press.
4. Folding problems due to brittleness of the paper.

Fineman et al. (1988) suggested that the drainability could be improved by minimal beating of the straw pulp and incorporation of retention aids. Although the wet web strength can be improved by beating, the simultaneous deterioration in the drainability seriously limits the effect of the beating. Addition of pine kraft pulp or cotton pulp and/or removal of epidermal and parenchyma cells from the stock improved the wet web strength by improvement of drainage and, thus, sheet dryness.

Table 6.25 Properties of Wheat Straw Soda–Oxygen Pulps Bleached under Various Conditions

	Bleaching yield, %		Breaking length, m		Burst Index kPa·m²/g		Tear Index mN·m²/g		Brightness (Elrepho), %		Optimum bleaching conditions
Variables	variables	$\bar{X}$	variables	$\bar{X}$	variables	$\bar{X}$	variables	$\bar{X}$	variables	$\bar{X}$	
NaOH; % based	8	94.1	4	5.8	4	3.4	4	5.3	12	70.1	
on O.D. fiber	12	92.5	8	5.0	8	3.1	12	5.1	4	55.4	12
Treatment time, min	30	82.5	40	6.3	20	3.3	20	5.9	20	71.0	
	20	80.0	20	5.8	30	3.3	40	5.5	30	70.1	20
Oxygen pressure, kg/cm^2	10	82.5	10	5.7	10	3.3	5	5.2	10	70.1	10
	5	81.5	5	5.0	5	3.3	10	5.1	5	67.4	
Consistency, %	20	88.9	12.5	6.5	12.5	3.8	12.5	5.1	12.5	62.2	
	12.5	87.2	20	5.6	20	3.4	20	4.9	20	59.9	12.5

Source: Eroglu, H. and Usta, M., Oxygen bleaching of soda–oxygen wheat straw pulp, Non-Wood Plant Fiber Pulping, Prog. Rept. No. 19, Tappi Press, Atlanta, 89, 1991.

Other amelioration steps suggested by these investigators included addition of a filler such as calcium carbonate, removal of hemicelluloses, use of an acidic environment, chemical treatment of the fraction enriched with epidermal and parenchyma cells, and surface sizing. Cheng et al. (1994) suggested that the high specific surface was the predominant cause of the poor drainage properties of straw pulps and found that removal of the fines fraction from wheat straw pulp significantly improved drainability.

Capretti and Marzetti (1991) evaluated the effect of steam explosion on the paper properties of wheat straw pulps. Steam explosion treatments were carried out in a laboratory pilot plant under different pressure/temperature/time conditions, both with and without pretreatment with sodium carbonate. The steam-exploded pulps exhibited paper properties similar to, or in some cases better than, the properties of straw pulps prepared by conventional batch techniques (Table 6.26). Most of the straw pulps had a high fines content, which was found to be important to the strength development of the sheets (Capretti and Marzetti, 1991).

Table 6.26 Paper Properties of Straw Pulps from Steam Explosion Treatment with or without Chemicals[1]

Temperature, °C	190		200	200		210		2300
Time, min	6		1	2		0.5		.5
Na_2CO_3 g/100 g	—	10	—	—	10	—	10	—
Apparent density, kg/m^3	550	745	480	530	625	540	725	645
Tensile index, Nm/g	24.0	53.0	20.5	27.0	40.5	23.0	49.5	33.0
Elastic modulus, MPa	1710	4250	1410	1680	2810	1580	3950	2250
Burst index, MN/kg	1.05	3.20	0.70	1.20	2.10	0.90	2.85	1.85
Scott bond, J/m^2	235	420	130	—	585	200	385	220
Intr. compres. index, Nm/g	18.0	28.5	17.0	20.0	22.5	18.5	27.5	20.0
CMT/30	130	180	115	155	165	130	185	130

[1] Values obtained at the same freeness SR 50.

Source: Capretti, G. and Marzetti, A., Steam explosion pulping of wheat straw, *Steam Explosion Techniques: Fundamentals and Industrial Applications*, Focher, B., Marzetti, A., and Crescenzi, V., Eds., Gordon & Breach, New York, 1991, p. 207.

Biodelignification of straw has also been briefly evaluated for possible biopulping approaches. However, most of the work has been directed on the growth, physiology, and delignification mechanisms of the white-rot fungus, *Panus conchatus*, and no pulping trials were reported (Yu, 1988).

4.3 Kenaf

In the extensive plant screening work carried out by the USDA, kenaf was identified as a highly desirable plant species for pulp and paper. This was based on its good productivity and positive agronomic characteristics. Intensive evaluation of kenaf in laboratory pulping trials followed at the USDA Northern Regional Research Center in Illinois (Clark et al., 1962; Clark and Wolff, 1962; Bagby et al., 1979; Cunningham et al., 1979; Touzinsky, 1987, 1980).

Kenaf, like jute, has both bast and core material that contain fibers with different characteristics. The bast comprises about 20% of the whole stem. The core fibers are only about 0.6 mm in length while the bast fibers are much longer, around 2.6 mm in length. As one would expect, the bast and core yield pulps with distinctly different properties. The bast gives higher yields and the pulp has better strength properties, even superior to softwood pulps. Pulp from the woody core contains relatively more pentosans and lignin and is very slow draining.

During the past five years, high value uses for the core stock have been developed. The woody core is ground into small particle sizes and used as an absorbent for oil spills, animal bedding, packing material, potting material, and a variety of other purposes. This leaves the bast for high quality paper. Indeed, a company is producing high quality writing and printing paper from the bast at a premium price. However, it has been demonstrated in pilot plant trials that the separation of the bast and core fibers is an expensive operation (Atchison, 1994).

Both whole and separated (bast and core) kenaf have been evaluated for pulp and paper properties. Green stalks are sometimes run through a crusher screw press to remove juices and other extraneous materials to reduce chemical consumption in pulping and bleaching. Whole chopped kenaf, when chemically pulped, tends to have slow draining characteristics that can limit paper machine speeds (Figure 6.18). Very good pulps have been obtained from kenaf from both mechanical and chemical processes (Misra, 1980; Clark, 1969; Touzinsky, 1987).

Figure 6.18 Chopped kenaf ready for charging to the digester in Thailand (Photo by R. A. Young).

4.3.1 Mechanical Pulping

Straight mechanical pulps from whole kenaf are usually weaker than the corresponding wood pulps, and conventional thermomechanical pulping of cubed kenaf yields a dark pulp of low strength. Although whole kenaf TMP is not adequate alone for newsprint grade paper, Touzinsky (1980) reported that paper with strength comparable to commercial newsprint could be prepared from furnishes containing bleached kenaf TMP and at least 12% kraft softwood pulp. Subsequent treatment of the TMP pulp with alkaline hydrogen peroxide (2% H_2O_2 and 2% NaOH) yields a pulp with properties equivalent or even superior to pine TMP as shown in Table 6.27 (Touzinsky, 1980, 1987). The final yield of the kenaf TMP was 80% with an optimum post-treatment pH of 8.5.

Table 6.27 Comparison of the Properties of Kenaf and Southern Pine TMP and CTMP

	TMP		CTMP	
	Kenaf	Southern Pine	Kenaf	Southern Pine
Brightness, %	67	54	70	54
Burst index, MN/kg	1.4	1.3	1.7	1.5
Tear index, N·m²/kg	8.1	8.5	8.0	9.7
Breaking length, km	3.5	3.0	4.3	3.4
Apparent density, kg/m³	318	330	388	386
Wet web tensile strength, N/m	30	21	34	24
Long fiber, %	36	55	33	53
Fines, %	49	34	53	37
Opacity, %	95	93	90	93
Scattering coefficient, m²/kg	66	48	66	50

Source: Touzinsky, G. F., *TAPPI*, 63(3), 109, 1980; Touzinsky, G. F., Kenaf, in *Pulp and Paper Manufacture, Vol. 3, Secondary Fibers and Non-Wood Pulping*, TAPPI Press, Atlanta, 106, 1987.

Chemithermomechanical pulping produces very good pulps from kenaf. In this case the alkaline hydrogen peroxide (2.5% H_2O_2 and 3.2% NaOH) is added to the kenaf just prior to the pressurized refining stage. These pulp properties are also compared to pine CTMP in Table 6.27. The kenaf CTMP has superior brightness, breaking length, and wet web strength (Touzinsky, 1987).

Shuhui et al. (1993) evaluated chemimechanical pulping of kenaf core material with alkaline hydrogen peroxide. The most effective treatment was with a mixture of 1.5–2% hydrogen peroxide and 3–5% caustic soda at a maximum temperature of 95°C. The pulp yield was 80%, the breaking length over 4 km, the brightness 60%, and the opacity 92%. Paper with strength comparable to commercial newsprint could be made from a furnish containing 90% of the kenaf CTMP and 10% of long chemical fiber.

Work sponsored by the U.S. Department of Agriculture in cooperation with Kenaf International demonstrated convincingly that kenaf CTMP is very suitable for production of newsprint (Kugler, 1988). Hammermilled kenaf was processed in a two-stage disk refining operation with 2.6% sodium hydroxide, 3.0% hydrogen peroxide, and a steam pressure of 2.1×10^{-5} Pa present in the first stage and atmospheric double disk refining to a freeness of 115 mL in the second stage (Kugler, 1988; Horn et al., 1992). The kenaf pulp compared favorably to southern pine and spruce/fir TMP. Energy requirements were lower; tear, tensile, brightness, and opacity were all in the acceptable range. A newsprint furnish of from 82–95% kenaf CTMP and 5–18% bleached kraft pulp proved suitable. The kenaf-based paper was run in the newspaper pressrooms of the *Bakersfield Californian*, the *Houston Chronicle*, the *Dallas Morning News* and the *St. Petersburg Times*. All runs were successful and the kenaf newsprint ran well on TKS offset presses yielding a sheet with very good visual impact. Analysis showed extremely good print contrast; reduced ruboff; and tear, tensile, and burst indices superior to southern pine newsprint (Kugler, 1988).

Horn et al. (1992) also produced acceptable newsprint from blends of 25% of the same kenaf CTMP and 75% deinked recycled newsprint. The mixture produced a paper with greater burst and tensile strength than the control furnish of 100% recycled newsprint. Other properties were not affected with the exception of opacity, which decreased with increasing kenaf content, especially beyond the 25% level. Despite the high brightness of kenaf CTMP, blends of kenaf and the deinked recycled newsprint did not shown a noticeable increase in brightness for blends containing less than 50% kenaf CTMP.

In very recent work, Sabharwal et al. (1994a,b, 1995) and Young et al. (1994b) obtained very good pulps via biomechanical pulping of kenaf. Cut kenaf bast strands were treated with a CZ-3 strain of the white-rot fungus *Ceriporiopsis subvermispora* for two weeks and then refined in a disc refiner under atmospheric conditions. Compared with RMP from kenaf, the energy consumption was substantially lower and the strength properties substantially higher for the fungal-treated kenaf. The burst, tensile and tear strengths increased by 65%, 58.3%, and 56.6%, respectively at about 200 mL CSF for the biomechanical kenaf pulp (Figures 6.19 and 6.20). The opacity and drainage properties were also superior for the fungal-treated pulps, but the brightness was lower. Scanning electron microscopy of fungal-treated kenaf bast strands after refining showed that the fibers appeared to separate more readily from adjacent fibers compared with the non-inoculated samples.

4.3.2 Chemical Pulping

Kenaf has been successfully pulped by the soda, kraft, and NSSC processes. Tumbling digesters are utilized to cook chopped and washed kenaf stalk. It has also been reported that removal of the juice from the kenaf plant by crushing is desirable to eliminate slime and reduce chemical consumption in both pulping and bleaching. Juice removal was especially beneficial when using a Pandia continuous digester, and cooking times as low as 10 minutes could be utilized for kraft pulping in this continuous system (Misra, 1980; Clark, 1969; Clark et al., 1962; 1971)

Ten percent caustic soda can be utilized to produce a bleachable soda pulp from kenaf by cooking for 3.5 h at 170°C. Clark et al. (Clark et al., 1962, Clark and Wolff, 1962) utilized chemical charges of 15–21% (dry matter basis) and a time of 2 hours at 170°C for chemical pulping of kenaf (soda, kraft, NSSC). Their results indicated that an active alkali charge of 16–16.6% was suitable for a bleachable grade pulp from both soda and kraft pulping. Low sulfidity in kraft cooking reduced

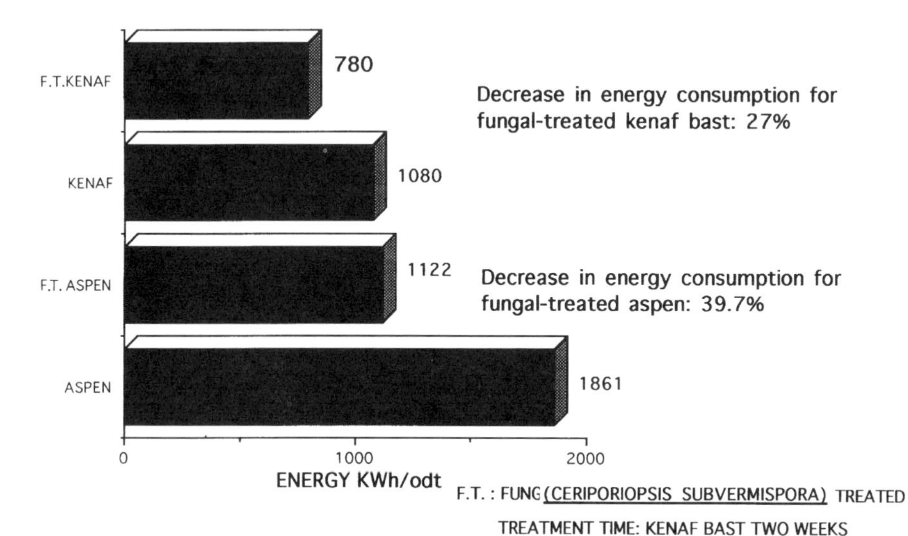

[a] kwh = kilowatt hours; odt = oven dry ton of pulp

Figure 6.19 Comparative energy consumption for refining of untreated and fungal-treated aspen and kenaf bast (Sabharwal et al., 1994a).

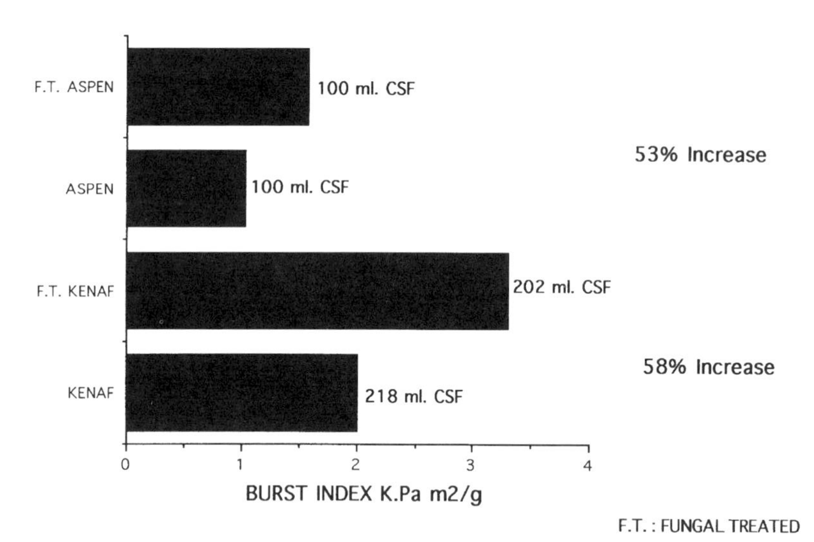

Figure 6.20 Comparative burst index of pulps from untreated and fungal-treated aspen and kenaf bast (Sabharwal et al., 1994a).

lignin values and bleach consumption and gave high bulk. Low sulfidity and high alkalinity gave higher strength properties, except for tear (Clark et al., 1962; Clark and Wolff, 1962; Bagby, 1978).

Contrary to pulping of wood, the kenaf soda pulps turned out to have strength characteristics that were equal or superior to the kraft pulps (Table 6.28) (Bagby,

Table 6.28 Kenaf Chemical Pulp Characteristics

	Soda[a]	Kraft[b]
Yield, %	62	55
Cellulose		
Crude, %	87	91
Alpha, %	68	71
Pentosan, %	19	20
Kappa number	45.4	27.3
Bleach consumed as Cl_2, %	10.2	6.5
Yield bleached, %[c]	53	48
Burst factor, g/cm²/gsm[d]	54	50
Tear factor, g/gsm[d]	102	93
Breaking length, m[d]	9,600	10,300

[a] Active alkali = 16.3%.
[b] Active alkali = 16.3%; Sulfidity = 10%.
[c] Basis digester charge; 75% brightness; CEH sequence.
[d] At 500 mL SR freeness

Source: Bagby, M. O., *Non-Wood Plant Fiber Pulping*, Prog. Rept. No. 9, TAPPI Press, Atlanta, 75, 1978.

1978). The yields were comparable but the soda pulps also had better drainage properties. The strength characteristics at the same freeness levels were superior to hardwood pulps and the tear was even better, comparable to a softwood pulp. It was also found that when kenaf chemical pulp was blended with wood chemical pulps there was a synergistic effect on the properties (Clark and Wolff, 1962). NSSC pulping of kenaf gave higher yields and less consumption of chemical.

Watson et al. (1976) also evaluated chemical pulping of kenaf in Australia. They carried out kraft pulping trials on both the bast and core stock from the kenaf plant and produced a series of different paper products from blends of the two pulps. They then determined the minimum tolerable proportions of bast fiber in the blends to reach minimal tolerable limits of drainage time, freeness, breaking length, and tear strength as shown in Table 6.29 (Watson et al., 1976). The results show that quite reasonable combinations of properties can be obtained, particularly when the bast pulp is unbeaten and the core stock pulp is either only lightly beaten or not at all.

Schroeter (1994) pulped both bast and core material using a soda–AQ process for use in linerboard. He found that the use of kenaf core fiber pulp in place of hardwood pulps improved sheet smoothness. The bast pulp required only minimal refining while the core pulp required no refining at all. The latter, when utilized as the top sheet in linerboard, thus resulted in reduced energy costs due to refining and improved sheet smoothness, without negatively impacting sheet strength properties.

Kenaf pulp can be bleached in a three-stage sequence (CEH) to a brightness of 80–85; and a brightness of 90 is possible in a four-stage sequence of CEHD or CdEHD. Some bleached yields, bleach consumption, and other pulp properties are shown in Table 6.28 (Clark and Wolff, 1962).

Mittal and Maheshwari (1994) recently reported on their experience with commercial kraft pulping of whole kenaf stalks in Thailand (Phoenix Paper Mill). They found that during the process of pulping, a higher proportion of the bast fibers remained with the final pulp, which resulted in a higher average fiber length and

Table 6.29 Tolerable Limits for Kenaf Core Pulp Proportions for Various Pulp Specifications

Specification[a]				Beating (PFI rev.)[b]		Acceptable range of woody core proportions (%)
Max. DT	Min. CSF	Min. BL	Min. tear	Stem	Bark	
20	200	6.5	9	0	0	5–70
				1,000	0	5–55
				0	1,000	0–30
				1,000	1,000	0–12
			12	0	0	5–40
				1,000	0	5–40
				0	1,000	0–30
				1,000	1,000	0–12
		7.5	9	0	0	0
				1,000	0	45–55
				0	1,000	0–30
				1,000	1,000	0–12
			12	0	0	0
				1,000	0	0
				0	1,000	0–30
				1,000	1,000	0–12
	300	6.5	9	0	0	5–60
				1,000	0	5–40
				0	1,000	0–10
				1,000	1,000	0
			12	0	0	5–40
				1,000	0	5–40
				0	1,000	0–10
				1,000	1,000	0
		7.5	9 or 12	0	0	0
				1,000	0	0
				0	1,000	0–10
				1,000	1,000	0
10	200	6.5	9 or 12	0	0	5–35
				1,000	0	5–25
				0	1,000	0–5
				1,000	1,000	0–1
		7.5	9 or 12	0	0	0
				1,000	0	0
				0	1,000	0–5
				1,000	1,000	0–1
	300	6.5	9 or 12	0	0	5–35
				1,000	0	5–25
				0	1,000	0–5
				1,000	1,000	0
		7.5	9 or 12	0	0	0
				1,000	0	0
				0	1,000	0–5
				1,000	1,000	0

[a] DT = drainage time; CSF = Canadian Standard Freeness; BL = breaking length/km; tear index, mNm^2/g.
[b] PFI rev. = revolutions in laboratory refiner.
Source: Watson, A. J. et al., The growing of kenaf in northern Australia and its potential for papermaking and food production, *D. V. Chem. Technol. Paper No. 7*, CSIRO, Australia, 1976.

good strength properties of the final pulp sheets. Strength properties of the unbleached pulps are shown in Figure 6.21. They also found that the black liquor from kenaf pulping did not cause scaling problems due to the low silica content of the plants. An overall recovery efficiency of 95 ± 0.5% was achieved.

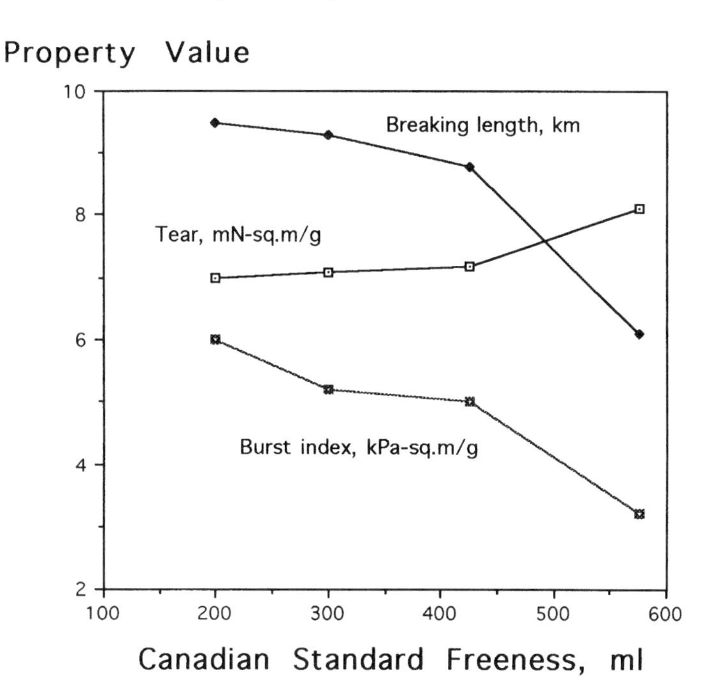

Figure 6.21 Physical properties of unbleached kenaf kraft pulp from Phoenix Pulp and Paper Mill (Mittal and Maheshwari, 1994).

However, a number of other problems were encountered in the commercialization of kenaf kraft pulps. Problems with procurement of the raw material were encountered because farmers were more interested in growing cassava, sugarcane, and other crops due to a better financial return. This resulted in a 50–60% increase in the cost of the kenaf raw material.

Handling and storage also presented problems since this was all done manually and proved to be cumbersome. Pitch was found to be a major problem at startup by accumulating on felts and the papermachine wire; however, incorporation of a pitch dispersant alleviated this problem. A large wire length fourdrinier table was also necessary to handle the slow drainage of the kenaf pulps and further problems were encountered in the press and drying sections. Efforts are continuing to solve these problems.

Similar to bagasse, very good steam explosion pulps can be produced from kenaf with an 8% sodium sulfite pretreatment. As shown in Figure 6.22, the tensile–tear relationship for kenaf SEP is far superior to these relationships for kenaf CMP and CTMP (Kokta and Ahmed, 1993). It was also found that less refining energy was required to reach a specific tensile strength for the kenaf SEP compared to the kenaf

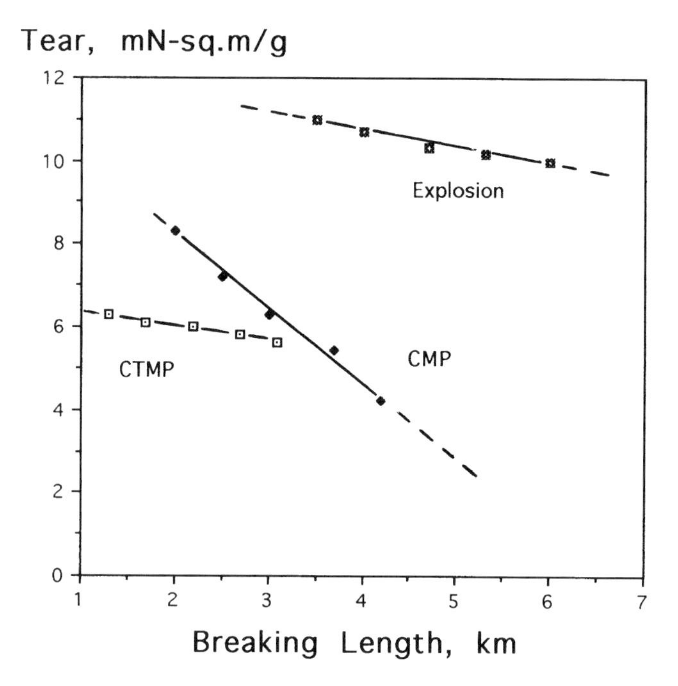

Figure 6.22 Variation of tear strength as a function of breaking length for kenaf pulps (SEP impregnation solution: 8% Na_2SO_3) (Kokta and Ahmed, 1993).

CMP and CTMP. The kenaf SEP can easily be bleached to the 77% level with a one-stage hydrogen peroxide treatment (Kokta and Ahmed, 1993). Properties of kenaf organosolv pulps (Alcell) are shown in Table 6.17.

4.4 Jute

Jute is not as widely grown and available as most of the other agro-based fibers discussed in this book. The high moisture and temperature requirements for growth have restricted production mainly to India, Bangladesh, Thailand, Brazil, China, and a few other countries. Jute has been a very important part of the economy of the West Bengal region of India and in Bangladesh. However substitution of synthetics for traditional jute applications such as bags (burlap) and carpet backings has devastated the export market for these products (Young, 1994a). A number of philanthropic organizations have recently been attempting to identify new, diverse applications to rectify the demand for jute and thus bolster the jute industry in all sectors, from the farmers in the agricultural regions to the mill processors and retailers. Increased utilization of jute for pulp and paper would therefore be a highly desirable pursuit.

Although jute has been utilized for pulp and paper for many years, rarely is whole jute utilized. Pulp mills generally acquire salvaged products such as old jute sacks, cuttings, and waste wrapping material. Care must be taken to remove extraneous substances such as tar, dirt, and metal contaminants to maintain the quality of the pulp. However, it is certainly possible to utilize the whole plant for paper production. Similar to kenaf, the stem consists of a woody core (60%) and the outer bast fibers. The bast fibers are likewise longer and more desirable for papermaking (Figure 6.23). The shorter "jute stick" core fibers provide greater opacity to sheets similar to hardwood fibers (Misra, 1980; Clark, 1969; McGovern et al., 1987).

Figure 6.23 Raw jute bast fiber after retting and separation (Photo by R. A. Young).

4.4.1 Mechanical Pulping

There is not extensive literature available on the use of mechanical processing methods for pulp production from jute. However, due to the similarities of the jute plant to kenaf, it should be possible to produce reasonable quality mechanical or enhanced mechanical pulps from jute. Indeed Sabharwal et al. (1995) and Young et al. (1994b) have demonstrated that a reasonable quality RMP can be produced from chopped jute as discussed below.

A comparative pulping study of jute bast, stick and the whole jute plant by chemimechanical, kraft, and soda processes was carried out by scientists at the Jute Technological Research Laboratory (JTRL) in Calcutta, India. Their results are shown in Table 6.30. Caustic soda (10–15%) was used for treatment of the jute material prior to mechanical disintegration in a disc refiner for the chemimechanical pulping trials. Fairly good strength properties were obtained, although not as good as the full chemical pulps.

Sabharwal et al. (1994b, 1995) and Young et al. (1994b) evaluated both RMP and biomechanical pulping of jute bast fibers and obtained very good pulp properties. Atmospheric mechanical refining of untreated, fungal-treated (*Ceriporiopsis sub-*

Table 6.30 Physical Properties of Jute Pulps

Raw Material	Pulping Process[1]	Yield, %[2]		Breaking Length, km	Burst Factor	Tear Factor	Fold
		UBLD	BLD				
Jute Bast	Chemi-Mech (Soda)	86–90	82–85	7.3	30	130	250
	Soda	65–67	62–64	8.5	38	135	900
	Kraft	68–70	63–65	8.8	40	150	1077
Jute Stick	Chemi–Mech (Soda)	76–78	68–70	5.0	20	51	78
	Soda	45	42	6.5	29	70	200
	Kraft	48	44	7.1	29	78	241
Whole Jute Plant	Chemi-Mech (Soda)	80–82	76–80	6.0	20	60	250
	Kraft	68	63	7.5	30	101	375

[1] Total chemicals charged for pulping = 10–20%.
[2] UBLD = unbleached, BLD = bleached by CEH sequence.
Source: Jute Technological Research Laboratory, Calcutta, India.

vermispora), and alkali-treated jute bast was explored in this investigation. Both hammermilled and cut jute bast strands were disintegrated in a two-stage laboratory single disc refiner. Although the energy consumption was higher for the cut jute, the resulting pulps had higher strength properties. Considerable energy savings in disc refining were realized with the fungal treatment of the jute, up to 33%, but the savings were somewhat less (25%) at 330 mL CSF where the optimum strength properties were obtained. However, the overall energy consumption was one-half that obtained for mesta TMP and 60% less than that obtained from fungal-treated aspen at 100 mL CSF (Figure 6.24). The energy consumption of 450 kWh/t for biomechanical pulp from jute bast, compared to an average of 1,600 kWh/t for pulp produced from wood, indicates that a savings of 1,150 kWh/t could be realized if wood was replaced by jute bast for pulp production by the biomechanical process.

The strength properties of the pulps obtained from fungal-treated cut jute bast were considerably improved; the burst, tensile, and tear strengths increased by 39%, 22%, and 33%, respectively, for the fungal-treated jute (Figures 6.25 and 6.26). Treatment of cut jute bast strands with sodium hydroxide solutions (5% and 10%) prior to refining gave pulps with higher strength properties but there was considerable loss in yield as shown in Table 6.31.

4.4.2 Chemical Pulping

Jute, like the other agro-based fibers, is generally cooked by alkaline processes. The lime, soda, and kraft processes have all proved successful for pulp production from jute bast and stick materials. It has been found that oil, waxes, and grease frequently contaminating the salvaged jute are removed in the alkali. Generally the bast material is digested in a rotary digester with alkali (10–20% chemical charge) for 3–4 hours, and sometimes for up to 8–10 h, at 140–150°C and pressures of

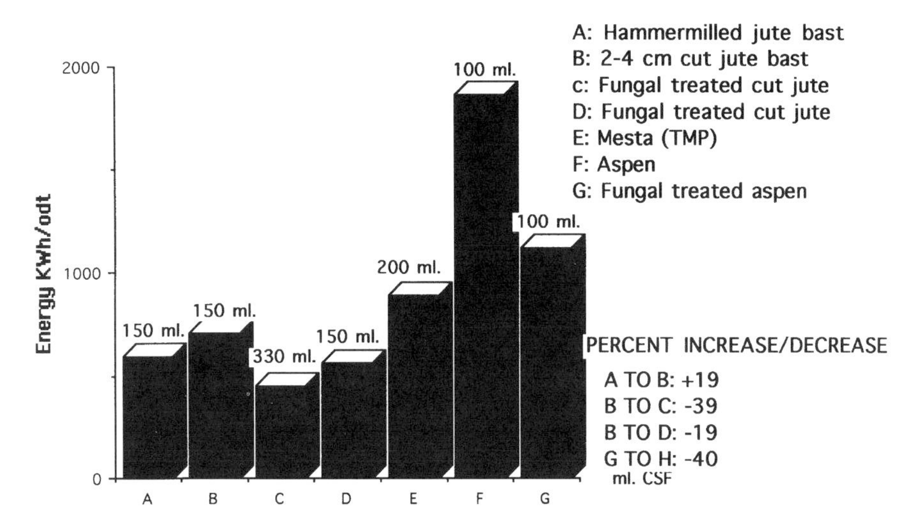

Figure 6.24 Specific energy consumption by various untreated and fungal-treated wood and agro-based materials (Sabharwal et al., 1995).

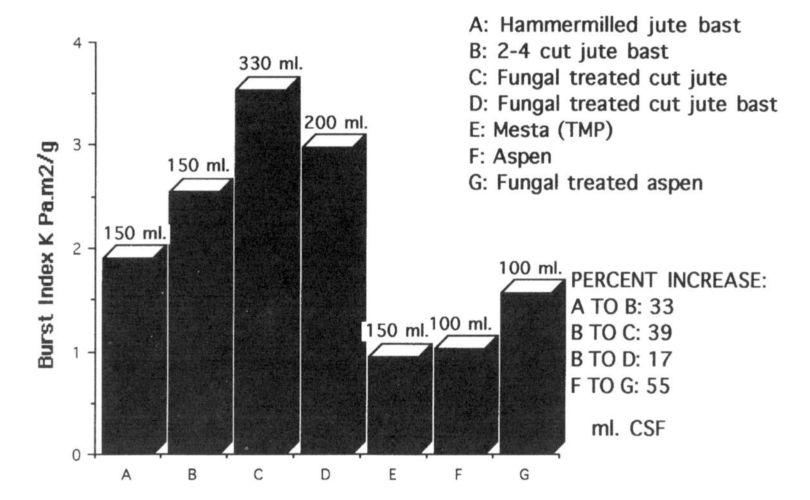

Figure 6.25 Comparative burst index of pulps from untreated and fungal-treated wood and agro-based materials (Sabharwal et al., 1995).

135–210 kPa. Some delignification takes place but, due to the mild conditions, it is insufficient for total fiber liberation and much of the pulp remains as fiber bundles as shown in Figure 6.27. This results in high yields, in the range of 60–70%, and distinctly different pulp properties (Table 6.30), as discussed below. To achieve better fiber liberation, one mill in India utilizes a two-stage kraft process with the first stage run at low pressure at 150°C for 8 h and the second stage at high pressure for 4 h at 165°C. The product is washed free of alkali and sent through breaker-beaters to give the unbleached pulp. Problems are frequently encountered with clumping and matting of the jute fibers, which cause difficulties in uniform cooking, blowing

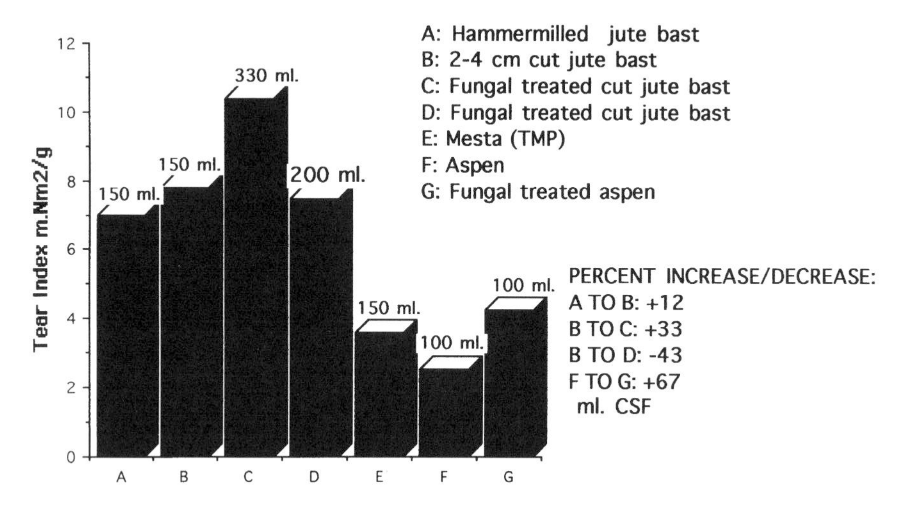

Figure 6.26 Comparative tear index of untreated and fungal-treated wood and agro-based materials (Sabharwal et al., 1995).

Table 6.31 Properties of Refined Pulps from Cut Jute Bast Treated with Alkali

Sodium Hydroxide %	Yield %	Freeness mL CSF	Burst Index kPa m²/g	Tensile Index N·m/g	Tear Index m·Nm²g
5	81.0	170	4.25	86.0	5.7
10	73.5	170	4.42	102.2	7.2
10	77.3	331	3.54	72.6	10.4

Source: Sabharwal, et al. *Holzforschung*, 1995.

of the digester, washing, screening, and disintegration. The continuous cooking methods previously described would be much better suited for jute pulping (Misra, 1980; Clark, 1969; Bandyopadhyay, 1970).

Alkaline sulfite and neutral sulfite anthraquinone (NS–AQ) pulping have also been applied to jute bast fibers. The conditions of the alkaline sulfite cook were: caustic soda, 12%; sodium sulfite, 14–16%; 165°C for 4 h. The unbleached yield was 44% with a kappa number of 40; bleaching with a CEH sequence gave 83.5% brightness. The strength properties of the pulp were equivalent to jute kraft pulp (Tribeni Tissue, 1992). NS–AQ pulping was performed on jute bast with a mixture of sodium sulfite and sodium carbonate. Higher yields and strength properties equivalent to softwood kraft properties were reported by Shafi et al. (1993) for the NS–AQ pulps.

Jute sticks (crushed and chipped) can also be pulped by the alkaline processes, but the yields are more typical of hardwoods, in the range of 45–50% (Clark, 1969; McGovern et al., 1987). The strength properties are lower than the bast fibers due to the much shorter fiber lengths (Table 6.30). Jute sticks can be easily pulped by soaking in a caustic solution for 8 h and then mechanically disintegrating in a disc refiner. Jute sticks are also sometimes cooked with magnesium bisulfite prepared by burning sulfur and passing the sulfur dioxide gas through an absorption tower

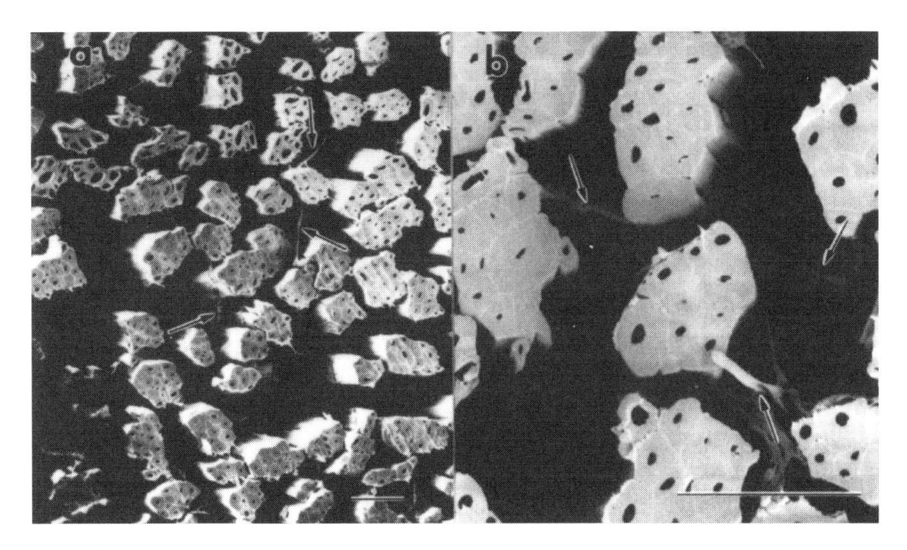

Figure 6.27 Scanning electron micrographs at two levels of magnification (a and b) of transverse sections of jute fibers that have been pretreated with *C. subvermispora*. Groups of fibers occur in loose aggregates allowing hyphae (arrows) to readily colonize the substrate from spaces between cells, Bar = 50 μm (Sabharwal et al., 1995).

containing a suspension of magnesia water. Direct or indirect pulping is then performed at 150–160°C. Pulps from jute sticks have a high freeness value (high wetness), drain slowly, and have a tendency to gel in refining, especially when prepared by semi-chemical or bisulfite pulping methods. Pulps refined in this manner yield dense, impervious sheets similar to grease proof paper. Paper produced from jute stick pulp is usually stronger than that from bamboo (McGovern, 1987). When the whole jute plant is pulped by either the chemical or chemimechanical processes, the pulp properties are intermediate between those from the bast and stick as shown in Table 6.30.

The pulps from both the jute bast and jute sticks are frequently bleached with solutions of 5–10% sodium or calcium hypochlorite (based on fiber). Two stages are utilized for the bast fiber to give a brightness of 50–60, while a one-stage hypochlorite bleach is usually sufficient for the brighter jute stick pulps. A CEH sequence is utilized to obtain brighter jute pulps. Jute pulps are difficult to bleach to high brightness (90%) and this may be related to insufficient delignification in the cooking stage or to a different mode of lignin distribution in the cell wall of the jute fiber (Misra, 1980; Clark, 1969; McGovern et al., 1987).

Because of the limited degree of delignification achieved when jute bast material is pulped under the specified mild conditions, the jute fibers show little collapse and are stiff, hard and durable. The result is a pulp which is very suitable for heavy duty tag stocks, manila or buff-colored drawing papers and wrapping papers. In blends with other pulps, however, jute can be used in a wide variety of paper grades, especially for cigarette papers where it imparts superior porosity properties to the final product. The percentage of jute pulp in cigarette papers is generally around

70% and ranges from 55–75% in tissue products. Printing, bond, and writing papers are also produced with about 55% of jute pulp in the product. Paper containing greater than 75% jute fibers is exempt from excise duties in India (Misra, 1980; Bagby, 1978; Tribeni, 1992).

Morgen (1964) reported that a dissolving pulp could be produced from jute by first prehydrolyzing with 0.5% sulfuric acid for one h at 130°C followed by soda cooking (159 g/L) at 170°C for one hour. The pulp was then bleached with a two stage hypochlorite sequence to give the final product. The production of dissolving pulps from jute stick has also been evaluated in India (Bandyopadhyay, 1970).

Further research is needed to realize the full papermaking potential of jute. A better understanding of the delignification and fiber liberation mechanisms is required to ascertain the effect on fiber properties of stiffness and strength. Also, alternate bleaching processes may produce brighter jute pulps for a broader market.

4.5 Bamboo

Bamboo is a very rapidly growing plant species that occurs as two types: the clump type is generally endemic to countries with warm tropical climates such as India, Brazil, New Guinea, Australia and Kenya to name a few; and the running type occurs in more temperate climates (Figure 6.28). Bamboo has been grown in the United States on a limited basis for more than 75 years and the USDA lab in Savannah, Georgia at one time had one of the most extensive collections of hardy bamboos in the world (Haun et al., 1966).

Figure 6.28 Bamboo "forest" in Japan (Photo by R. A. Young).

Bamboo is a perennial grass which flowers at irregular intervals. The plant hs only one culm but new culms spring up every year to form a clump (clump type). The bamboo is cut every 4–6 years in India with a yield of 2–4 t/ha/yr (Misra, 1980). However there is wide variation in reported yields, with one study reporting 50 t/ha in a single harvest after 5–6 years (Bhargava, 1987). With properly managed plantations of bamboo it should be possible to obtain yields of 10–20 t/ha/yr.

Two species of bamboo, *Bambusa arundinacea* and *Dendrocalamus strictus*, are typically used for pulp and paper in India. The average length of the fibers compares favorably with coniferous fibers but the length varies considerably, from 1.5–4.4 mm. The average length is usually in the range of 2.2–2.6 mm. The average fiber width is about 14 μ. Bamboo fibers are generally stiff due to a thick cell wall and narrow lumen. Philippine bamboos have higher ash and silica contents compared to the other Asian species. Haun et al. (1966) found that the tropical bamboo, *Bambusa vulgaris*, had longer fibers and somewhat superior papermaking characteristics compared to the other bamboos that they studied.

Although bamboo was considered unsuitable for papermaking at one time in the western nations, the Chinese have utilized bamboo to make paper for centuries. The cut green bamboo was soaked in lime milk for days, washed, ground and the rejects were picked out by hand. The bamboo paper sheets were then bleached in the sunshine (Ying, 1985). Today, bamboo is utilized for papermaking in Bangladesh, China, Kenya, the Philippines, Taiwan, and Thailand, and it is a major raw material for papermaking in India (1.7 million t/yr). However, Pant et al. (1991) have reported that demand for bamboo is exceeding supply and the Indian industry is becoming increasingly dependent on hardwoods.

Bhargava (1987) has listed the advantages and disadvantages of bamboo as a raw material for papermaking. The advantages include: (a) fast growth rate, (b) debarking not required, (c) high yields from kraft pulping, (d) low chemical consumption, (e) easily bleachable kraft pulps, and (f) good strength pulp, stronger than most tropical hardwood pulps. The disadvantages of bamboo include: (a) gregarious and sporadic flowering which upsets and disrupts regular supply, (b) difficult to chip because of hollow stems, (c) difficult to handle mechanically because of variable diameter and crookedness of stems, (d) dense nodes which are highly lignified and difficult to pulp, (e) high silica which causes scale formation and problems with lime mud reburning, and (f) high percentage of parenchyma cells.

The bamboo stems are washed with showers to remove external dirt and sand and carried on chain conveyors to the chipper. The bamboo is converted to chip form with multiknife (modern drum) chippers for charging to the digester. It has been shown that crushing of the stems improves penetration of the cooking liquor and makes chipping easier, especially for the denser nodes; however, power consumption is high for the crushing operation and the additional step reduces output with the crushed chips of irregular size.

4.5.1 Mechanical Pulping

Bamboo mechanical pulps have been utilized for centuries to produce Chinese ceremonial papers. Green bamboo stalks can be disintegrated directly or, if dry, soaked before mechanical disintegration. It has been reported that pulp from green culms produces a stronger and finer-textured mechanical paper. The grinder pockets are first charged with rice straw prior to introduction of the cut bamboo culms. Bhargava (1946) also evaluated groundwood mechanical pulping of bamboo but was not able to produce a satisfactory pulp. He found that the groundwood bamboo pulp was like a fine flour rather than fibrous in nature. However, a semi-chemical cold soda process was found to produce a satisfactory pulp from bamboo. After treatment with a 26% caustic solution for 16 hours and disintegration in a disc refiner, a pulp with a breaking length of 2,200 m, a burst factor of 9.9, and a tear factor of 45.2 was obtained.

4.5.2 Chemical Pulping

The earliest work on bamboo pulping was done by the Chinese. In the 1870s Routledge (1879) documented his work on bamboo pulping from juvenile culms (4–6 months). Raitt (1925) later applied wood pulping techniques to bamboo and demonstrated the suitability of this approach for production of high quality pulps. Today the Indian industry commonly produces paper grades with up to 60% bamboo pulp in the furnish (Pant et al., 1991).

Alkaline processes have been shown to be advantageous for bamboo pulping, and the kraft process produces better pulps than the soda process. However, mills in Taiwan utilize the soda process (13–20% for 8–10 h) to avoid the odors associated with kraft pulping. Sulfite pulping has not been utilized for bamboo because of the lower pulp strength, the high cost of sulfur, the instability of bisulfite liquor in warm climates, and the lack of suitable recovery systems (Misra, 1980).

A problem with bamboo pulping is to achieve good penetration of the bamboo tissue. The difficulties are caused by the presence of an impenetrable epidermis and strong bundle sheaths, limited areas of conducting channels and the complete absence of rays (Rydholm, 1965). Therefore, Raitt (1925) developed a two-stage kraft digestion system specifically for bamboo. A mild first stage (pectin digestion) is employed to achieve good liquor penetration and to remove the easily dissolved and degraded materials such as gums, tannins, starch, sugars, etc. The second stage (lignin digestion) is a more drastic cook to complete the delignification process. The spent liquor from the second stage is used for the first stage, while a fresh solution of caustic soda/sodium sulfide is used to charge the second stage. The first stage runs for 2 h at 115–125°C, with 6–10% alkali based on oven-dry bamboo. The second stage is run for 3 h at 140–155°C at 16–18% active alkali. Although there is good delignification (permanganate number = 10–12) and chemical consumption is low, the two-stage cook is more time consuming, which gives a lower output, and there is greater steam consumption. A comparison of the properties of bamboo pulps from the two-stage vs. the one-stage process is shown in Table 6.32 (Haun, 1966).

Most pulp mills utilize a single stage kraft (overhead pulping) digestion process. The conditions for this cook are: active alkali = 20–22%; sulfidity = 20–25%; cooking time = 2–2.5 h at 170°C; and a liquor-to-wood ratio = 2.5–3:1. After the bamboo chips are charged to the digester, the liquor is directly steam-heated and circulated to the digester. An impregnation stage at 120°C is performed for 1 h, followed by a 2–2.5 h cook. The total cycle takes about 4 h. Yields of 44–46%, with

Table 6.32 Physical Properties of Bamboo Pulps from Raitt Two-Stage and Kraft Single-Stage Processes[a]

Pulping method and species tested	Pulp yield %	Beating time Min	Burst factor g/cm²/gsm	Tear factor g/gsm
Raitt two-stage				
Bambusa longispiculata	41	62	22.0	132
B. tuldoides	45	53	22.5	103
Guadua angustifolia	44	60	18.6	135
Phyllostachys bambusoides	44	40	20.3	85
P. purpurata "Solidatem"	44	48	31.1	151
Kraft single-stage				
P. viridis	49	47	22.2	110
P. flexuosa	50	28	31.1	93
P. nidularia	54	48	22.5	77
P. purpurata "Straightstem"	53	43	41.5	108
Pine chips	57	55	43.6	106

[a] 400 mL CSF.

Source: Haun, J. R. et al., *Fiber and Papermaking Characteristics of Bamboo*, Tech. Bull. No. 1361, ARS/USDA, Washington, D.C., 1966.

permanganate numbers of 13–16, have been reported for the single-stage cook (Misra, 1980). Semana et al. (1967) utilized a single-stage kraft cook for pulping a variety of different types of bamboos from the Philippines and compared their results with the properties of kraft pulps from tropical hardwoods and temperate softwoods. The Philippine bamboos were easily digested giving yields in the range of 41–48%. In general the bamboo pulps had higher tearing strengths but lower tensile, burst and fold strengths compared to the hardwood and softwood pulps as shown in Table 6.33. Semana et al. (1967) explained the differences between the strength properties of the bamboo and wood pulps based on differences in the relative rigidity and length of the bamboo fibers.

Neutral sulfite pulping has also been utilized with bamboo. Naffziger et al. (1961) reported a yield of 61% from NSSC pulping of *Phyllostachys bambusoides*, while Chen and Lin (1973) reported yields of 55–69% for NSSC pulping of bamboo in Taiwan. The latter investigators reported considerable losses in bleaching, with yields down to 48–53%. Both investigators found that the bleached bamboo pulps had strength properties equal to or superior to bleached softwood pulps, and the pulp was a suitable substitute for long-fiber pulp in various paper grades. The strength properties of an experimental newsprint furnish comprised of 20% NSSC and 80% cold soda bamboo pulps were comparable to a control sheet from 80% aspen groundwood and 20% unbeaten softwood sulfite pulp (Misra, 1980).

Continuous pulping processes have proven to be very effective for pulping of bamboo. In India, bamboo and bagasse are cooked separately but in the same horizontal-tube digester at different times. Uniform quality bamboo pulps are produced in both Kaymr and Pandia continuous digesters. Lungren (1993) has described the use of two types of Kaymr digesters (ASTHMA and Steam Phase) for continuous pulping of bamboo. Such modified digesters are in use in Thailand (Phoenix Mill) and China (Yibin Paper Mill) (Figure 6.29).

Table 6.33 Physical Properties of Pulps from Bamboo and Other Fibrous Raw Materials[a]

Species	Density g/cm^2	Burst factor	Tear factor	MIT folding endurance	Breaking length, m	Opacity, %
Bolo	0.57	70	170	1,040	8,750	98.8
Buho	0.63	66	153	700	8,150	97.8
Giant bamboo	0.60	63	195	780	8,250	97.2
Kauayan-kiling	0.60	64	183	1,050	9,300	98.4
Kauayan-tinik	0.61	64	150	850	8,500	98.5
Yellow bamboo	0.65	66	101	610	9,300	99.0
Mean of above species	0.61	65.5	159	838	8,708	—
Range	0.57–0.65	63–70	101–195	610–1,050	8,150–9,300	—
Mean of 10 Philippine hardwood kraft pulps	0.64	73.6	125	1,186	10,205	—
Range	0.52–0.72	44–94	103–172	90–2,050	8,000–12,350	—
Mean of 5 softwood kraft pulps	0.67	82.6	126	1,124	9,640	—
Range	0.61–0.72	69–94	107–148	1,000–1,300	8,500–10,500	—

[a] 300 ml CSF.

Source: Semana, J. A. et al. *TAPPI*, 50(8), 416, 1967.

Figure 6.29 The modern Phoenix Pulp and Paper Mill in Thailand (Photo by R. A. Young).

Bamboo pulps are suitable for a broad range of paper grades. Haun et al. (1966) evaluated a series of bamboo pulps for a variety of paper grades as shown in Table 6.34. The bamboo pulps were produced by both the one- and two-stage kraft processes, bleached with a CEH sequence, and mixed with the following pulps: conventional bleached kraft softwood, bleached soda hardwood, oak groundwood, flax pulp, and cotton rag pulp. The bamboo pulps have proven especially suitable for absorbent products such as paper towels (Haun et al., 1966; Bharvaga, 1987).

It has been found to be more difficult to bleach bamboo pulps compared to other agro-based fibers. Typically a DEH or a CEHH sequence has been utilized to bleach bamboo pulped by the kraft process to a κ number of 11–16. The CEHH sequence involves application of 7% chlorine in chlorination, 2.5% NaOH in extraction and 3% and 2% chlorine, respectively, in the two calcium hypochlorite stages to give a brightness of 68–78%. The total chlorine consumption was thus 12% (McGovern, 1967). High κ number pulps require additional steps in the bleach sequence to achieve high brightness levels (80–85%). CEHEH, CEHED, and CCEHD have been suggested, with the latter sequence of two chlorination stages costing less. However, chlorination causes excessive pollution and alternate, less polluting bleaching processes need to be explored for bamboo and other agro-based fibers, such as introduction of oxygen (O) stages for chlorination (C). Bleaching additives such as sulfamic acid, diammonium hydrazine phosphate, and sodium silicate have been found useful for controlling viscosity, preserving strength properties, and improving brightness of bamboo pulps.

Table 6.34 Composition and Physical Characteristics of Bamboo Furnish Papers Made on Pilot Fourdrinier Paper Machine

Paper Guide	Furnish	Basis Weight g/m²	Burst Factor g/cm²/gsm	Tear Factor g/gsm	Breaking Length m	Fold (MIT)	Porosity		Double Smoothness 8-ply	Brightness %
Bond	Bamboo (R),[a] 80%; SP kraft,[b] 20%	85.1	28.1	60.1	5,870	15	211.0	211.4	70.8	68
Saturating	Bamboo (R), 90%; SP kraft, 10%	166.8	25.3	12.0	5,680	189	10.5	11.0	—	—
Tissue	Bamboo (R), 90%; SP kraft, 10%	37.3	22.6	43.0	5,000	1	30.8	34.2	77.6	70
Wet strength	Bamboo (R), 100%	93.2	20.4	78.3	4,580	50	2.7	2.7	46.4	—
Book	Bamboo (R), 66%; soda,[c] 34%	72.7	12.6	48.1	2,940	1	9.3	9.3	116.4	76
Magazine	Bamboo (R), 56%; SP kraft, 12%, and soda, 32%	86.7	13.8	50.8	3,690	4	36.4	36.4	117.2	68
Saturating	Bamboo (R), 100%	155.2	21.3	79.2	5,280	13	1.6	21.6	26.8	—
Bond	Bamboo (R), 100%	258.1	11.2	24.8	1,700	1	154.9	155.0	104.8	74
Book	Bamboo (R), 50%; soda, 25%; and SP kraft, 25%	81.2	19.9	72.6	3,690	5	24.5	27.7	168.4	70
Book	Bamboo (R), 100%	96.0	19.0	47.9	4,510	10	410.7	410.8	286.8	—
Bond	Bamboo (R), 50%; rag pulp, 50%	66.2	27.6	10.4	6,340	69	80.3	80.2	86.4	—
Tissue	Bamboo (R), 100%	38.2	16.6	41.0	3,840	1	92	92	184.4	—
Food container board	Bamboo (R), 100%	22.7	18.0	86.2	4,570	66	11.5	11.5	54.0	—
Bond	Bamboo (R), 50%; flax pulp, 50%	95.4	16.2	67.0	4,610	7	155	155	104.8	74
Bible	Bamboo, 50%; flax pulp, 50%	44.6	11.0	47.1	3,440	1	68.0	68.1	143.2	—

Corrugating medium	Bamboo (K),[d] 100%	12.4	17.0	62.6	4,440	10	3.3	3.2	47.2	Not bleached
Colored wrapper	Bamboo (K), 100% and 31%	45.5	15.5	77.1	4,700	1	2.9	2.9	74.0	Too red for reading
Magazine	Bamboo (K), 22%; bamboo (R), 46%; oak groundwood, 16%; and soda, 16%	69.5	22.3	61.9	5,470	1	62.4	62.4	106.8	68
Stationary	Bamboo (R), 100%	79.1	23.1	63.2	5,300	15	65.8	65.8	90.0	73

[a] (R) – bamboo kraft pulp from two-stage Raitt process.
[b] SP – kraft-bleached southern pine kraft pulp.
[c] Soda – bleached soda hardwood pulp.
[d] (K) – bamboo kraft pulp from kraft single-stage process.

Source: Haun, J. R. et al., *Fiber and Papermaking Characteristics of Bamboo*, Tech. Bull. No. 1361, ARS/USDA, Washington, D.C., 1966.

The Phoenix Pulp and Paper Company in Thailand utilizes bamboo for production of commercial grade kraft pulp. Due to environmental pressures, the mill has introduced an oxygen delignification stage prior to bleaching. This has resulted in a 30% reduction in κ number that, in turn, has reduced chemical consumption in bleaching, particularly chlorine and hypochlorite, and significantly reduced effluent load as COD, color, and dissolved solids. The pulp quality was found to be comparable in terms of strength properties (Mittal and Maheshwari, 1994).

A number of investigators have evaluated the production of dissolving pulps from bamboo (Naffziger et al., 1960). Both clump and running types of bamboo yield satisfactory dissolving pulps. Joglekar and Donofrio (1951) used sulfuric acid for pre-hydrolysis kraft digestion and bleaching to produce high quality bamboo-dissolving pulps, although the yields were low and the ash contents greater than commercial pulps. Naffziger et al. (1960) evaluated nitric acid, water, and steam as prehydrolytic agents prior to kraft pulping. Commercial quality dissolving pulps were obtained from all three approaches in the yield range of 23–32%. Dissolving-grade pulp is bleached in a CEHEDH sequence with a final sulfur dioxide treatment to a brightness exceeding 90% and a viscosity of 16–20 cp (Misra, 1980). These pulps are used to produce rayon fibers as filaments and staple for yarns and tire cord applications.

4.6 Reeds

Reeds are a grass-type plant, such as *Phragmites communis*, that generally grow in swamplands, river bottom land, and delta areas such as in Russia, Romania, Egypt, China, North Korea, Iraq, and Turkey. The stalks may grow to 15 feet and the fiber length ranges from about 1.5–3.0 mm. The "rice culture" of China and Korea, with drainage and dam systems, provides greater ease of harvesting of reeds in these countries, while harvesting in the delta regions of Eastern Europe is more problematic and often restricted to the months of November and December, when the ground is frozen, or through the use of specially designed Caterpillars with wide treaded rubber tracks. Small boats are used in Egypt and Iraq for harvesting the reeds from the salt marshes and deltas of the Nile, Tigris, and Euphrates rivers.

Losses in storage and handling are particularly problematic with reeds. The many steps involved in preparation of the reeds for pulping are shown in Figure 6.30 (Wiedermann, 1987). Discounting moisture loss, the loss of dry material can be as much as 30% through these stages if proper care is not exercised. For optimum harvesting and transportation, the reeds should be biologically mature, contain a minimum amount of moisture, and have lost the majority of the leaves. The decrease in the moisture content reduces biological degradation and transportation costs, and the leaves have high silica contents and low fiber content (Clark, 1969; Wiedermann, 1987).

Like many other agro-based plant material, cold maceration of reeds in lime milk for 2–3 weeks, followed by mechanical defibration, was utilized to produce a pulp prior to the advent of modern technology. Also similar to the other agro-based plants, the alkaline processes appear to be the most favorable for pulping of the reeds. Rapid pulping is realized by the kraft process both in batch and continuous

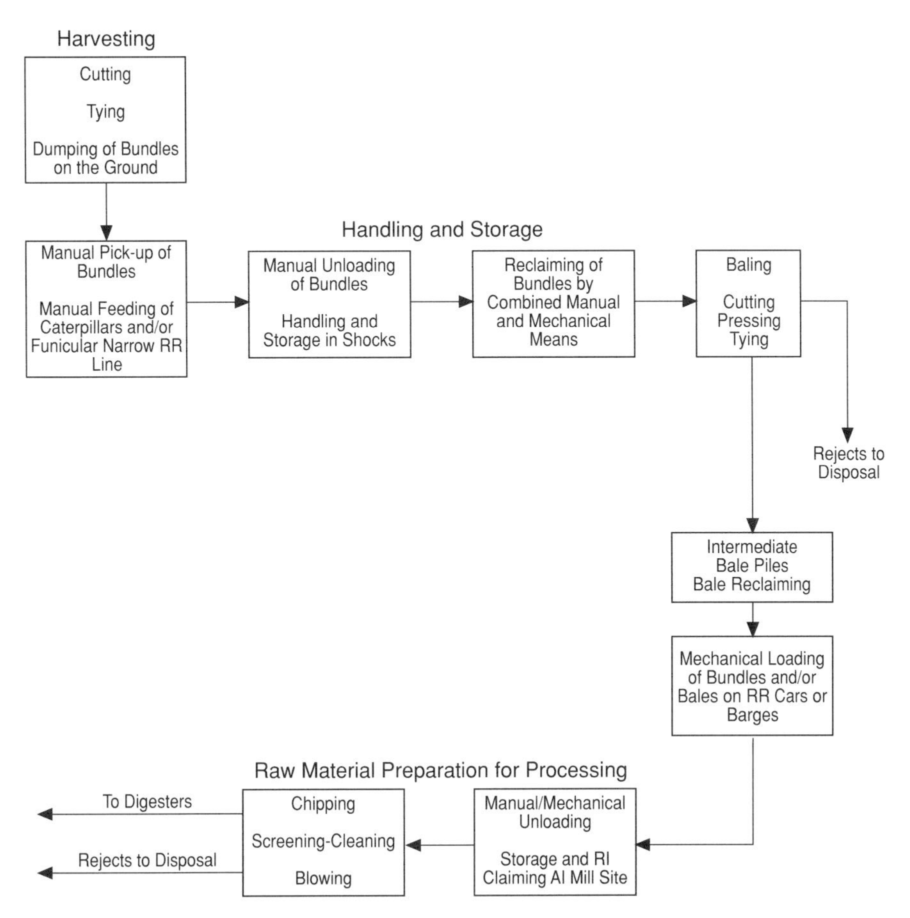

Figure 6.30 Preparation of reeds for pulping — schematic sequence and type of operation (Wiedermann, 1987).

(Pandia) operations with the following conditions: active alkali, 15–20%; sulfidity, 15%; cooking time, 20 min; temperature, 170°C. Reed pulps show good burst and tensile strengths. Typical pulp properties are given in Table 6.35 (Wiedermann, 1987).

Chemimechanical, NSSC and magnesium bisulfite processes have also been commonly utilized for pulping of reeds. The chemimechanical process involves treatment with 7–9% caustic soda at 100°C for 45–75 min, followed by mechanical disintegration to give a yield of 65–70% (10–14% lignin). For NSSC pulp the reeds are cooked at 155–170°C, for 75–120 min with 15% sodium sulfite and 5% sodium bicarbonate to give a final yield of 58–63% (4–5% lignin). The high brightness of the unbleached NSSC pulp was notable. Properties of the pulps from these processes are summarized in Table 6.35 (Wiedermann, 1987).

Both CEH and CEHH sequences have been utilized to bleach reed pulps to brightness ranges of 80–85 and final yields of 38–42% (Misra, 1980). The high ash content of reed pulps mandates an acid pre-treatment for production of dissolving

Table 6.35 Properties of Reed Pulps[a]

	Semichemical Soda	Neutral Sulfite Semi-chemical	Kraft
Yield, %	66–71	58–63	50–52
Permanganate No.	—	11–14	12–13
Lignin, %	10.8–14.0	3.7–5.0	1.7–2.1
Breaking length, km	3,900–4,400	4,200–4,800	7,200
Burst, kg/cm^2	1.5–2.3	1.4–2.3	2.2
Tear, g/gsm	76–82	62–71	—
Double fold, no.	20–42	6–14	330
Brightness, %	32–38	44–53	—

[a] 250 ml CSF.

Source: Wiedermann, A., Reeds, in *Pulp and Paper Manufacture, Vol. 3, Secondary Fibers and Non-Wood Pulping*, TAPPI, Press Atlanta, 99, 1987.

pulps. Although some grades of paper and board can be manufactured entirely from reed pulp, a level of 35–70% of reed pulp is the recommended maximum for high quality writing and printing papers (Table 6.36) (Wiedermann, 1987). The bleached pulp contributes considerably to the bulk and opacity of the paper but very high contents of reed pulp in the furnish have been reported to cause frequent breaks on the paper machine due to poor wet sheet strength properties. Chengwu et al. (1988) suggested that these problems could be alleviated by proper selection of pulping and refining conditions and by removal of non-fibrous cells.

Table 6.36 Characteristics of Papers Containing 50–100% Reed Pulp[a]

Type of Paper	Reed Pulp, % in Furnish	Breaking Length, km	Double Fold	Tear Strength, g/gsm	Burst Strength, kg/cm^2
Writing	50	4–4.5	12–16	55–60	1.4–1.6
Printing	50	3.6–4.0	8–10	50–60	—
	80	3.5–4.0	6–8	40–48	1.3–1.6
	100	2.7	3–5	40–45	1.1–1.2

[a] Other component in furnish is bleached sulfite pulp.

Source: Wiedermann, A., Reeds, in *Pulp and Paper Manufacture, Vol. 3, Secondary Fibers and Non-Wood Pulping*, TAPPI Press, Atlanta, 99, 1987.

4.7 Esparto

Esparto grass (*Lygeum spartum*) grows wild over large areas of North Africa and is found in Spain and Mexico. The plant grows in clumps to heights of 1–1.25 m from seed. It is harvested manually twice a year and can be stored up to one year prior to pulping. The plant, which contains about 38-43% fibrous material, is first threshed to remove the outer layer of wax and cut to the desirable length for pulping. The length of the esparto fiber varies from 0.25–2.5 mm with an average length of 1.5 mm and an average thickness of 7 μ. The ultimate fibers of esparto are among the most slender of the agro-based fibers (Misra, 1980; Clark, 1969; Rydholm, 1965; Young, 1994a; McGovern et al., 1987).

Esparto pulping was initiated in England in 1865 as a substitute for rags, which were in short supply. Coal freighters from England would return from the Mediterranean areas with loads of esparto grass in the 1800s. After World War II when esparto supplies were cut off and coal shipments decreased, esparto pulping in England disappeared; now the grass is pulped mainly in Algeria, Morocco, Tunisia, Spain, and France. Esparto pulps are sold at a premium price because of the distinctive properties. As well as good strength properties, the pulps exhibit high bulk, opacity, dimensional stability, absorbency, and softness (Clark, 1969).

Soda pulping of esparto is carried out by both batch and continuous processes. The batch technique is based on the methods utilized in England in the 1800s and involves dusting, boiling, and washing prior to pulping and bleaching. The dusting is carried out with conical screens that remove about 1–6% of the weight of the plant material in the form of sand and dust. The uncut grass is then fed through the top of a specially-designed digester (Figure 6.31), which holds about 6 tons of grass. Caustic solution (19–25 1b/100 lb grass) is added through a manifold to wilt and pack the esparto, followed by further addition of fresh grass until the digester is fully charged (Misra, 1980; Clark, 1969).

After the digester cover is secured, steam is introduced at the bottom of the digester until a pressure of 3–3.5 kg/cm^2 (6–7 kPa) is reached, which takes about 45 min. The cooking time is about 2.5 h with the blow down covering about 30 min. The liquor is circulated through discharge pipes at opposite sides of the digester. The spent liquor is drained through the perforated screen and the pulp is washed with hot water. These solutions are recovered for reuse. The pulp is discharged from the side of the digester or from the bottom in modified units. A high pressure water jet forces the cooked esparto through the discharge orifice which causes further disintegration of the pulp. Final washing and bleaching is performed in washer-beaters and the temperature is increased to 100°F by direct steaming just prior to bleaching. The pulp is bleached with calcium hypochlorite (Clark, 1969).

A variety of continuous pulping processes have been utilized for esparto, including the continuous soda–chlorine process, a Celdecor–Kaymr process and Pandia continuous processing. Ouchtati and Carney (1981) described the use of a continuous cooking of esparto in the Pandia digester by the soda process in Tunisia. Typical cooking conditions were 15% active alkali, 165°C for 18–20 min. A CEH bleaching sequence yielded a pulp of 80–82 brightness, while CEHD was utilized for higher brightness levels (90%). A sulfur dioxide treatment was necessary to control color reversion. Spanish varieties of esparto give higher bleached pulp yields (45%) compared to the African varieties (38%).

The properties of esparto pulps are given in Table 6.37 (Misra, 1980). As previously mentioned, paper made from esparto pulp is noted for its high bulk, opacity and dimensional stability. After processing, esparto pulps have an apparent specific gravity of 0.45 g/cm^3 compared with 0.80–0.95 g/cm^3 for wood pulps. The exceptional dimensional stability is demonstrated by only 0.8% shrinkage on drying, or expansion on wetting, compared with twice this amount for wood pulp. This results in sheets that remain very flat in use. Esparto pulps are also extremely

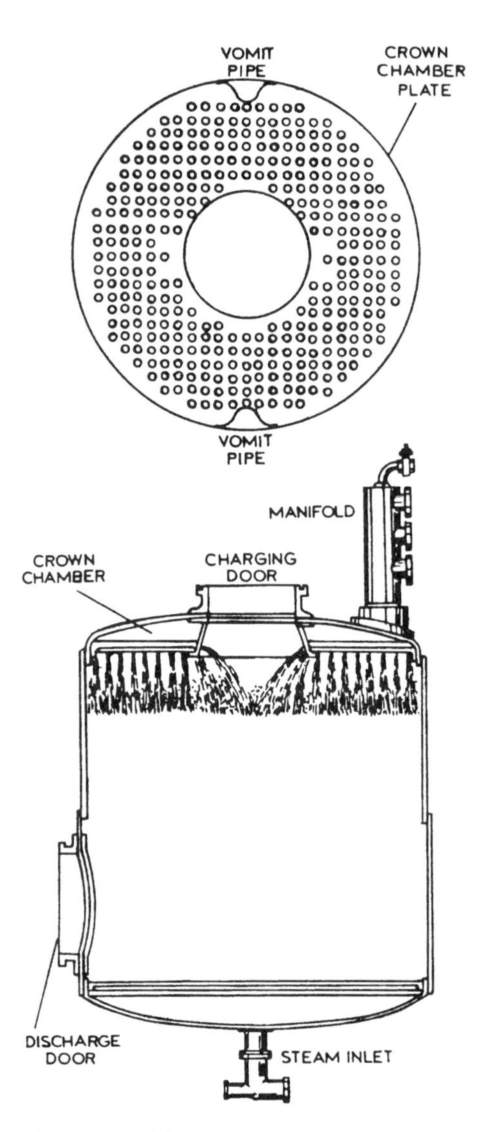

Figure 6.31 Grass-type digester used for cooking of esparto (Clark, 1969).

receptive to ink, and render exceptionally sharp images. In addition to book, printing and writing papers, Ouchtati and Carney (1981) identified several grades of paper where esparto pulps are especially suited:

1. Cigarette filter and wrapper papers where high porosity and good opacity are essential.
2. Filter papers where good porosity is also desired.
3. Dielectric and condenser paper where high opacity, suppleness, and purity are essential.

Table 6.37 Strength Properties of Unbleached and Bleached Esparto Compared with Wood Pulps

Canadian Standard Freeness (mL)	Burst Index (kPa/gsm)	Tear Index (mN/gsm)	Folding Endurance (MIT double fold)	Breaking length (km)	Sheet Density (g/cm³)
		Unbleached esparto pulp			
450	3.08	12.36	17	4.14	0.42
330	3.73	12.26	59	5.74	0.48
230	4.25	11.37	64	6.65	0.52
		Bleached esparto pulp			
450[a]	2.94	10.79	24	5.40	0.43
325	4.14	11.08	88	6.69	0.50
210	4.31	10.44	352	6.87	0.55
		Mixed hardwood (aspen, maple, beech) pulp[b]			
450[a]	3.53	8.43	40	7.0	0.69
		Mixed softwood (jack pine, hemlock, fir) pulp[b]			
450[a]	7.36	10.39[c]	1,600	11.4	0.79

[a] These values at Canadian Standard freeness 450 mL are interpolated.
[b] USDA Forest Service Forest Products Laboratory, Madison, WI.
[c] Tear index for unbeaten pulp.

Source: Misra, D. K., Pulping and bleaching of non-wood fibers, *Pulp and Paper Chemistry and Chemical Technology*, Vol. 1, 3rd edition, J. P. Casey, Ed., Wiley–Interscience, New York, 504, 1980.

4.8 Sabai Grass

Sabai grass (*Eulaliopsis binata*) was once an important source of fiber for papermaking in India; in 1952 it comprised 22% of the fibrous material pulped in this country. The quality of the pulps were considered by the Indian papermakers to be roughly equivalent to esparto pulps. The use of this material, however, has diminished considerably in recent years due to difficulties in procurement. Some plantations have been established to provide a more consistent supply of the raw material. Many small mills in India continue to use Sabai grass for production of wrapping, writing, and printing papers (Clark, 1969; McGovern et al., 1987).

The fibers range in length between 0.5–4.9 mm with an average of about 4.1 mm. The fiber widths are in the range of 9–16 μm. The fibers are thick-walled with a narrow lumen and produce very good book and printing papers. Before pulping the grass is chopped to 2.5–5 cm sections and dusted on rotary screens. The material is then pulped by the soda process with 10–15% caustic soda (5:1 liquor/grass ratio) in rotary spherical digesters or esparto–type digesters as shown in Figure 6.31. The washed pulp is bleached with calcium hypochlorite in one stage to about 70% brightness or in two stages to 75–80%

brightness. Unbleached paper from Sabai grass can be produced more simply, by hot soda pulping in a hydrapulper. The liquor is heated by direct steaming to a temperature of 98°C and the pulping time is 1 h. Good quality paper is obtained from the Sabai grass pulp alone or in mixtures (50:50) with rice straw pulp (Clark, 1969; McGovern et al., 1987).

Jeyasingam (1991) described some of the special problems–considerations related to Sabai grass as follows:

1. High cost of harvesting, collecting, transporting, and storage of the material by traditional methods with hand labor.
2. High cost compared to straw, an agricultural residue. Cost for straw harvesting and collection are offset by grain production.
3. High cost compared to bamboo. Local bamboo is a denser material and therefore more economical to harvest, handle, and transport.

4.9 True Hemp

The source of true hemp is the bast fiber of *Cannabis sativa*, and sunn hemp, *Crotalaria juncea,* also produces a bast fiber. *Cannabis sativa* originated in China and is now grown in Central Asia and Eastern Europe. The stem is used for fiber, the seeds for oil, and the flower and leaves for drugs, among them marijuana. Sunn hemp is native to India and is now widely grown in temperate and tropical climates. Hemp fibers have been traditionally used to produce cordage, yarns, nets, etc. (Young, 1994a).

Like the other bast fibers, jute and kenaf, the hemps are retted to obtain the bast fibers. Water, dew, or snow retting is utilized depending on the geographical location. The fibers are relatively thick-walled with a broad lumen. The source of the raw material for pulping is frequently waste rope, cordage, etc., although both virgin and waste material are viable sources of the bast fiber (Young, 1994a).

Pulping of hemp ropes involves first chopping the rope strands to 5 mm lengths and opening of the bundles on a bolting machine. The cut ropes are then cleansed by leaching in a solution of lime or lime and caustic soda in tanks similar to those used for leaching rags. The material is then pulped by the soda process in a rotary digester. The charge of caustic soda is 20% (O.D. basis) at a liquor-to-fiber ratio of 5:1. The agro-material is then cooked for 4 h at 145°C. The pulp is extracted for 3 h in a washer-beater and then bleached in a three-stage sequence, CEH, to a brightness of near 80% (Misra, 1980; Clark, 1969; McGovern et al., 1987).

Sunn hemp is also an excellent source of papermaking fibers. Cunningham et al. (1978) evaluated kraft pulping of sunn hemp stalks by the kraft process. The pulping conditions in the rotary digester were: active alkali, 16.4%; sulfidity, 33.9%; liquor to solid ratio, 7:1; temperature, 170°C, for 2 h. The screened pulp yield was 54% and the strength tests (Table 6.38) indicated that the sunn hemp pulp had burst and tearing strengths similar to commercial southern mixed hardwood pulp and had considerably higher folding endurance strength. The pulp would be suitable for a variety of grades of paper and a mill in India produces cigarette paper from sunn hemp pulp.

Table 6.38 Characteristics of Sunn Hemp Pulps[a]

Characteristic	Georgia	Indiana
Density, g/cc	0.720	0.710
Burst factor, g/cm²/g/m²	47.8	58.0
Breaking length, m	9,100	10,100
Zero-span breaking length, m	10,900	10,200
Tear factor, g/g/m²	94.0	96.5
Folding endurance, M.I.T. No.	750	700
Porosity, per 100 mL sec	665	2,520
Stiffness. Gurley, mg	112	113

[a] Freeness = 600 °SR.

Source: Cunningham, R. L. et al., TAPPI, 61(2), 37, 1978.

Tjeerdsma and Zomers (1993) evaluated organosolv pulping of the woody core of hemp (*Cannabis sativa*) as a potential source of papermaking fiber for the Netherlands. The stem wood was chipped and pulped autocatalytically in 60% aqueous ethanol. The optimum conditions were found to be 2.5 h at 195°C to give a yield of 48.6% and a κ number of 24.8. The degree of polymerization of the pulp cellulose remained relatively high (2,752 from viscosity) which indicated that the strength properties should be reasonably good. Chemi-mechanical (soda) pulps from the woody core of hemp also gave reasonable strength properties as shown in Table 6.39 (deGroot et al., 1988). Hemp has been seriously considered as source of papermaking fiber in the Netherlands.

Table 6.39 Strength Properties of Chemimechanical Pulps from Hemp Woody Core

% NaOH	0	9	9	16	20
Freeness, °SR	58	38	73	38	58
Tensile strength, km	1.6	8.3	7.1	8.8	7.7
Burst factor, kPa·m²/g	0.4	4.1	4.1	3.5	5.0
Tear factor, mN·m²/g	2.0	3.9	3.6	5.4	2.9
Brightness, % GE	68	49	64	50	42

Source: de Groot, B. et al., The use of non-wood fibers for pulping and papermaking in the Netherlands, *Int'l. Non-Wood Fiber Pulping and Papermaking Conference*, Beijing, P. R. China, July, 216, 1988.

4.10 Leaf Fibers

The important leaf fibers for papermaking include abaca (*Musa textilis*) and sisal (*Agave sisilana*). Abaca is also called Manila hemp from the port of its first shipment, but it bears no relationship to the true hemps which are bast fibers. The plant is grown in the Philippines, and Central and South America. Sisal originated in the tropical western hemisphere and has been transplanted to East Africa, Indonesia, Brazil, and the Philippines. Sisal is named after the port in the Yucatan from which it was first exported. The traditional uses for the leaf fibers have been for cordage, ropes, twines, etc. (Young, 1994a).

4.10.1 Abaca

The abaca stalk consists of a central stem enclosed by leaf stalks or sheaths. The fiber is obtained from the outer sheaths of the stem by separating the outer layers (called tuxies) with a knife, pulling the layers free and final cleaning from parenchyma cells and other extraneous materials by drawing under a knife. This can be done by hand or mechanically. In the latter case a spindle stripper provides more constant tension on the knife to give a more uniform fibrous material for papermaking. Although fibers obtained in this manner are not suitable for porous tissue, they are excellent raw material for dense "onionskin" type tissues with high tear strengths (Misra, 1980; Clark, 1969; Rydholm, 1965; Young, 1994a).

The abaca fibers are then pulped in spherical or globe digesters by the soda or alkaline sulfite processes. The yield from soda pulping is 50% under the following conditions: caustic soda, 12–16% (O.D. basis); temperature, 140–150°C for 5–6 h. For alkaline sulfite pulping a charge of 16–18% sodium sulfite is used in a cook at 150–170°C. The yield is variable depending on the method used for stripping the plant stalk, 58–62% for hand stripped stalks and 67–71% for mechanically stripped stalks (Misra, 1980).

Mita et al. (1988) reported on a non-sulfur process for pulping abaca. These investigators utilized a hydrogen peroxide alkaline process to produce a high brightness pulp with a yield of 70.9%. The cooking liquor was a mixture of hydrogen peroxide, sodium hydroxide, and/or sodium carbonate and small quantities of anthraquinone plus a chelating agent. After cooking for 1 h at 140°C, the pulp was bleached in a single stage (alkaline hydrogen peroxide or calcium hypochlorite) to a brightness of 85–89%.

Abaca stems can also be pulped directly, without separating the tuxies, by the Gocellin process (Misra, 1980; Gomez, 1966, 1967). In this case the green stalk is cut, crushed and sliced in a specially designed chipper which removes water and produces 5 cm fragments. The material is then pressure-cooked at 90–100°C and mechanically defibrated at 120–130°C for 30 min in a hydrapulper. The defibrated stock is screened and sent to the pulp mill for final pulping.

Rope material is first cut to 5 cm lengths and dusted, cleaned, and opened. The alkaline cook is carried out in a rotary rag-type boiler with 10% lime or a 10% lime, 5% sodium carbonate solution for 8–10 h at 25 psi. The yield is about 50–65%. In another process the cut rope is cooked in an open tub with 10–14% lime and a small amount of sodium hydroxide (0.1%) (Misra, 1980; Bidwell, 1925).

A rag beater-breaker is used for both washing and bleaching of the abaca pulps. The pulps can be bleached to 80% brightness with hypochlorite in one or two stages, with the two stages requiring less chlorine. A CED bleaching sequence has also been used to produce pulps with excellent strength as well as high brightness. The strength properties of abaca pulps are shown on Table 6.40 (da Silva and Pereira, 1985).

Abaca pulps have been used to produce porous types of tissue grades such as tea bag paper, electrolytic capacitor tissues, filter paper, etc. Due to the high tensile and tear strength, abaca pulps in blends with other pulps would have applicability for use in hard papers such as currency paper (Jeyasingam, 1991). Table 6.41 shows

Table 6.40 Properties of Abaca, Sisal, and Cotton Pulps

Quality	Sisal	Abaca	Cotton
Freeness, CSF	662	665	664
	261	186	211
Basis weight, g/m^2	59.7	61.5	61.1
	58.2	63.3	60.8
Thickness, mm	0.163	0.162	0.209
	0.118	0.120	0.132
Density, g/cm^3	0.37	0.38	0.29
	0.49	0.52	0.46
Porosity, BN mL/min	2,800	2,650	3,000
	380	60	620
Traction Index, N·M/g	17.4	50.4	4.22
	65.3	89.7	25.7
Breaking length, m	1,778	5,140	430
	6,663	9,144	2,623
Elongation, %	1.7	2.9	—
	3.8	3.9	2.3
Tear Index, mN/gsm	14.7	41.4	4.9
	23.3	24.7	8.5
Burst Index, kPa/gsm	1.31	4.37	0.53
	5.40	8.82	1.63
Double Fold, MIT	2	2,030	—
	1,290	5,820	10
Opacity, %	77.0	81.3	75.3
	75.7	77.9	81.7

Source: da Silva, N. M. and Pereira, A. D., Experience of pioneer sisal simultaneous resource for pulp and energy, *Non-Wood Plant Fiber Pulping*, Prog. Rept. No. 15, TAPPI Press, Atlanta, 63, 1985.

the specific properties of abaca which fit with the variety of paper grades suitable for this pulp (Franco et al., 1981). The high price of abaca pulps, 4–5 times that of coniferous pulps, is due to the high cost of labor for fiber extraction and the low yields from the plant. Paper from abaca pulp maintains a position in the market due to its unique qualities of high permeability and tensile strength at low basis weights.

4.10.2 Sisal

The sisal plant leaves grow from a central bud and are 0.6–2 m long with a thorn-like tip. The leaves are harvested twice a year. The fibers are embedded longitudinally in the leaves, which are cut and defibrated in a hammermill. The juice is extracted from the mass in a crushing machine and is sold as a valuable byproduct to the chemical and pharmaceutical industries. The fiber is metered to the pre-impregnation vessel at the bottom of a Kaymr continuous digester and after series impregnation is cooked by a conventional soda process (Misra, 1980; da Silva and Pereira, 1985). The pulp is then screened, defibered, washed, and bleached (Figure 6.32). A brightness of about 92 is achieved in a multistage bleaching sequence of CEDED. The tear strength increases progressively with increased beating time as

Table 6.41 Application/Uses of Abaca Paper and Nonwovens

Product Requirement
Strong, thin, porous and wettable
Tea bags, coffee brewing bags, meat casings, stencils, paper towel, alcohol and absorbent sheets, sanitary napkin, liquid filter, medical plaster, incontinent pads, baby diapers, etc.
Strong, thin and porous
Air filter, cigarette plug wrap, hospital gowns and masks, bed sheets, medical plaster, throw-away disposable clothing
Strong and thin
Bible papers, lens tissues, cigarette column wrap paper, carbon paper, wax paper, metallizing paper
Strong and pliable
Currency notes, masking tapes, security paper, vacuum bags, containers (cement bags, seed bags), leatherette base
Strong and firm binder
Wall boards, ceiling boards, roof tiles, asphalt coated sheet, wallpaper sheets

Source: Franco, P. T. et al., Abaca pulp and paper industry in the Philippines, *Proc. Pulping Conf.*, TAPPI Press, Atlanta, 133, 1981.

shown in Figure 6.33 (da Silva and Pereira, 1985). The pulp has excellent strength properties, comparable to softwood pulps, as shown in Table 6.40. The table also contains data for abaca and cotton pulps for comparison. Table 6.42 shows the variety of paper grades suitable for incorporation of both unbleached and bleached sisal pulps according to da Silva and Pereira (1985). The first market pulp mill based on sisal is operating in Camacari, Brazil.

4.11 Cornstalks

Corn, or maize (*Zea Mays*), is produced mainly for food and feed from the grain or for use of the entire plant as silage. However, substantial portions of the plant remain unused or underused. Over the years this has prompted scientists to explore the papermaking potential of cornstalks. As early as 1837, a patent was issued to Bouchet in France for paper from cornstalks, and wrapping paper and paperboard were produced in Austria in the 1880s. The Cornstalks Products Company, formerly of Danville, IL, produced 45 t/day of cornstalk pulp in 1927. The pulp was blended with wood pulps to produce various grades of paper of acceptable market value and limited editions of more than 100 newspapers and magazines were printed on these papers. Problems of continued supply and preservation of the cornstalks eventually led to the demise of the mill. The Israelis also collected and pulped cornstalks in the 1950s, but abandoned the operation due to economic factors similar to those encountered by the American firm. The use of picker-shellers which discharge the broken stalks to the ground has further limited utilization of the stalks (Bagby and Widstrom, 1987).

The cornstalk stem is surrounded by leaves, but without the leaves accounts for about 50% of the plant (by wt.), with nodes (28%) and pith (21%) accounting for the rest of the stem. Therefore, the total stalk contains about 50% fibers and 50% parenchyma cells and vessels and only about 1% epidermis cells (Rydholm, 1965). As with bagasse, clean removal of the pith material improves the properties of the

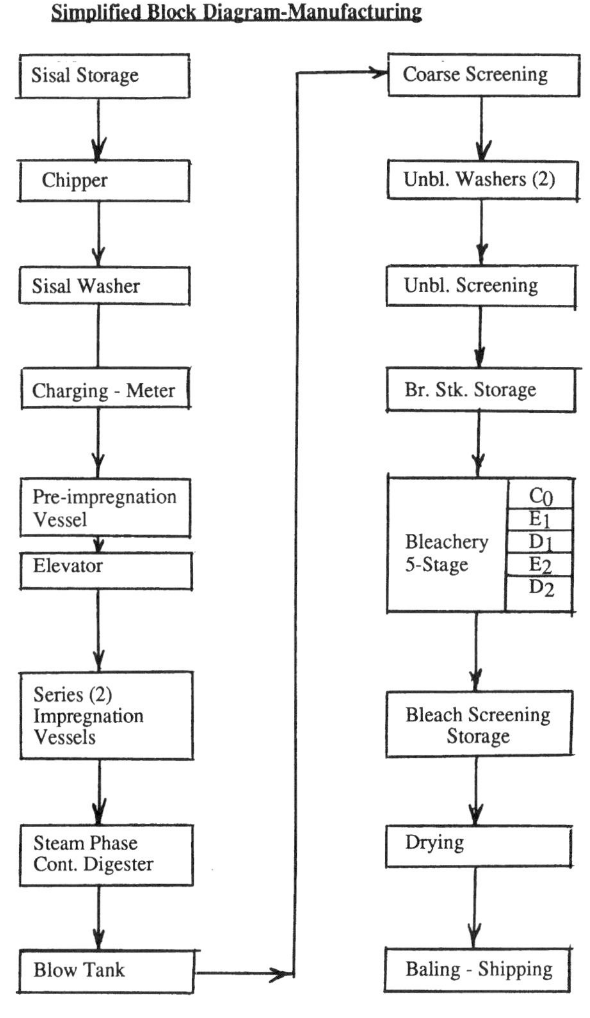

Figure 6.32 Block diagram for production of sisal pulp (da Silva and Pereira, 1985).

pulp and the efficient use of the pulping chemicals, although this has not been a common practice with cornstalks. The fibers are somewhat shorter compared to straw and reed plants, with an average fiber length of 1.2–1.4 mm and average width of 15–17 mm. If the pith is not removed, the pulp contains considerable nonfibrous elements, up to 50% for kraft pulps and 60–70% for semi-chemical pulps (McGovern et al., 1987).

Cornstalks traditionally have been pulped by alkaline processes, including the use of milk of lime (cold and hot), soda, and kraft processes. Both batch (rotary spherical) and continuous pulping at atmospheric and elevated pressures have been utilized to produce pulps from cornstalks. For soda pulping the cornstalks are charged with 8–11% caustic soda and cooked at 150–160°C for about 2 h to give rather low yields of about 37–40%. Higher temperatures in soda pulping of cornstalks have a

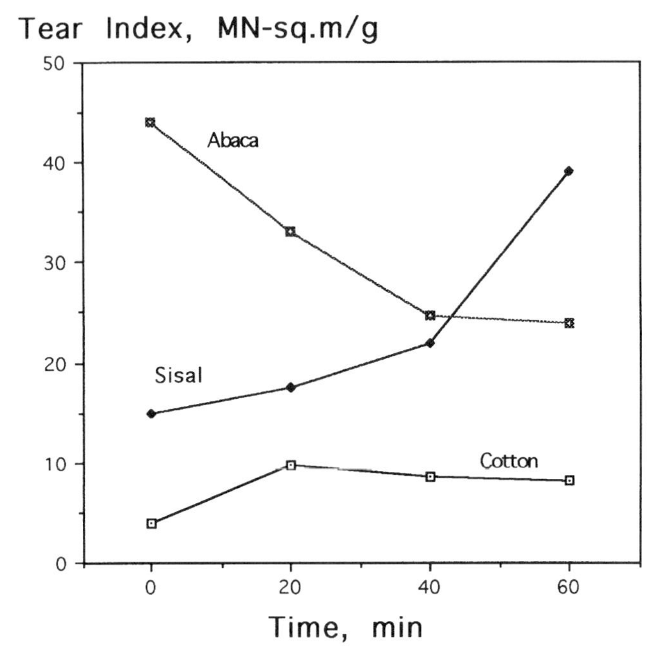

Figure 6.33 Tear index comparisons for agro-based kraft pulps as a function of beating times (da Silva and Pereira, 1985).

Table 6.42 Applications of Sisal Pulps

Unbleached Sisal Pulp	Bleached Sisal Pulp
(1) Paper for Dielectric Purposes	(1) Filtering Papers
- Insulation of Electrical Cables	- Coffee Makers
- Insulation of Telephone Cables	- Automotive Types
- Printed Circuits	- Vacuum Oil and Air Cleaners
Properties: Low Conductivity, Resistance to Traction, Cleanliness, Elongation	Properties: Porosity, High Quality, Cleanliness
(2) Wrapping Papers	(2) Specialties
- Cement Bags	- Security
- Sacks	- Cigarettes (Plug Wrap, Tipping)
- Pressboards	Properties: Tear, Porosity, Double Fold
Properties: Tearing Strength	
	(3) Disposables
	(a) Fluff Types – Tissues, Diapers
	Properties: Absorbency, Cleanliness, Fluffy
	(b) Non-Woven Types – Medical Clothing – Covers, Tea Bags, Other
	Properties: Porosity, Cleanliness, Fiber Type

Source: da Silva, N. M. and Pereira, A. D., Experience of pioneer sisal simultaneous resource for pulp and energy, *Non-Wood Plant Fiber Pulping*, Prog. Rept. No. 15, TAPPI Press, Atlanta, 63, 1985.

negative effect on strength properties. A higher alkali charge, 15–16% as sodium oxide, and a higher temperature, 160–165°C (2.5 h) are necessary for kraft cooking of the stalks. A high liquid-to-solids ratio (greater than 5:1) is recommended for pulping these materials, otherwise the liquid is totally absorbed in the fiber-pith mass resulting in wide variations in the pulp properties. Strength properties of soda and kraft pulps from cornstalks are shown in Table 6.43 (McGovern et al., 1987).

Table 6.43 Physical Properties of Cornstalk Pulps

Test	Soda	Kraft
Refining Index (interpolated), °SR	45	45
Breaking Length, m	5.3	5.8
Folding Endurance	60	81
Burst Strength, kg/cm^2	2.21	2.50
Tear Strength, g/gsm	87.5	92.0

Source: McGovern, J. N. et al., Other fibers, *Pulp and Paper Manufacture, Vol. 3, Secondary Fibers and Non-Wood Pulping*, TAPPI Press, Atlanta, 110, 1987, Chap. IX.

Usta et al. (1991) evaluated the use of soda–oxygen pulping with cornstalks. The loose, open structure of the cornstalks promoted the use of short cooking times at low temperatures with this more environmentally sound process. They found the optimum cooking conditions to be: alkali charge, 18%; temperature, 120°C; time at maximum temperature, 40 min; oxygen pressure, 10 kg/cm^2 and liquor-to-wood ratio, 5:1. The screened yields were found to be lower than those from wheat and rice straw and were ascribed to the lower holocellulose content and the greater amount of leaf and pith materials in cornstalks. The maximum strength properties reported for the soda–oxygen pulps were: breaking length, 7.5; burst index, 4.2; tear index, 6.2; and brightness, 46.3%.

Washing of cornstalk pulps requires relatively large washers due to the jellylike nature of the pith and the large amount of fines in the pulps. This also results in reduced drainage rates on the fourdrinier wire of the paper machine with concomitant reduced paper production rates.

Cornstalks have also been processed to dissolving pulps with purity greater than 95% (Bagby and Widstrom, 1987; Abou-Salem et al., 1983). The pulps were produced in a multi-step process involving hydration for one hour in boiling water, followed by prehydrolysis in 0.75% sulfuric acid for 6 h at 100°C (20:1 liquor-to-solids ratio) and then kraft or soda cooking for 5 h at 100°C. The degree of polymerization of the cornstalk dissolving pulp (DP = 970–1140) was comparable to dissolving pulps from bagasse (DP = 920) and softwoods (DP = 860).

4.12 Flax

Flax (*Linum usitatissimum*) is grown in many parts of the world, mainly for the production of linen from the fiber and linseed oil from the seed. Different strains of flax are grown for fiber or oil, with the latter emphasized in the United States. Traditionally the flax plant has been retted to remove the woody core, which accounts for 70% of the plant, from the bast fibers. Stream retting is believed to produce the highest quality linen fibers and the river Lys in Belgium has been noted to produce particularly desirable flax bast fibrous material. The bast strands, which may ultimately represent only 10% of the original plant material, are utilized to produce the high quality linen fabrics in Europe, especially Ireland (Misra, 1980; Young, 1994a; Tucker, 1969).

Flax material becomes available for papermaking as wastes from the linen industry in the form of fibers and fabrics from processing in the spinning and textile mills and from waste linen rags. The rags and fiber waste from the textile mills give the purest form of flax fiber since the material has been previously retted to remove the short fiber core stock. The fiber is further scrutched, hackled, and combed to yield a "flax tow" for pulp and paper (Tucker, 1969).

In contrast, the whole plant material remaining after removal of the seed, known as "seed flax straw," is also available as a by-product for use in pulp and paper. However, the whole plant cannot be used directly for pulp and paper because the core material of the bast-fiber plant contains a considerable amount of non-fibrous cells which must first be removed through mechanical decortication. This involves passing the raw stalk through breakers with fluted rollers and screening. The "seed flax tow" is quite different than the "flax tow" obtained by retting and has a shive content of gummy bast material as high as 30% (Misra, 1980; Clark, 1969; McGovern et al., 1987).

Although the bast fiber strands produced from seed flax are very short and unsuitable for textile uses, the ultimate fibers are almost identical to the ultimate fibers produced by the flax used for linen production, and therefore, both are very suitable for pulp and paper. The long, slender ultimate fibers have an average length of 28–30 mm and an average diameter of 20–22 mm to give a length-to-diameter (L/D) ratio of 1300:1. This L/D ratio is one of the highest for fibers utilized for papermaking, exceeded only by cotton fibers (Misra, 1980; McGovern et al., 1987).

Because of the different composition of "flax tow" and "seed flax tow," it is necessary to pulp these two materials under different conditions, the seed flax tow requiring a harsher environment to produce a good pulp. Flax tow is pulped in a rotary digester with a charge of 10–15% sodium hydroxide at 150°C for 4 h, while seed flax tow is treated with 15–25% caustic soda at 160°C for 6 h. Kraft pulping also can be utilized for pulping flax (Abou-Salem et al., 1983). The unbleached and bleached yields for the flax tow are 60–65% and 56–62%, respectively, while these yields for the seed flax tow are 33–40% and 30–37%. The pulp is washed in a beater-breaker and bleached in CEH, CEP, or CED sequences to a brightness of around 80%. The flax tow may be cut early in the process to facilitate processing or later on through the action of the refining equipment (Abou-Salem et al., 1983).

The quality of the paper from the seed flax tow is dependent on the extent of cleaning and removal of the shive type material which is further dependent on the economics and end-use of the product. Since the seed flax tow often contains a greater amount of shives, the paper is usually of lower quality, especially in terms of tear and fold strength. The paper from flax tow is generally more durable and resistant to deterioration. However, Wong (1987) found very similar properties for seed flax tow and flax tow when pulped by the neutral sulfite process. As shown in

Figure 6.34, the two pulps and their blends showed almost identical tensile-tear property relationships. Flax tow pulp, however, gave much better opacity (Figure 6.35), while the seed flax tow was much easier to bleach, even when the pulp had a higher kappa number (Tucker, 1969).

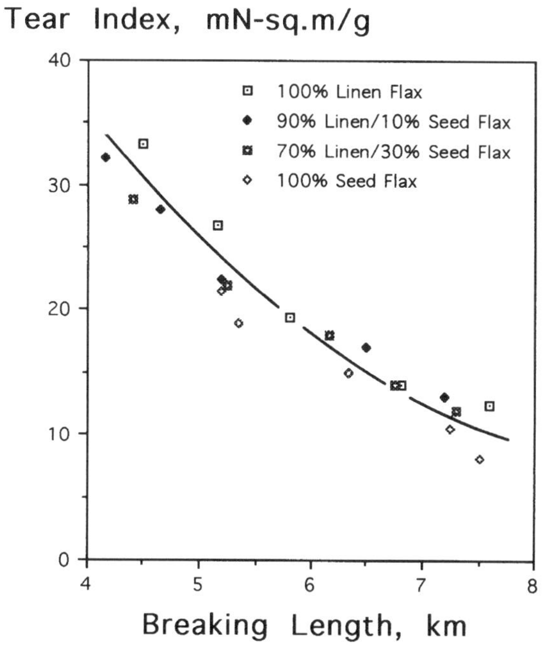

Figure 6.34 Tear strength of sheets from "flax tow" (linen flax) and "seed flax tow" (seed flax) pulps at different breaking lengths (Wong, 1987).

In general, paper from flax pulps is known for high strength (especially tear), toughness, durability, and an impression of outstanding quality. However flax pulps are expensive to produce and have been primarily used for production of specialty papers such as cigarette and other expensive thin papers, currency paper, and condenser tissue.

There have also been a considerable number of studies performed on steam explosion pulping of flax. Although the vast majority of these investigations have been directed to the production of staple fiber for textile applications (Kessler et al., 1994; Forcher et al., 1988), Kokta and Ahmed (1993) demonstrated that good papermaking pulps could be obtained by SEP of flax. As shown in Figure 6.36 the tensile–tear relationship for flax SEP is superior to those of flax RMP and CTMP when the SEP was prepared by pretreatment with 8% sodium sulfite and 1% sodium hydroxide. Explosion flax pulp with a yield of 75% can also be bleached to a level of 70% in one stage with 4% hydrogen peroxide (Kokta and Ahmed, 1993).

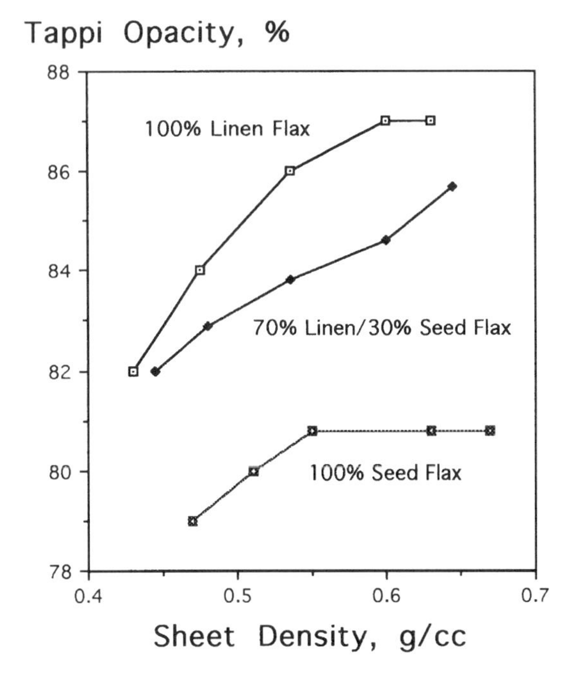

Figure 6.35 Opacity values for sheets prepared from "flax tow" (linen flax) and "seed flax tow" (seed flax) at different sheet densities (Wong, 1987).

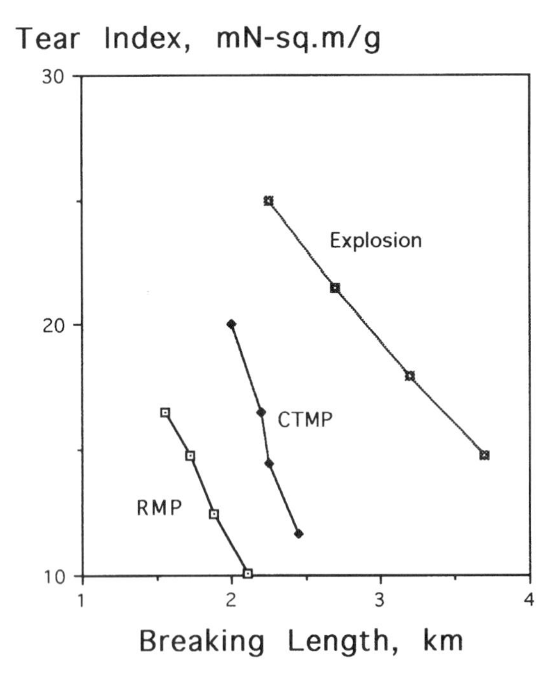

Figure 6.36 Tear and breaking length relationship of RMP, CTMP and explosion pulp flax (SEP pretreatment solution: 8% Na_2SO_3 + 1% NaOH) (Kokta and Ahmed, 1993).

4.13 Other Plants Utilized for Pulp and Paper

A number of other plant materials have been evaluated for use for pulp and paper most of which are listed in Tables 6.3–6.5 as evaluated by the USDA. Those which have received additional attention are discussed below.

4.13.1 Sorghum

This plant is also known as Indian millet and is used for a variety of purposes. There are a number of different varieties used for grain, syrup, and forage. Broom corn is used for the manufacture of brooms, baskets, and mats. Grain sorghum was introduced into the U.S. from the Sudan and is therefore sometimes referred to as Sudan grass (Clark, 1969; Jeyasingam, 1991a). Grain sorghum is grown in 20 states in the U.S. and, at 3 tons per acre, a total of 28 million tons of stalk is available in the U.S. The pulp from sorghum has potential for use in corrugating medium (Atchison, 1994).

Clark et al. (1973) evaluated 5 species of sorghum as raw materials for pulp and paper and found considerable variation according to year, species, and line. Crude cellulose contents ranged from 42–56% and α-cellulose from 26–36%. Several varieties of *Sorghum bicolor* appeared to be have promising characteristics. Jeyasingam (1991a) evaluated sorghums for semi-chemical pulp production and found that the physical properties were nearly equal to wheat straw pulps. Xie et al. (1988) suggested the following conditions for alkaline sulfite pulping of sorghum: alkali charge (as NaOH) 12–13%, with sodium monosulfite amounting to 60%, liquor-to-material ratio 1:3, maximum temperature 165°C and maximum pressure of 6 kg/cm^2. Sorghum was rated as one of the top five potential agro-based materials for pulp and paper in the U.S. by Atchison (1994).

4.13.2 Giant Cane

This is a perennial grass plant, *Arundo donax*, native to the Mediterranean region but now widely grown. The culms grow 2–8 m high with a diameter of 1–4 cm. The cellulose content is similar to wood and bamboo, while the pentosan content is greater for the giant cane. The cane has been pulped in Italy by first extracting chipped material with water (100°C) to remove the sugars (14%), which are fermented to alcohol (230 l/t pulp). The extracted cane is then cooked with calcium bisulfite in stationary digesters, screened, washed, and bleached in a three-stage sequence (CEH) to yield a pulp (36%) suitable for rayon production (Clark, 1969; Perdue, 1958).

4.13.3 Miscanthus sp.

One *Mercanthus* species, *Miscanthus sacchariflous* or Amur silver grass, is utilized in China for pulp and paper. Based on laboratory studies, Li and Pen (1988) suggested that Amur silver grass CMP pulps would make good substitutes for wood pulp when high brightness was not required. Wu and Qin (1988) found that Amur silver grass has different properties and behaves differently than other plants in kraft pulping; for example, the grass has a higher content of holocellulose (greater than 75%), and it has a higher lignin content (about 20%) and a lower ash content than straw. They suggested mild cooking of the grass under kraft pulping conditions, thus, a slow gradual rise in the pulping temperature to 100°C for the bulk delignification stage and then cooking at 150–165°C for 20 min. When run on the paper machine, bleached kraft pulps with high contents of Amur silver grass resulted in sticking to the press rolls and frequent wet web breaks. The running could be considerably improved by addition of pine groundwood pulp to the Amur silver grass-bleached kraft pulp; 20–25% of a pine groundwood pulp proved suitable for printing papers. For machine speeds greater than 400 m/min about 15–18% of a semi-bleached pine kraft pulp is needed as reinforcement to produce printing papers (Hua and Tu, 1988). The addition of the Amur silver grass pulps to the pulp furnish improved bulk and opacity of the printing papers.

Recent work in Germany has shown the suitability of *Miscanthus x giganteus* for plantation growth and pulp and paper in that country (Neuman, et al., 1994). Plantations of the *Mercanthus* were grown for five years, harvested and pulped by both alkaline and acidic processes. The properties of the soda–AQ pulps from *Mercanthus* were very similar to the properties of a spruce sulfite pulp. Similar properties were obtained for ASAM and ASA pulping of *Miscanthus*. These investigators also demonstrated that the solvent based Acetocell process was very suitable for pulping *Miscanthus*. The strength properties of the *Miscanthus* Acetocell pulps were of the same order as the soda–AQ pulps as shown in Figure 6.37 (Neuman et al., 1994). This solvent-based system offers a much less polluting alternative to conventional systems.

4.13.4 Palms

Various palm species have been evaluated for papermaking such as coconut palm trunks (*Cocos nucifera*), Palmyra palm leaves (*Borassus flabellifer*), and date palm leaves and stalks (*Phonrix dactylifera*). Laboratory work done in the Philippines indicated that various grades of paper such as onion skin, bag, and corrugating medium could be produced by blending coconut palm trunk pulps with kraft wood pulps. Native to India, the Palymra palm leaves were utilized in antiquity for recording ancient literature using a sharp, pointed steel pencil. Although further work is necessary, initial laboratory trials have indicated that the leaf stalk could be used as a papermaking material. The date palm grows mainly in the Middle East and North Africa. The use of date palm stalk for papermaking was evaluated in Iraq both in laboratory and mill scale trials. Although a low yield was obtained from kraft pulping of the stalk (28–30%), a satisfactory grade of paper was obtained (Sabharwal et al., 1981; Jeyasingam, 1991a).

4.13.5 Banana and Other Musa Species

There are hundreds of varieties of banana that are related to abaca. The stems of the plantain *Musa paradisiaca*, for example, contain about 70% fibers and 30%

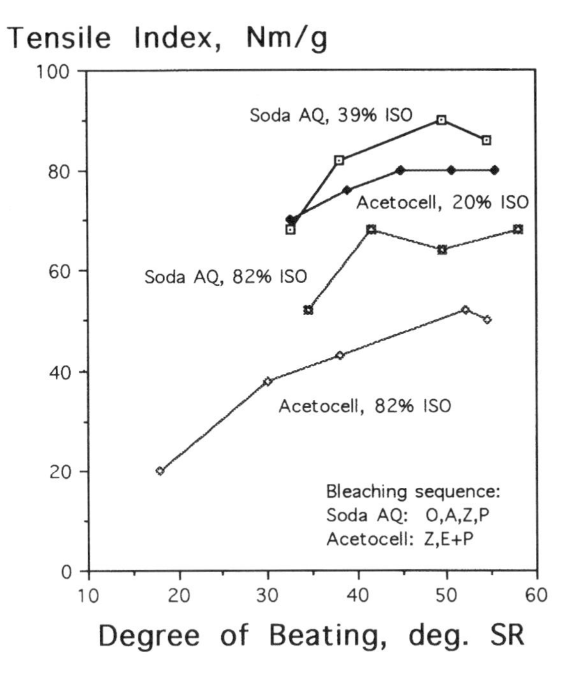

Figure 6.37 Comparison of strength properties of Soda–AQ and Acetocell *Miscanthus* pulps (Bleaching stages, O = oxygen; A = acid wash; Z = ozone; P = peroxide; E + P = caustic attraction + peroxide) (Neumann et al., 1994).

pith, the latter of which must be removed before pulping. The average fiber length is about 4 mm, which compares very favorably with softwood fibers. The stem has a lower Klason lignin and holocellulose content than both bagasse and wood, but higher ash and extractives. Pulping experiments indicated that a good quality pulp can be produced from the stems by both the kraft and soda processes with a low active alkali content (9% on wood). The kraft pulps had superior strength to the soda pulps. The strength properties of the banana stem pulps were about the same as kraft pulps from southern pines (Darkwa, 1988).

Escolano et al. (1988) evaluated the pulp and papermaking properties of eight Musa species related to abaca. Based on the proximate chemical analysis, the Musa fibers exhibited superior properties to woods due to their high holocellulose and pentosan contents and lower lignin contents. The materials responded favorably to pulping by soda, kraft, and alkaline sulfite pulping processes. Yields for the soda process ranged from 53.6% for latundan to 65.1% for giant cavendish, while for the kraft process butuhan yielded 39.0% and giant cavendish, 63.7%. This latter species gave a yield of 72.3% with the alkaline sulfite process. With the exception of alinsanay and giant cavendish, all the pulps had lower strength properties compared to abaca pulps; however, all the pulps were comparable to commercial kraft wood pulps. Thus, the pulps from the various Musa species ranked as follows: giant cavendish, alinsanay, lagkitan, saging-matsing, saba, latundan, pakol, and butuhan.

4.13.6 Cotton

Rags, linters, and stalks from cotton have all been utilized for pulp and paper. The use of cotton rags, of course, has a very long history, dating to the first century A.D. in China and was perfected by the Arabs and later, the Europeans. Since cotton rags do not contain any lignin, the cooking process is employed to mainly purify the cotton. Lime or caustic soda, or a combination of the two, is utilized to remove dirt, grease, starch and other loading materials added to the cotton fabrics. Lime is especially good for removing colored dyes from the fabrics. The rags are cut into small pieces before the cooking to prevent "roping" in the boiler. Bleaching is generally carried out with calcium or sodium hypochlorite. The rag stock imparts excellent strength properties to pulp furnishes, and the rag pulps are in high demand for high quality bond and writing papers (Table 6.40). Paper containing cotton fibers is distinguished by the characteristic watermark, at usually 10–25% cotton (formerly "rag") (Tucker, 1969).

Cotton linters also provide good pulps but have not been utilized to the extent of cotton rags, even though considerable research effort has been expended to improve the processes (Tucker, 1969). Pulps from cotton stalks have not been extensively exploited. Yang et al. (1988) pulped whole cotton stalks by a variety of semi-chemical and chemical processes, including neutral ammonia sulfite, soda and soda–AQ pulping and reported reasonable properties for the pulps; however, at least 20% of a wood kraft pulp would be necessary in the furnish to meet general utilization demands.

4.13.7 Other Grasses

Lemon grass and canary grass have also been considered as potential raw materials for pulp and paper (Patel, 1985; Paavilainen and Torgilsson, 1994). There have been recent efforts in Scandinavia (Finnish/Swedish agro fibers project) to further exploit canary grass. A variety of pulping processes were evaluated for canary grass and straw, including alcohol and other solvent processes, soda–oxygen, lime–soda–oxygen, phosphate pulping, steam explosion treatment, and biological treatment. It was found that Reed canary grass and Tall fescue have short narrow fibers and produce a pulp with a greater number of fibers than most hardwoods. Thus these pulps have been suggested for substitution of birch pulp in fine paper furnishes to impart improved printing properties (Paavilainen and Torgilsson, 1994).

Investigators at Oregon State University have shown the seed grass straw yields a pulp equivalent to wheat straw. In the Willamette Valley of Oregon alone, more than one million tons of straw residues from grass seed production are burned annually. The Weyerhaeuser Company is carrying out a comprehensive study of the potential of this material for pulp and paper (Atchison, 1994).

4.13.8 Additional Plants

As mentioned, many other plant materials have been evaluated for pulp and paper. Patel et al. (1985) reported on the pulping characteristics of lemon grass, lantana, impomea, tobacco stalks, kendu leaves, water hyacinth, *Urena lobata* and *Sesvania sp*. Alcaide et al. (1993) obtained promising paper strength results for handsheets prepared from five agricultural residues cooked by the soda process; namely, olive tree fellings, wheat straw, sunflower stalks, vine shoots and cotton stalks. Also evaluated for pulp and paper have been *Cajanus cajan* (Upadhyaya and Singh, 1988), Di (An and Zhu, 1988), pineapple (Jeyasingham, 1988), Illuk grass (Jeyasingam, 1991b) and forage rape (Morrison, 1994). Good strength properties were obtained from the forage rape pulps as shown in Table 6.44.

Table 6.44 Physical Properties of Paper from Forage Rape Pulps

Cultivar	Burst Index $kPa \cdot m^2 g^{-1}$	Tear Index $mN \cdot m^2 g^{-1}$	Tensile Strength Index $N \cdot mg^{-1}$	Density $g \cdot cm^{-3}$
Crack	3.15	6.7	59.2	0.66
Bonar	2.49	4.3	41.3	0.56
Winifred	2.45	2.3	41.0	0.71
Eucalyptus	6.80	11.8	89.0	0.70
Mechanical	1.22	4.1	24.5	0.46

Source: Dr. Ian Morrison, Scottish Crop Research Institute, Invergowrie, Dundee, Scotland.

REFERENCES

Abou-Salem, M. A., Abd El-Megeed, F. F., and Nesseem, R. I., Dissolving pulps from *Zea maize* by alkaline sulfide and sulfite pulping, *Ind. Eng. Chem. Prod. Res. Dev.*, 22, 506, 1983.

Abril, A., Aguero, C., and Hdez, A., The pulp and paperboard production in Cuba with bagasse pulps, *Int'l. Non-Wood Fiber and Papermaking Conf.*, Beijing, P.R. China, July, 572, 1988.

Alcaide, L. J., Baldovin, F. L., and Herranz, J. L. F., Evaluation of agricultural residues for paper manufacture, TAPPI, 76(3), 169, 1993.

Annus, S. and Szekeres, M., Experiences with the reconstruction of straw pulp mill, *Int'l. Non-Wood Fiber and Papermaking Conf.*, Beijing, P.R. China, July, 159, 1988.

Assumpcao, R. M. V., Nonwood fiber utilization in pulping and papermaking – UNIDO's activities, *Non-Wood Plant Fiber Pulping*, Prog. Rept. No. 20, TAPPI Press, Atlanta, 191, 1991.

Atchison, J. F., Bagasse, *Pulp and Paper Manufacture, Vol. 3, Secondary Fibers and Non-Wood Pulping*, TAPPI, Atlanta, Chap IV, 1987a.

Atchison, J. E., Critical factors to be considered in production of bagasse newsprint and the necessary high yield mechanical pulp required, *Non-Wood Plant Fibers Pulping*, Prog. Rept. No. 17, TAPPI Press, Atlanta, 63, 1987b.

Atchison, J. E., Worldwide capacities for non-wood plant fiber pulping—increasing faster than wood pulping capacities, *Non-Wood Plant Fiber Pulping*, Prog. Rept. No. 19, TAPPI Press, Atlanta, 1, 1991.

Atchison, J. E., Making the right choices for successful bagasse newsprint production, Part I, *TAPPI*, 75(12), 63,1992.

Atchison, J. E., Making the right choices for successful bagasse newsprint production, Part II, *TAPPI*, 76(1), 187, 1993.

Atchison, J. E., *Present Status and Future Prospects for Use of Non-Wood Plant Fibers for Paper Grade Pulps,* paper presented at AF&PA Pulp and Fiber Fall Seminar, Tucson, AZ, November, 1994.

Atchison, J. E. and McGovern, J. N., History of paper and the importance of non-wood plant fibers, *Pulp and Paper Manufacture*, Vol. 3, *Secondary Fibers and Non-Wood Pulping*, TAPPI Press, Atlanta, GA, 1987, Chap. 1.

Aziz, S. and McDonough, T. J., Ester pulping – A brief evaluation, *TAPPI*, 70(3), 137, 1987.

Aziz, S. and Sarkanen, K., Organosolv pulping – A review, *TAPPI*, 72(3), 169, 1989.

Bagby, M. O., *Non-Wood Plant Fiber Pulping,* Prog. Rept. No. 9, TAPPI Press, Atlanta, 75, 1978.

Bagby, M. O., Cunningham, R. L., Touzinsky, G. F., Hemerstrand, G. E., Curtis, E. L., and Hofreiter, B. T., *Non-Wood Plant Fiber Pulping Conf.,* Prog. Rept. No. 10, TAPPI Press, Atlanta, 111, 1979.

Bagby, M. O. and Widstrom, N. W., Biomass uses and conversion, *Corn, Chemistry and Technology*, Watson, S. A. and Ramstad, P. E. Eds., Am. Assoc. Cereal Chem., St. Paul, MN, 575, 1987.

Bambanaste, R., Fernandez, W., Sabatier, T., Aguero, C., and Garcia, O. L., High yield pulps from sugar cane bagasse, development perspectives of the Cuban technology, *Int'l. Non-Wood Fiber and Papermaking Conf.*, Beijing, P.R. China, July, 170, 1988.

Bandyopadhyay, S. B., Advances in bleaching of bagasse pulp, *Ann. Rept. Jute Technol. Res. Lab.*, Indian Council Ag. Res., Calcutta, 1970.

Battista, O. A., Ed., *Synthetic Fibers in Papermaking*, Wiley-Interscience, New York, 1964.

Benitez, A. B. and Bottan, G. L., Oxidative extractions: The future of bagasse bleaching, *Non-Wood Plant Fiber Pulping*, Prog. Rept. No. 15, TAPPI Press, Atlanta, 29, 1985.

Bhargava, R. L., *Indian Forest Bull.*, No. 129, Forest Research Inst., Dehra Dunn, 1946.

Bhargava, R. L., Bamboo, *Pulp and Paper Manufacture*, *Vol. 3, Secondary Fibers and Non-Wood Pulping* Chap. 5, TAPPI Press, Atlanta, 71, 1987.

Bidwell, G. L., *U.S. Patent No.* 1531728, 1925.

Biermann, C. H., *Essentials of Pulping and Papermaking*, Academic Press, New York, 1993.

Bleier, P. , UNIDO/SIDA desilication–Fact and basic principles, *Proc. Int'l. Seminar and Workshop in Desilication*, UNIDO, Vienna, 13, 1991.

Brink, D. L., Merriman, M. M., Radakrishna, R., Berndt, H., Reddy, M., and Yang, Y. S., Rice straw pulping and bleaching, *Non-Wood Plant Fiber Pulping,* Prog. Rept. No. 18, TAPPI Press, Atlanta, 1, 1988.

Britt, K. W., *Pulp and Paper Technology*, Van Nostrand Reinhold, New York, 1970.

Cai, J., Dazhen, Z., Wu, Y., Weng, J., Zhang, R., and Shen, Z., A preliminary study of organosolv-pulping of wheat straw with acid-acetate process, *Int'l. Non-Wood Fiber and Papermaking Conf.*, Beijing, P.R. China, July, 180, 1988.

Capretti, G. and Marzetti, A., Steam explosion pulping of wheat straw, *Steam Explosion Techniques: Fundamentals and Industrial Applications*, Focher, B., Marzetti, A. and Crescenzi, V., Eds., Gordon & Breach, Philadelphia, 1991, p. 207.

Captein, A. A., Knapp, S. B., Watt, R. A., and Wethern, J. D., Sugarcane bagasse as a fibrous papermaking material, V, Neutral sulfite and bisulfite pulping of Hawaiian bagasse, TAPPI, 40(8), 620, 1957.

Carvelo, A. A. S., Alaburda, J., Botaro, V. R., Lechat, J. R., and deGroote, R. A. M. C., Acetosolv pulping of sugarcane bagasse, *TAPPI*, 73(11), 217, 1990.

Casey, J. P., *Pulp and Paper Chemistry and Chemical Technology,* Vol. 1, 3rd edition, Wiley-Interscience, New York, 1980.

Chen, S. C. and Lin, S. J., *Coop. Bull. Taiwan Forestry Res. Inst.*, No. 21, 18, 1973.

Cheng, Z., Leminen, J., Ala-Kaila, K., and Paulapuro, H., The basic drainage properties of Chinese wheat straw pulp, *Proc. Pulping Conference*, TAPPI Press, San Diego, 735, 1994.

Chengwu, Z., Liji, J., Yongcheng, D., and Nan, W., Industrial experiences and problems involved in the manufacture of sulfite reed printing paper, *Int'l. Non-Wood Fiber Pulping and Papermaking Conf.*, Beijing, P.R. China, July, 778, 1988.

Chornet, F. and Overend, R., Phenomenological kinetics and reaction engineering aspects of steam/aqueous treatments, *Steam Explosion Techniques: Fundamentals and Industrial Applications*, Focher, B., Marzetti, A., and Crescenzi, V., Eds., Gordon and Breach New York, 1991, p. 21.

Chin, Y. C., Seger, G. E., and Chang, H. M., Low pollution bleaching of Hawaiian bagasse soda–AQ pulps, *Int'l. Non-Wood Fiber and Papermaking Conf.*, Beijing, P.R. China, July, 509, 1988.

Clark, T. F., Annual crop fibers and the bamboos, *Pulp and Paper Manufacture*, Vol. II, 2nd edition, R. O. MacDonald, Ed., McGraw-Hill, New York, 1969, p. 1.

Clark, T. F. and Wolff, I. A., A search for new fiber crops, Part VI, Kenaf and wood pulp blends, TAPPI, 45(10), 786, 1962.

Clark, T. F., Nelson, G. H., Nieschlag, H. J., and Wolff, I. A., A search for new fiber crops, Part V, Pulping studies on kenaf, TAPPI, 45(10), 780, 1962.

Clark, T. F., Cunningham, R. L., and Wolff, I. A., A search for new fiber crops, Part XIV, Bond paper containing continuously pulped kenaf, TAPPI, 54(1), 63, 1971.

Clark, T. F., Nelson, G. A., Cunningham, R. L., Kwolek, W. F., and Wolff, I. A., A search for new fiber crops, Potential of sorghums for pulp and paper, TAPPI, 56(3), 107, 1973.

Cunningham, R. L., Clark, T. F., Kwolck, W. F., Wolff, I. A., and Jones, Q., A search for new fiber crops, Part XIII, Laboratory scale pulping studies continued, TAPPI, 53(9), 1697, 1970.

Cunningham, R. L., Clark, T. F., and Bagley, M. O., *Crotalaria juncea* — Annual source of papermaking fiber, TAPPI, 61 (2), 37, 1978.

Cunningham, R. L., Touzinsky, G. F., and Bagby, M. O., Brightening of kenaf thermomechanical pulp, TAPPI, 62(4), 69, 1979.

da Silva, N. M. and Pereira, A. D., Experience of pioneer sisal simultaneous resource for pulp and energy, *Non-Wood Plant Fiber Pulping*, Prog. Rept. No. 15, TAPPI Press, Atlanta, 63, 1985.

Darkwa, N. A., Pulping characteristics of plantain pseudostems, *Int'l. Non-Wood Fiber and Papermaking Conf.*, Beijing, P.R. China, July, 973, 1988.

deGroot, B., van Zuilichem, D. J., and van der Zwan, R. P., The use of non-wood fibers for pulping and papermaking in the Netherlands, *Int'l. Non-Wood Fiber Pulping and Papermaking Conf.*, Beijing, P.R. China, July, 216, 1988.

Dou, D. Z., The study of the principal parameters of the hydrogen peroxide bleaching of bagasse kraft pulp, *Int'l. Non-Wood Fiber and Papermaking Conf.*, Beijing, P.R. China, July, 516, 1988.

Duenar, R. S., Prado, J. J. R., Romero, N., and Irulegui, A., Low pollution alternative process to obtain bleached pulp from sugarcane bagasse, *Proc. Seventh Int'l. Symp. on Wood and Pulping Chem.*, Vol. 3, Beijing, P.R. China, May, 280, 1993.

El-Taraboulsi, M. A. and Abou-Salem, A. H., Rice straw for fine papers, Part II, Bleaching of rice straw soda pulps of high reverted brightness, TAPPI, 50(11), 115A, 1967.

Eroglu, H., Soda–oxygen anthraquinone pulping of wheat straw, *Non-Wood Plant Fiber Pulping*, Prog. Rept. No. 17, TAPPI Press, Atlanta, 133, 1987.

Eroglu, H. and Usta, M., Oxygen bleaching of soda–oxygen wheat straw pulp, *Non-Wood Plant Fiber Pulping*, Prog. Rept. No. 19, TAPPI Press, Atlanta, 89, 1991.

Escolano, J. O., Visperas, R. V., and Ballon, C. H., The pulping characteristics of some *Musa* sp. fibers other than abaca, *Int'l. Non-Wood Fiber and Papermaking Conf.*, Beijing, P.R. China, July, 197, 1988.

Fernandez, N., Naranjo, M. E., and Alvarez, J., Advances in the bleaching of bagasse pulp, *Non-Wood Plant Fiber Pulping*, Prog. Rept. No. 15, TAPPI Press, Atlanta, 43, 1985.

Fineman, I., Wagberg, L., Xiaoping, Z., Kuang, S. J., and Nian, L. F., Straw pulp as a raw material for offset paper – Problems and solutions, *Int'l. Non-Wood Fiber and Papermaking Conf.*, Beijing, P.R. China, July, 619, 1988.

Forcher, B., Marzetti, A., and Crescenzi, V., *Proc. Int'l. Workshop on Steam Explosion Techniques: Fundamental and Industrial Applications*, Oct., 1988.

Franco, P. T, Cruz, O. J., and Tabora, P. C., Abaca pulp and paper industry in the Philippines, *Proc. Pulping Conf.*, TAPPI Press, Atlanta, 133, 1981.

Fuentes, E. F., Bagasse chemithermomechanical pulps and their utilization for newsprint, *Proc. Pulping Conf*, TAPPI Press, Atlanta, 233, 1981.

Gohre, O., Semichemical straw pulp process for the manufacture of fiber raw materials in fluting liner wrapping paper and solid cardboard, *Int'l. Non-Wood Fiber and Papermaking Conf.*, Beijing, P.R. China, July, 206, 1988.

Gomez, F. M., *Philippine Patent*, Nos. 2699 (1966); 2812 (1966); 3661 and 3921 (1967).

Goyal, S. K., Ray, A. K., Bhardwaj, N. K., and Gupta, A., Pulping studies of rice straw using soda and soda–AQ processes, *Non-Wood Plant Fiber Pulping*, Prog. Rept. No. 19, TAPPI Press, Atlanta, 71, 1991.

Granfeldt, T., Nordin, S., Danielson, O., and Ryrberg, K. G., Towards effluent tree production of bagasse and eucalyptus pulps for newsprint, *Non-Wood Plant Fiber Pulping*, Prog. Rept. No. 19, TAPPI Press, Atlanta, 39, 1991.

Haun, J. R., Clark, T. F., and White, G. A., *Fiber and Papermaking Characteristics of Bamboo*, Tech. Bull. No. 1361, ARS/USDA, Washington, D.C., 1966.

Henderson, J. T. and Knapp, S. D., *U.S. Patent* 3,013,935.

Hinrich, D. D., Knapp, S. B., Milliken, J. H., and Wethern, J. D., Sugarcane bagasse as a fibrous papermaking material, VI, Modified alkaline and hydrotropic pulping of bagasse, TAPPI, 40(8), 626, 1957.

Horn, R. A., Wegner, T. H., and Kugler, D. E., Newsprint from blends of kenaf CTMP and deinked recyled newsprint, TAPPI, 75(12), 69, 1992.

Hua, M. and Tu, H., Experiments and practices in the manufacture of printing papers with Amur Silver Grass BKP, *Int'l. Non-Wood Fiber and Papermaking Conf.*, Beijing, P.R. China, July, 639, 1988.

Hurter, A. M., Utilization of annual plants and agricultural residues for the production of pulp and paper, *Non-Wood Plant Fiber Pulping*, Prog. Rept. No. 19, TAPPI Press, Atlanta, 49, 1991.

Iyengar, R. S., Expereinces in refining bagasse and other non-wood fibers, *Proc. Pulping Conf.*, TAPPI Press, Atlanta, 111, 1981.

Jeyasingam, J. T., Industrial experience in the semi-chemical pulping of straw, *Non-Wood Plant Fiber Pulping*, Prog. Rept. No. 17, TAPPI Press, Atlanta, 127, 1987.

Jeyasingam, J. T., Critical analysis of straw pulping methods – worldwide, *Non-Wood Plant Fiber Pulping*, Prog. Rept. No. 18, TAPPI Press, Atlanta, 103, 1988a.

Jeyasingam, J. T., Critical analysis of straw pulping methods worldwide, *Int'l. Non-Wood Fiber and Papermaking Conf.*, Beijing, P.R. China, July, 223, 1988b.

Jeyasingam, J. T., A summary of special problems and considerations related to non-wood fiber pulping worldwide, *Non-Wood Plant Fiber Pulping*, Prog. Rept. No. 19, TAPPI Press, Atlanta, 149, 1991a.

Jeyasingam, J. T., Mill experience in the application of non-wood fiber for papermaking, *Non-Wood Plant Fiber Pulping,* Prog. Rept. No. 20, TAPPI Press, Atlanta, 39, 1991b.

Jin, B., Hasuike, M., and Murakami, K., Differences of papermaking properties between rice and wheat straw fibers, *J. Japan Wood Res. Soc.,* 34(11), 923, 1988.

Joglekar, M. H. and Donofrio, C. P., TAPPI, 34(6), 254, 1951.

Kaldor, A. F., Kenaf, An alternate fiber for the pulp and paper industries in developing and developed countries, *TAPPI*, 75(10), 141, 1992.

Kessler, R. W., Becker, V., and Goth, B., Stream explosion of flax – A superior technique for upgrading fiber value, paper presented at *207th Am. Chem. Soc. Mtg.*, San Diego, CA, March, 1994.

Knapp, S. B., Milliken, J. H., and Wethern, J. D., Sugarcane bagasse as a fibrous papermaking material, III, Effects of depithing and bagasse variables in pulping characteristics, TAPPI, 40(8), 602, 1957.

Knapp, S. B. and Wethern, J. D., Sugarcane bagasse as a fibrous papermaking material. IV. Alkaline pulping of Hawaiian bagasse, TAPPI, 40(8), 609, 1957.

Kokta, B. and Ahmed, A., Steam explosion pulping (SEP) of bagasse, flax and kenaf: Comparison with conventional processes, *Proc. Seventh Int'l. Symp. on Wood and Pulping Chem.*, Beijing, P.R. China, May, 410, 1993.

Krishnamachari, K. S., Rangan, S. G., and Ravindranathan, N., Experiences of bagasse pulping with rapid continuous digester, *Proc. Pulping Conference*, TAPPI Press, Atlanta, 61, 1981.

Kuang, S. J. and Fineman, I., How to improve the runnability of wheat straw pulp, *Int'l. Non-Wood Fiber and Papermaking Conf.*, Beijing, P.R. China, July, 649, 1988.

Kugler, D. E., Kenaf newsprint: Realizing commercialization of a new crop after four decades of research and development, *A Report on the Kenaf Demonstration Project, Special Projects and Program System*, USDA/CSRS, May, 1988.

Kulkarni, A. G., Pant, R., and Panda, A., Controlled carbonation–A prelude to selective separation of silcia from black liquors, in *Proc. Int'l. Seminar & Workshop on Desilication*, UNIDO, Vienna, 58, 1991.

Lai, Y. Z., Siter, W., and Guo, X., Chemimechanical pulping of bagasse, *Int'l. Non-Wood Fiber and Papermaking Conf.*, Beijing, P.R. China, July, 233, 1988.

Lathrop, E. C., Fiber shipping containers from sugarcane bagasse, TAPPI, 40(10), 788, 1957.

Lathrop, E. C. and Aronovsky, S. I., TAPPI, 32(4), 145, 1949.

Li, Y. and Pen, Z., Preliminary study of Amur Silver Grass CMP during high consistency refining, *Int'l. Non-Wood Fiber and Papermaking Conf.*, Beijing, P.R. China, July, 659, 1988.

Liang, W. Z., Zhou, F. C., Xiao, X. R., and Wang, Z. H., Organosolv pulping characteristics of wheat straw, *Int'l. Non-Wood Fiber and Papermaking Conf.*, Beijing, P.R. China, July, 271, 1988.

Lin, W., Zao, H., Chang, F., and Gao, J., Organosov pulping of bagasse with ethanol–water system, *Int'l. Non-Wood Fiber and Papermaking Conf.*, Beijing, P.R. China, July, 281, 1988.

Liu, S. D., Studies on the brittleness of bagasse paper, *Int'l. Non-Wood Fiber and Papermaking Conf.,* Beijing, P.R. China, July, 681, 1988.

Liu, M. G., Zhao, J. N., Zhou, G. Z., and Cao, P. F., Neutral ammonium sulfite production of wheat straws and the utilization of its black liquor, *Int'l. Non-Wood Fiber and Papermaking Conf.*, Beijing, P.R. China, July, 299, 1988.

Lora, J. H. and Pye, E. K., The ALCELL Process, *Proc. Solvent Pulping Symp.*, TAPPI Press, Atlanta, 1992.

Lora, J. H. and Pye, E. K., The ALCELL process, an environmentally sound approach to annual fibers pulping, in *Proc. Non-Wood Fibers for Industry Conf.*, Vol. II, Pira/Silsoe Res. Inst., Pira, Leatherhead, Surrey, U.K, 1994.

Luna, G. V. and Torres, C. A., High yield pulp from bagasse, *Proc. Pulping Conf.*, TAPPI Press, Atlanta, 69, 1981.

Lungren, S., Production of high quality bleached bamboo pulp, *Proc. Seventh Int'l. Symp. on Wood and Pulping Chem.*, Beijing, P.R. China, May, 318, 1993.

Mannar, G. R., Bagasse pulping in a continuous digester for bleachable grades, *Non-Wood Plant Fiber Pulping*, Prog. Rept. No. 18, TAPPI Press, Atlanta, 39, 1988.

Marchessault, R. A., Steam explosion: A refining process for lignocellulosics, in *Steam Explosion Techniques: Fundamentals and Industrial Applications*, Focher, B., Marzetti, A., and Crescenzi, V., Eds., Gordon & Breach, New York, 1991, 1.

Marmers, H., Yuritta, J. P., and Menz, D. J., The siropulpers: An explosive alternative for non-wood pulping, in *Proc. Pulping Conf.*, Atlanta, 261, 1981.

McGovern, J. N., Bleaching of non-wood pulps, *TAPPI*, 50(11), 63A, 1967.

McGovern, J. N., Coffelt, D. E., Hurten, A. M., Ahuja, N. K., and Wiedermann, A., Other fibers, *Pulp and Paper Manufacture, Vol. 3, Secondary Fibers and Non-Wood Pulping*, TAPPI, Atlanta, 110, 1987, Chap. IX.

Misra, D. K., Cereal straw, *Pulp and Paper Manufacture, Vol. 3, Secondary Fibers and Non-Wood Pulping*, TAPPI Press, Atlanta, Chap. VI, 1987.

Misra, D. K., Pulping and bleaching of non-wood fibers, *Pulp and Paper, Chemistry and Chemical Technology,* Vol. 1, 3rd edition, J.P. Casey, (ed.), Wiley-Interscience, New York, 504, 1980.

Mita, A., Kasiwabara, S., and Pono, R. Q., Preparation of hydrogen peroxide – alkaline pulp (PAP) from abaca, *Proc. Pulping Conf.*, TAPPI Press, 133, 1988.

Mittal, K. C., Optimization of prehydrolysis – Kraft delignification of bagasse, *Non-Wood Plant Fiber Pulping*, Proj. Rept. No. 19, TAPPI Press, Atlanta, 23, 1991.

Mittal, S. K. and Maheshwari, S., A few technical aspects of pulp production from kenaf, *Proc. Pulping Conference*, TAPPI, San Diego, 105, 1994b.

Mittal, S. K. and Maheshwari, S., Extended delignification of bamboo pulp by oxygen – A mill experience, *Proc. Pulping Conference*, TAPPI Press, San Diego, (6), 243, 1994.

Morgen, L. M., *Khim i Fiz.-Khim. Prirodnil Sintetich. Polimerov*, Akad. Nank Uz. SSR, Inst. Khim. Polimerov, No. 2, 93, 1964.

Morrison, I. M., Range, Provision and processing of non-wood fibres from temperate crops, *Proc. Non-Wood Fibers for Industry Conf.*, Vol. I, Pira/Silsoe Res. Inst., Pira, Leatherhead, Surrey, U.K, 1994.

Naffziger, T. R., Clark, T. F., and Wolff, I. A., Newsprint from domestic timber bamboo, TAPPI, 44(7), 472, 1961.

Naffziger, T. R., Matuszewski, R. S., Clark, T. F., and Wolff, I. A., Dissolving pulps from domestic timber bamboo, TAPPI, 43(6), 591, 1960.

Nelson, G. H., Heschlag, H. J., Daxenbichler, M. E., Wolff, I. A., and Perdue, R. E., A search for new fiber crops, Part III, Laboratory scale pulping studies, TAPPI, 44(5), 319, 1961.

Neuman, W., Gottleib, K., Rupp, M., and Meckel, J. F., Acetocell and *Miscanthus*: New facts for the green production of pulp, *Proc. Pulping Conference*, TAPPI Press, San Diego, 314a, 1994.

Nieschlag, H. J., Nelson, G. H., Wolff, J. A., and Perdue, R. E., A search for new fiber crops, *TAPPI*, 43(3), 193, 1960.

Orgill, B., Current status of bagasse newsprint technology, *Int'l. Non-Wood Fiber and Papermaking Conf.*, Beijing, P.R. China, July, 590, 1988.

Ouchtati, B. and Carney, R. J., Esparto pulp, its manufacture and use in specialty papers, *Proc. Pulping Conf.*, TAPPI Press, 131, 1981.

Paavilainen, L. and Torgilsson, R., Reed canary grass – A new nordic papermaking fiber, *Proc. Pulping Conference*, TAPPI Press, San Diego, 611, 1994.

Panda, A., Operational problems in pulping and chemical recovery plants of silica rich fibers; Raw materials and earlier desilication work carried out in India, in *Proc. Int'l. Seminar & Workshop on Desilication*, UNIDO, Vienna, 36, 1991.

Pant, R., Panda, A., Bansal, M., and Naithani, N. K., Pulp and paper research in India, TAPPI, 74(8), 69, 1991.

Patel, R. J., Angadujavar, C. S., and Rao, Y. S., Non-wood fiber plants for papermaking, *Non-Wood Plant Fiber Pulping*, Prog. Rept. No. 15, TAPPI Press, Atlanta, 77, 1985.

Perdue, R. E., *Arundo donax* – Source of musical reeds and industrial cellulose, *Economic Botany*, 12(4), 368, 1958.

Qian, Z. L. and Tang, Z. C., The utilization of AQ in straw material digestion, *Int'l. Non-Wood Fiber and Papermaking Conf.*, Beijing, P.R. China, July, 349, 1988.

Quirarte, J. R., Alkaline peroxide pretreatment in sugarcane bagasse chemimechanical pulping, *Non-Wood Plant Fiber Pulping*, Prog. Rept. No. 20, TAPPI Press, Atlanta, 163, 1991.

Raitt, W., *World's Paper Trade Rev.*, 84, 562, 618, 1925.

Ranqamannar, G., Michelsen, J., and Silver, L., Developments in bagasse newsprint technology, *Non-Wood Plant Fiber Pulping*, Prog. Rept. No. 20, TAPPI Press, Atlanta, 1, 1991.

Rao, Y. S. and Maheshwari, S., Bleaching characteristics of wheat straw pulp, *Non-Wood Plant Fiber Pulping*, Prog. Rept. No. 17, TAPPI Press, Atlanta, 111, 1987.

Rao, A. R. K., Rao, P. N., Jena, S. C., and Jayasayee, V., Comparative web responses of bagasse pulps to refining, *Int'l. Non-Wood Fiber and Papermaking Conf.*, Beijing, P.R. China, July, 696, 1988.

Rao, N. J., Ray, A. K., Kumar, N., Arora, A. K., and Mehrotra, A., Soda–oxygen delignification of rice straw, *Non-Wood Plant Fiber Pulping*, Proj. Rept. No. 20, TAPPI Press, Atlanta, 83, 1991.

Routledge, T., *Bamboo and Its Treatment*, Sunderland, 1879.

Rydholm, S. A., *Pulping Processes*, Wiley-Interscience, New York, 1965.

Saavedra, F., Triana, O., Diaz, R., Acevedo, R., and Pecez, M., Fluff pulp: bagasse as a raw material for disposable products, *Int'l. Non-Wood Fiber and Papermaking Conf.*, Beijing, P.R. China, July, 358, 1988.

Sabharwal, H. S., Akhtar, M., Blanchette, R. M., and Young, R. A., Biomechanical pulping of kenaf, *TAPPI*, 77(12) 1, 1994a.

Sabharwal, H. S., Akhtar, M., Blanchette, R. M., and Young, R. A., Bio-refiner mechanical pulping of bast type fibers, *Proc. Pulping Çonference*, TAPPI Press, San Diego, 623, 1994b.

Sabharwal, H. S., Akhtar, M., Blanchette, R. M., and Young, R. A., Refiner and biochemical pulping of jute, *Holsforschung*, 49, 537 (1995).

Sabharwal, H. S., Kasmoula, T., Zedan, K. M., Rahman, S. A., and Abas, R. F., Influence of prehydrolysis on the alkaline pulping characteristics of mid-rib and frond of date palm leaves: A potential source of pulp and paper, *Proc. Second Latin-American Congress on Pulp and Paper*, Torromolinos, Spain, June, 5, 1981.

Sadawarte, N. S., Dharwadkar, A. R., and Veeramani, H., Soda–AQ pulping of bagasse, *Proc. Pulping Conf.*, TAPPI Press, Atlanta, 117, 1981.

Sawhney, R. S., Newsprint manufacture from bagasse – present status, *Int'l. Non-Wood Fiber and Papermaking Conf.*, Beijing, P.R. China, July, 714, 1988.

Schroeter, M. C., Use of kenaf for linerboard quality enhancement, *Proc. Pulping Conference,* TAPPI Press, San Diego, 95, 1994.

Semana, J. A., Escolano, J. O., and Monsalud, M. R., The kraft pulping qualities of some Philippine bamboos, *TAPPI*, 50(8), 416, 1967.

Shafi, M., Akhtaruzzamarn, A. F. M., and Mian, A. J., Pulping of whole length jute by neutral sulfite AQ process, *Holzforschung*, 47, 83, 1993.

Shuhui, Y., Zhrivei, H., and Yorsen, L., Kenaf core alkaline hydrogen peroxide chemimechanical pulping, *Proc. Seventh Int'l. Symp. on Wood and Pulping Chem.*, Vol. 3, Beijing, P.R. China, 317, 1993.

Tandon, R., Gupta, A., Kulkarni, A. G., and Panda, A., Properties of black liquors from pulping of non-woody raw materials, in *Proc. Int'l. Seminar & Workshop on Desilication*, UNIDO, Vienna, 23, 1991.

Tapanes, E, Naranjo, M. E., and Aguero, C., Soda–AQ pulping of bagasse, *Non-Wood Plant Fiber Pulping*, Prog. Rept. No. 15, TAPPI Press, Atlanta, 19, 1985.

Tjerdsma, B. F. and Zomers, F. H. A., Organosolv pulping of hemp, *Proc. Seventh Int'l. Symp. on Wood and Pulping Chem.*, May, Beijing, P.R. China, 514, 1993.

Touzinsky, G. F., Laboratory papermachine runs with kenaf thermomechanical pulp, TAPPI, 63(3), 109, 1980.

Touzinsky, G. F., Kenaf, in *Pulp and Paper Manufacture, Vol. 3. Secondary Fibers and Non-Wood Pulping*, Chap. VIII TAPPI Press, Atlanta, 106, 1987.

Touzinsky, G. F., Cunningham, R. L., and Bagby, M. O., Papermaking properties of kenaf thermomechanical pulp, TAPPI, 63(1), 53, 1980.

Tribeni Tissue Company, Calcutta, India, personal communication, 1992.

Tucker, L. B., Pulping of rags and other fibers, *Pulp and Paper Manufacture*, Vol. II, 2nd edition, MacDonanld, R.O., Ed., McGraw-Hill, New York, 75, 1969.

Upadhyaya, J. S., Chinnapadasan, S., Rajan, S., Singh, R., and Dutt, D., Studies on alkali–oxygen–AQ delignification of bagasse, *Non-Wood Fiber Plant Pulping*, Prog. Rept. No. 20, TAPPI Press, Atlanta, 55, 1991.

Usta, M. and Eroglu, H., Soda–oxygen pulping of rye straw, *Non-Wood Plant Fiber Pulping*, Prog. Rept. No. 18, TAPPI Press, 113, 1988.

Usta, M., Kirci, A., and Eroglu, H., Soda–oxygen pulping of cornstalks, *Non-Wood Plant Fiber Pulping*, Prog. Rept. No. 20, TAPPI Press, Atlanta, 25, 1991.

Valladares, J., Rolz, C., Bermudez, M. E., Batres, F. R., and Custodio, M. A., Pulping of sugarcane bagasse with a mixture of ethanol–water solution in presence of sodium hydroxide and AQ, *Non-Wood Plant Fiber Pulping*, Prog. Rept. No. 15, TAPPI Press, Atlanta, 23, 1985.

Venkataraman, T. S., Manufacture of bagasse-based newsprint – Issues and options, *Int'l. Non-Wood Fiber and Papermaking Conf.*, Beijing, P.R. China, July, 739, 1988.

Venkataraman, T.S. and Torza, S., Beloit–SPB process achieves the breakthrough in bagasse newsprint, *Non-Wood Plant Fiber Pulping*, Prog. Rept. No. 17, TAPPI Press, Atlanta, 1, 1987.

Watson, A. J., Gartside, G., Weiss, D. F., Higgins, H. G., Mamers, H., Davies, G. W., Irvine, G. M., Wood, I. M., Manderson, A., and Crane, E. J. F., The growing of kenaf in northern Australia and its potential for papermaking and food production, *Div. Chem. Technol. Paper No. 7*, CSIRO, Australia, 1976.

Wethern, J. D. and Captein, H. A., *U.S. Patent*, 3,013,931.

Wiedermann, A., Reeds, in *Pulp and Paper Manufacture, Vol. 3, Secondary Fibers and Non-Wood Pulping*, Chap. VIII TAPPI Press, Atlanta, 99, 1987.

Winner, S. R., Goyal, G. C., Pye, E. K., and Lora, J. H., Pulping of agricultural residues by the Alcell process, *Non-Wood Plant Fiber Pulping Conf.*, Prog. Rept. No. 20, TAPPI Press, Atlanta, 171, 1991.

Wong, A., Neutral sulfite pulping of oilseed flax and textile linen flax fibers, *Non-Wood Plant Fiber Pulping*, Prog. Rept. No. 17, TAPPI Press, Atlanta, 29, 1987.

Wu, Y. M. and Qin, Y., The characteristics of *Miscanthus sacchariflorus* kraft cooking and using for process control, *Int'l. Non-Wood Fiber and Papermaking Conf.*, Beijing, P.R. China, 425, 1988.

Xie, B. Z., Chen, H. C., Yao, S. Y., and Zhao, G. Y., Experience in sorghum stalk alkaline sulfite pulping, *Int'l. Non-Wood Fiber and Papermaking Conf.*, Beijing, P.R. China, 436, 1988.

Yang, S. H., Li, Y. S., and Gao, Y., Research and characterization of cotton stalk as a papermaking raw material, *Int'l. Non-Wood Fiber and Papermaking Conf.*, Beijing, P.R. China, July, 139, 1988.

Ying, T., Pulping of bagasse and non-wood fibers in Taiwan, *Non-Wood Plant Fiber Pulping*, Prog. Rept. No. 15, TAPPI Press, Atlanta, 103, 1985.

Youchang, L., Making super-standard writing paper by blending bleached straw pulp, *Int'l. Non-Wood Fiber and Papermaking Conf.*, Beijing, P.R. China, July, 349, 1988.

Young, J. E., Kenaf newsprint is a proven commodity, *TAPPI* , 70(11) 81, 1987.

Young, R. A., Structure, swelling and bonding of cellulose fibers, *Cellulose: Structure, Modification and Hydrolysis*, Young, R. A. and Rowell, R. M., Eds., Wiley-Interscience, New York, 1986.

Young, R. A., Ester pulping: A status report, *TAPPI* , 72(4), 195, 1989.

Young, R. A., Acetic acid based pulping processes, *Proc. Symp. on Solvent Pulping*, TAPPI, Atlanta, 1992.

Young, R. A., Vegetable fibers, in *Kirk-Othmer Encyclopedia of Chemical Technology,* Vol 10, 4th ed., 1994a.

Young, R. A., Sabharwal, H., Akhtar, M., and Blanchette, R., Biomechanical pulping of non-wood species, *Proc. Non-Wood Fibers for Industry Conf.*, Vol. II, Pira/Silsoe Res. Inst., Pira, Leatherhead, Surrey, U.K, 1994b.

Young, R. A., Comparison of the Properties of chemical cellulose pulps, *Cellulose*, 1, 107, 1994c.

Yu, H. S., *Conchatus* and its prospects of application in biopulping, *Int'l. Non-Wood Fiber and Papermaking Conf.*, Beijing, P.R. China, July, 439, 1988.

Zanuttini, M. and Christensen, P. K., Effects of alkali charge in bagasse chemimechanical pulping, *Non-Wood Plant Fiber Pulping*, Prog. Rept. No. 19, TAPPI Press, Atlanta, 231, 1991.

COMPOSITES

CHAPTER 7

Opportunities for Composites from Agro-Based Resources

Roger M. Rowell

CONTENTS

1. INTRODUCTION

Until about 1920, the Western world greatly depended on the use of agro-based resources for materials. With the coming of plastics, high performance metals, ceramics, and other synthetics, the use of agro-based derived materials lost its market share. We are now aware that our landfills are filling up, our resources are being depleted, and our planet is becoming polluted. Because of this, there is renewed interest in technologies that are considered to be environmentally friendly and products that are biodegradable, and recyclable.

There is a greater awareness of the need for materials in an expanding world population with increasing affluence. It took all of recorded history for the world population to reach 1 billion in 1830. In 1930, it had doubled to 2 billion. While it took 100 years for the population to increase to 1 billion people by 1930, at the present rate, the population will add 1 billion people every 11 years.

It is estimated that by the year 2000, 25% of the population of China will become "middle class." This represents more people than the entire population of the United States. If the desire for materials in this growing segment of China equals the middle class of other countries, there will be a great need for new materials.

China is only one example of the large new markets that will open up for new materials. Asia, Mexico, South America, and Eastern Europe are also "emerging" as industrial consumers that will seek new materials.

2. FIBER SUPPLY

In any commercial development, there must be a long-term guaranteed supply of resources. In the case of agro-based composites, these resources include fiber, labor, water, energy, and processing equipment. In order to insure a continuous fiber supply, management of the agricultural producing land should be under a proactive system of land management whose goal is both sustainable agriculture and the promotion of healthy ecosystems. Ecosystem management is not a euphemism for preservation, which might imply benign neglect. Sustainable agriculture denotes a balance between conservation and utilization of agricultural lands to serve both social and economic needs, from local, national, and global vantage points. Sustainable agriculture does not represent exploitation but rather is aimed at meeting all the needs of the present generation without compromising the ability of future generations to meet their needs. It encompasses, in the present case, a continuous production of fiber, considerations of multi-land use, and conservation of the total ecosystem.

There is a wide variety of agro-based fibers to consider for utilization. All of them should be considered for composites to take advantage of unique fiber properties each plant type has to offer, not just because we have a desire to promote one fiber over another. Unless one particular fiber has some advantage in the market, it will be replaced with whatever resource has the market advantage. That market advantage can be based on many elements such as availability, price, or performance. Desire does not drive markets! Producers and manufacturers of agro-fibers must explore common interests and, where possible, prepare an enterprise-driven long range strategic plan for development and promotion of an agro-fiber industry (Rowell, 1994b).

Chapter 1 gives the inventory of some of the larger sources of agricultural crop fiber that could be utilized for composites. The traditional source of agro-based composites has been wood, and for many countries, this will continue to be the major source. Wood has a higher density than annual plants so there will be more bulk when using agricultural crop fiber. There are also concerns about the seasonality of annual crops, which requires considerations of harvesting, separating, drying, storing, cleaning, handling, and shipping. In the present system of using wood, storage costs can be reduced by letting the tree stand alive until needed. With any annual crop, harvesting must be done at a certain time and storage/drying/cleaning/separating will be required. This will almost certainly increase costs of using agro-based resources over wood depending on land and labor costs; however, in those countries where there is little or no wood resource left, or where restrictions are in place to restrict the use of wood, alternate sources of fiber are needed if there is to be a natural fiber industry in those countries.

As world population increases, there will be a greater need to grow grains to feed people. For every ton of grain harvested, there is 3 to 5 times that weight in stalk fiber. So, as the need for grain increases, the supply of stalk fiber also increases.

Other large sources of fiber can come from recycling agro-fiber based products, such as paper (Figure 7.1), waste wood, and point source agricultural residues such as rice hulls from a rice processing plant, sunflower seed hulls from an oil processing unit, and bagasse from a sugar mill (Rowell et al., 1993). Approximately 60% of the volume of solid waste in an average municipal waste stream consists of agro-based resources such as paper and paper-based products, wood, and yard wastes. There are also potentially millions of tons of wood fiber in timber thinnings, industrial wood waste, demolition waste, pallets, and pulp mill sludges. Recycling paper products back into paper requires wet processing and removal of inks, inorganics, and adhesives. Recycling these same products into composites can be done using dry processing (thus eliminating a waste water stream) and all co-existing resources can be incorporated into the composite.

Figure 7.1 Agro-based resources ready for recycling into composites.

3. POTENTIAL COMPOSITES FROM AGRO-RESOURCES

For this chapter, a composite will be defined as any combination of two or more resources held together by some type of mastic or matrix. Composites can be classified in many ways: by their densities, by their uses, by their manufacturing methods, or by other systems. For this paper, they will be classified by their uses. Eight different use classes will be covered: (1) geotextiles, (2) filters, (3) sorbents, (4) structural composites, (5) non-structural composites, (6) molded products, (7) packaging, and (8) combinations with other materials. In some cases, one type of composite can be used for more than one use. For example, once a fiber web has been made, it can be directly applied as a geotextile, filter, or sorbent, or it can be further processed into a structural or non-structural composite, molded product, used in packaging, or combined with other resources.

3.1 Geotextiles

Geotextiles mean the use of fabrics in association with the earth (see Chapter 13). The long bast or leaf fibers can be formed into flexible fiber mats, which can be made by physical entanglement, nonwoven needling, or thermoplastic fiber melt matrix technologies (Figure 7.2, Rowell, 1993). The two most common types are carded and needle-punched mats. In carding, the fibers are combed, mixed, and physically entangled into a felted mat. These are usually of high density but can be made at almost any density. A needle-punched mat is produced in a machine that passes a randomly formed machine-made web through a needle board that produces a mat in which the fibers are mechanically entangled (Figure 7.3). The density of this type of mat can be controlled by the amount of fiber going through the needle board or by overlapping needled mats.

Numerous articles and technical papers have been written and several patents have been issued on both the manufacture and use of nonwoven fiber mats containing combinations of textile and lignocellulosic fibers or lignocellulosic fibers alone. Combinations of long and short fibers can be used to make geotextiles, which can reduce the total cost of these composites.

Geotextiles have a large variety of uses. These can be used for mulch around newly planted seedlings (Figure 7.4). The mats provide the benefits of natural mulch; in addition, controlled-release fertilizers, repellents, insecticides, and herbicides can be added to the mats as needed. Research results on the combination of mulch and pesticides in agronomic crops have been promising.

The addition of such chemicals could be based on silvicultural prescriptions to ensure seedling survival and early development on planting sites where severe nutritional deficiencies, animal damage, insect attack, and weed problems are anticipated. Medium density fiber mats can also be used to replace dirt or sod for grass seeding around new homesites or along highway embankments (Figure 7.5). Grass or other type of seed can be incorporated in the fiber mat. Fiber mats promote seed germination and good moisture retention. Low and medium density fiber mats can be used for soil stabilization around new or existing construction sites. Steep slopes, without root stabilization, lead to erosion and loss of top soil.

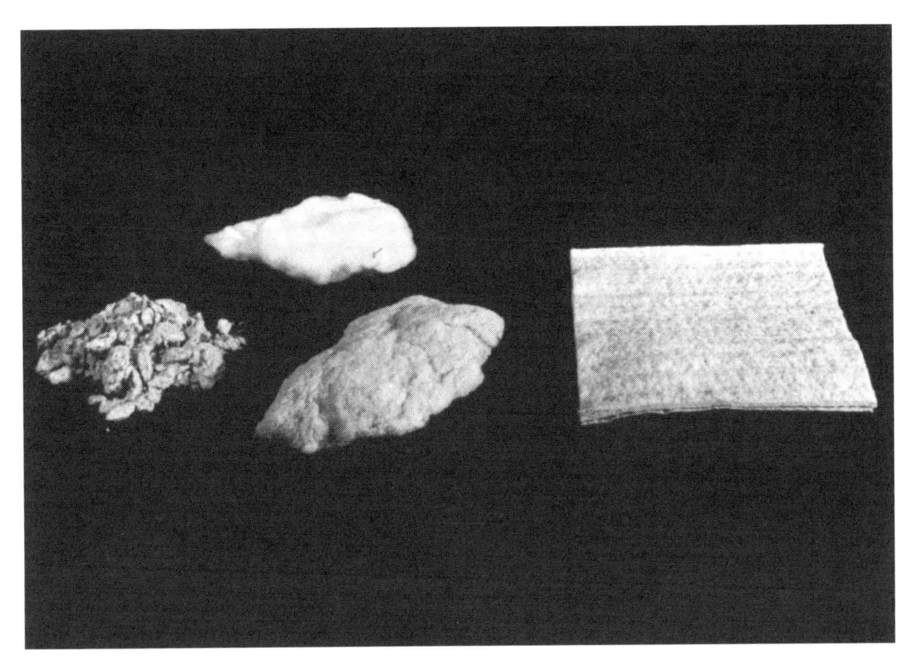

Figure 7.2 Mat-forming process combining short fibers with long binder fibers.

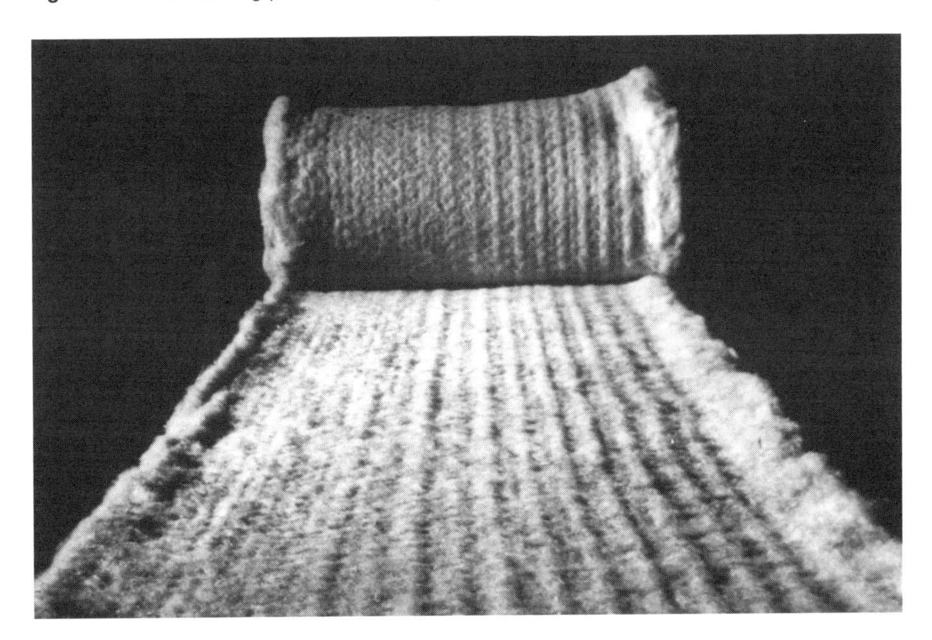

Figure 7.3 Continuous needled mat ready for pressing into complex shapes.

Medium and high density fiber mats can also be used below ground in road and other types of construction as a natural separator between different materials in the layering of the back fill (Figure 7.6). It is important to restrain slippage and mixing

Figure 7.4 Mulch mat in place to control the growth of weeds around a newly planted tree sapling.

Figure 7.5 Agro-based geotextile in use to help prevent erosion on a steep slope.

of the different layers by placing separators between the various layers. Jute and kenaf geotextiles have been shown to work very well in these applications but the potential exists for any of the long agro-based fibers.

Figure 7.6 Examples of agro-based geotextiles.

It has been estimated that the global market for geotextiles is about 800 million square meters, but this estimate has not been broken down into use categories, so it is impossible to determine the portion that is available for natural geotextiles.

3.2 Filters

Medium and high density fiber mats can be used for air filters. The density of the mats can be varied depending on the size and quantity of material being filtered and the volume of air required to pass through the filter per unit of time (Figure 7.7). Air filters can be made to remove particulates and/or can be impregnated or reacted with various chemicals as an air freshener or cleanser. Medium to high density mats can also be used as filtering aids to take particulates out of waste and drinking water or solvents. The fiber can be modified to selectively remove desired contaminates (see Chapter 13).

3.3 Sorbents

Tests are presently underway for use of agro-based sorbents to remove heavy metals, pesticides, and oil from rain water runoff in several cities in the United States (Figure 7.8). Medium and high density mats can also be used for oil spill clean up pillows. It has been shown that agro-based core material from kenaf or jute will

Figure 7.7 Agro-based fiber mats can also be used as air, water, or solvent filters.

preferentially sorb oil out of sea water when saturated with sea water. There are many other potential sorbent applications of agro-fiber/core resources such as removal of dyes and trace chemicals in solvents and in the purification of solvents (see Chapter 13).

Figure 7.8 Kenaf pith (core) used for oil spill cleanup.

The fiber can be modified and used as a chromatographic solid phase support to selectively separate organic and inorganic chemicals. It is also possible to use core fiber materials as sorbents in cleaning aids such as floor sweep. While this is not a composite as such, it is another way agro-based resources can be used as sorbents.

Lignocellulosic fiber can also be compressed and converted to activated carbon for use as a sorbent. Again, it is not a composite as defined in this report but another outlet for agro-resources in the area of sorbents.

3.4 Structural Composites

A structural composite is defined as one that is required to carry a load in use

Figure 7.9 Examples of many types of structural composites made using thermosetting resins.

(Figure 7.9). In the housing industry, for example, these represent load bearing walls, roof systems, subflooring, stairs, framing components, furniture, etc. In most, if not all cases, performance requirements of these composites are spelled out in codes and/or in specifications set forth by local or national organizations.

Structural composites can range widely in performance from high performance materials used in the aerospace industry down to wood-based composites which have lower performance requirements. Within the wood-based composites, performance varies from multi-layered plywood and laminated lumber to low cost particleboard. Structural wood-based composites, intended for indoor use, are usually made with a low cost adhesive which is not stable to moisture while exterior grade composites contain a thermosetting resin that is higher in cost but stable to moisture. Performance can be improved in wood-based as well as agro-based composites by using chemical modification techniques to modify fiber properties such as dimensional stability, biological and ultraviolet resistance, and stability to acids and bases, or treated with conventional fire retardant and/or decay control chemicals (see Chapter 11).

3.5 Non-Structural Composites

As the name implies, non-structural composites are not intended to carry a load in use. These can be made from a variety of materials such as thermoplastics, textiles, and wood particles, and are used for such products as doors, windows, furniture, gaskets, ceiling tiles, automotive interior parts, molding, etc. (Figure 7.10). These are generally lower in cost than structural composites and have fewer codes and specifications associated with them.

Non-structural composites can be produced by a variety of processes including extrusion, thermo-pressing, pultrusion, sheet molding, and injection molding.

Figure 7.10 Examples of non-structural composites made using a thermoplastic matrix.

3.6 Molded Products

The present wood-based composite industry mainly produces two-dimensional (flat) sheet products. In some cases, these flat sheets are cut into pieces and glued/fastened together to make shaped products such as drawers, boxes, and packaging. Flat sheet wood fiber composite products are made by making a gravity formed mat of fibers with an adhesive and then pressing (Figure 7.11). If the final shape can be produced during the pressing step, then the secondary manufacturing profits can be realized by the primary board producer. Instead of making low cost flat sheet type composites, it is possible to make complex shaped composites directly using agro-based long bast fibers.

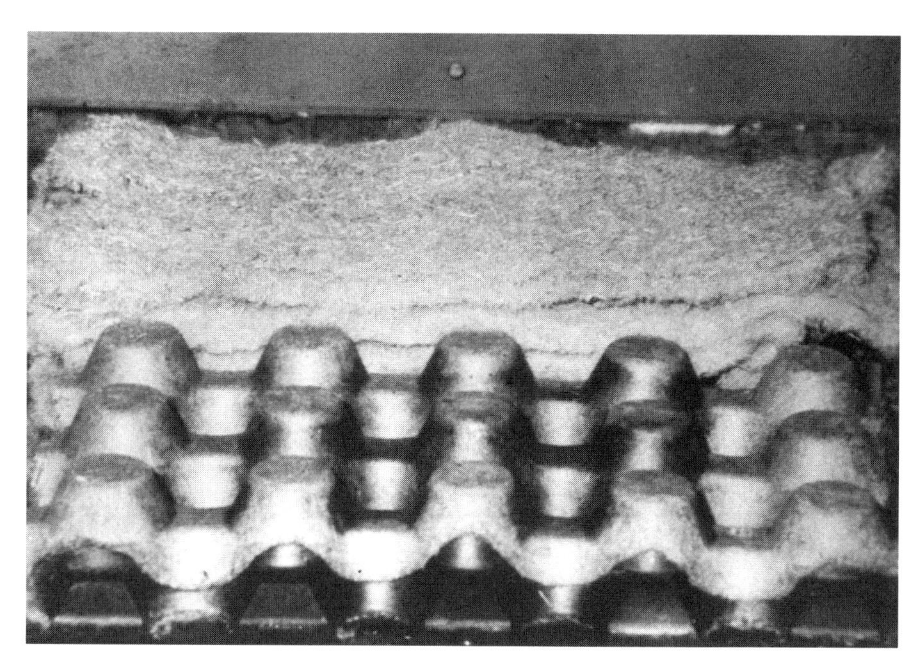

Figure 7.11 Agro-based mat being pressed into a complex shape.

In this technology, fiber mats are made similar to the ones described for use as geotextiles, except during mat formation an adhesive is added by dipping or spraying of the fiber before mat formation or added as a powder during mat formation. The mat is then shaped and densified by a thermoforming step. Within certain limits, any size, shape, thickness, and density are possible (Figure 7.12). These molded composites can be used for structural or non-structural applications as well as packaging, and can be combined with other materials to form new classes of composites. This technology will be described later.

3.7 Packaging

"Gunny" bags made from jute have been used as sacking for products such as coffee, cocoa, nuts, cereals, dried fruits, and vegetables for many years. While there are still many applications for long fiber for sacking, most of the commodity goods are now shipped in containers. These containers are not made of agro-fibers today but there is no reason why they cannot be (Rowell, 1994a). Medium and high density agro-based fiber composites can be used for small containers, for example, in the tea industry and for large sea-going containers for commodity goods (Figure 7.13). These composites can be shaped to suit the product by using the molding technology described previously or formed into low cost, flat sheets and made into containers.

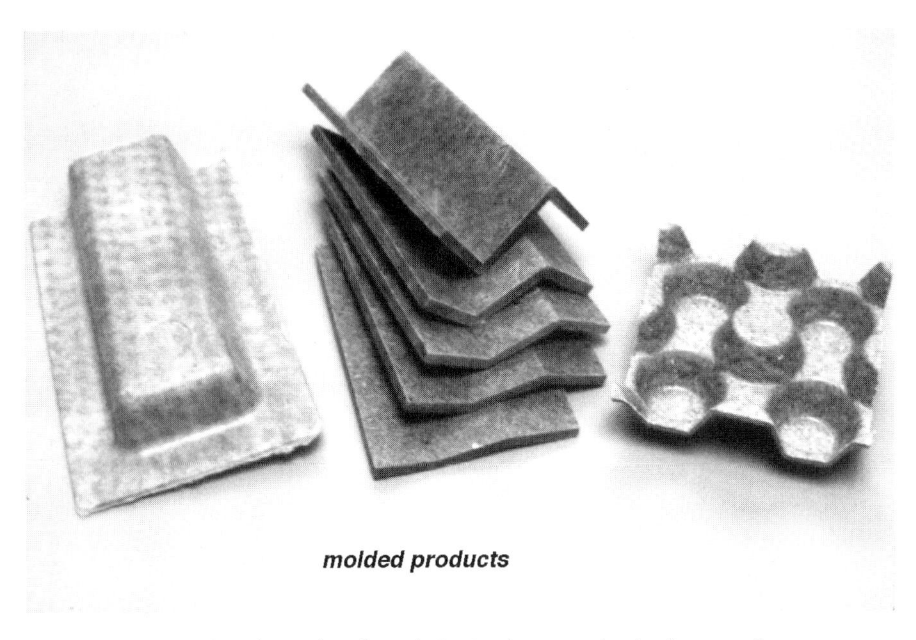

Figure 7.12 Examples of complex shaped structural or non-structural composites.

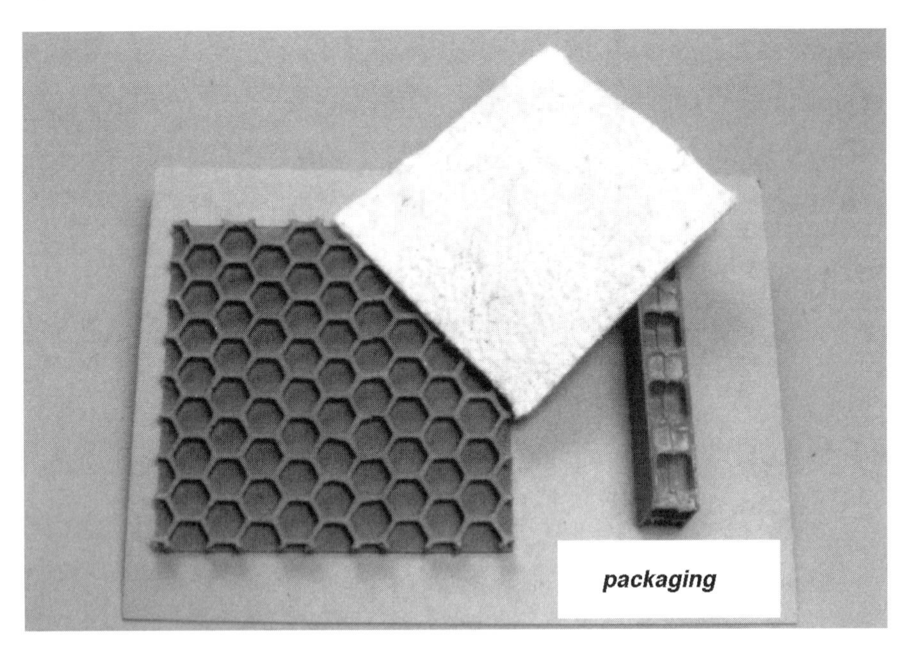

Figure 7.13 Lightweight packaging using "Spaceboard" (see Chapter 10) or mats for nesting-type packaging.

Agro-based fiber composites can also be incorporated into returnable containers where the product is reused several times. These containers can range from simple crease-fold types to more solid, even nestable, types. Long agro-fiber fabrics and mats can be overlayed with thermoplastic films such as polyethylene or polypropylene to package such products as concrete, foods, chemicals, and fertilizer. Corrosive chemicals require the plastic film to provide water resistance and reduce degradation of the agro-based fiber. There are also many applications for agro-based fiber as paper sheet products for packaging. These vary from simple paper wrappers to corrugated, multi-folded, multi-layered packaging.

3.8 Combinations with Other Resources

It is possible to make completely new types of composites by combining different resources. Leaf, bast, and/or stick fiber can be combined, blended, or alloyed with other materials such as glass, metals, plastics, and synthetics to produce new classes of materials (Figures 7.14 and 7.15). The objective is to combine two or more materials in such a way that a synergism between the components results in a new material with much better properties than the individual components.

Figure 7.14 Many different types of fibers can be combined in a mat to enhance the performance of a composite.

Agro-based fiber/glass fiber composites can be produced with glass as a surface material or combined as a fiber with lignocellulosic fiber (Figure 7.16). Composites of this type often have a very high stiffness-to-weight ratio. The long bast fibers can also be used in place of glass fiber in many different types of liquid composite molding (LCM) systems such as resin transfer molding (RTM), resin injection molding (RIM), structural reaction injection molding (SRIM), and sheet molding compounding (SMC) (Figure 7.17). All of these techniques include a fibrous mat mixed with a liquid resin, which is polymerized to form a reinforced fiber composite.

Figure 7.15 Combinations of agro-based fibers with thermoplastics, rubber, and metals.

In almost all present industrial applications, the fiber used is glass and the resins include polyesters, vinyl ester, polyurethane, phenolics, melamines, phthalates, alkyds, isocyanates, epoxides, acrylates, and silicones. Agro-fibers are lower in specific gravity, higher in specific tensile strength, lower in cost, and less energy intensive to process, so they are well suited to these types of technologies. The biggest single application of these technologies today is in automotive parts but there are growing markets in medical, sporting, space, tank, tub, boat, and corrosion-resistant equipment. Problems of dimensional stability and compatibility with the resin must be addressed but this could also lead to major new markets for agro-based resources.

Metal films can be overlayed onto smooth, dimensionally-stabilized fiber composite surfaces or applied through cold plasma technology to produce durable coatings. Such products could be used in exterior construction to replace all aluminum or vinyl siding markets, where agro-based resources have lost market share.

Metal fibers can also be combined with stabilized fiber in a matrix configuration in the same way metal fibers are added to rubber to produce wear-resistant aircraft tires. A metal matrix offers excellent temperature resistance and improved strength properties, and the ductility of the metal lends toughness to the resulting composite. Application for metal matrix composites could be in the cooler parts of the skin of ultra-high-speed aircraft. Technology also exists for making molded products using perforated metal plates embedded in a phenolic-coated fiber mat, which is then pressed into various shaped sections.

Bast or leaf fiber can also be combined in an inorganic matrix. Such composites are dimensionally and thermally stable, and they can be used as substitutes for

Figure 7.16 Combinations of agro-based fibers with fiberglass.

Figure 7.17 Resin transfer molding using (left) jute, (center) kenaf, or (right) fiberglass.

asbestos composites. Inorganic-bonded bast fiber composites can also be made with variable densities that can be used for structural applications.

One of the biggest new areas of research in the value added area is in combining natural fibers with thermoplastics (Sanadi et al., 1994 a,b,c, see Chapter 12). Since prices for plastics have risen sharply over the past few years, addition of a natural powder or fiber to plastics provides a cost reduction to the plastic industry (and in some cases increases performance as well); but to the agro-based industry, this represents an increased value for the agro-based component. Most of the research on these types of composites has been concentrated on the use of compatibilizers to make the hydrophobe (plastic) mix better with the hydrophil (lignocellulosic). The two components remain as separate phases, but if delamination and/or void formation can be avoided, properties can be improved over those of either separate phase. These types of materials are usually referred to as natural fiber/thermoplastic blends.

Another combination of fiber and thermoplastics is in products that can best be described as fiber–plastic alloys (Rowell et al., 1994a,b). In this case the fiber and plastic become one material and it is not possible to separate the two components. Fiber–plastic alloys are possible through fiber modification and grafting research. This can be done by considering that agro-based fibers consist of a thermoset polymer (cellulose) in a thermoplastic matrix (lignin and the hemicelluloses). The glass transition temperature (T_g), however, of the dry natural thermoplastic matrix is higher than the decomposition temperature of the fiber. If the T_g were lowered through chemical modification, it should be possible to thermoplasticize the lignin and, possibly, the hemicelluloses for thermo-molding at temperatures below decomposition. If a reactive thermoplastic is then reacted with the modified bio-based fiber, it should be possible to form fiber/thermoplastic alloys. It is also possible to thermoplasticize cellulose but then the strong, stiff, reinforcing thermoset structure is lost. If the cellulose were also plasticized, it would be possible to produce agro-based films and foams.

4. INTEGRATED PROCESSING METHODOLOGIES

4.1 Multi-Fiber Processing Technology

One way to optimize the yield of any agro-based resource is to make use of all parts of the plant. The scheme shown in Figure 7.18 gives possible processing pathways that lead to the composite products identified in this chapter that can come from each fraction of the plant. The entire plant (leaves, stock, pith, roots) can be used directly to produce structural and non-structural composites. By using the entire plant, processes such as retting, fiber separation, fraction purification, etc. can be eliminated, which increases the total yield of plant material and reduces the costs associated with fraction isolation. This also gives the farmer a different option for crop utilization; that is, transporting the entire plant to a central processing center rather than getting involved in plant processing.

Another option is to separate the higher value long fiber from the other types of shorter fiber for use in value-added products. When the long fiber is separated, the by-product is a large amount of short fiber and pith material that can be used for such products as sorbents, packing, lightweight composites, and insulation. By

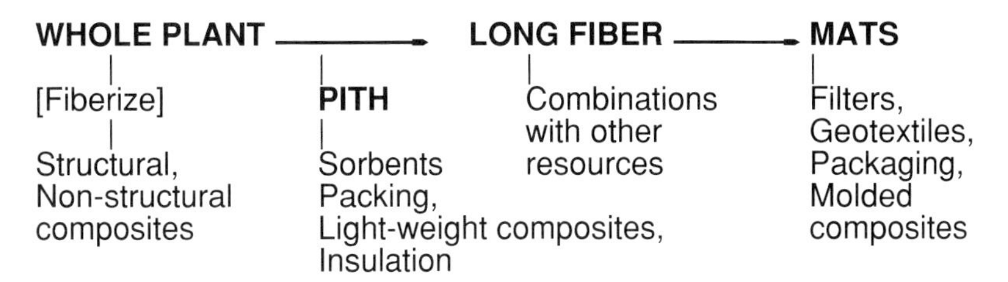

Figure 7.18 Multi-functional fiber separation scheme.

utilizing the by-product from the long fiber isolation process, the overall cost of long fiber utilization is reduced. The isolated long fiber can then be used to make mats that have value-added applications in filters, geotextiles, packaging, molded composites, and structural and non-structural composites.

More than one plant type can be processed at the same time using this scheme. Without the need to separate different agricultural resources, the costs of sorting and storage are reduced. Recycled natural fiber can also be used in this scheme. Desired product output and resource costs will drive this type of multi-fiber use technology.

4.2 Multi-Product Processing Technology

The key issues in the agro-fiber processing arena are what to collect, how to collect it, and what to make from what has been collected. In the preceding section, we discussed how eight different types of composites can be made utilizing different parts of the plant. We now offer a system for collection and processing of agro-based resources to produce a wide variety of products based on the same principle as a petroleum refinery. In this scheme, fiber for composites is just one of the product options available.

In the petroleum refinery, crude oil is brought into the refinery and management uses the computer to decide what products will be made, based on market demand. The same program could be applied to the agro-based industry. Our present paradigm of processing agriculture resources is to compartmentalize each segment of the industry rather than integrate them into one encompassing system.

Figure 7.19 shows the basic concept of an agro-refinery. In any geographic area, all agro-based resources are brought to the refinery. No predetermined markets are assumed and, as with the petroleum refinery, decisions as to processing and final product selection are made by market demands. From the cities within the geographic area come recycled paper products, waste wood, grass, leaves, and other agricultural wastes (Rowell et al., 1993). From the rural area come these same resources along with all agricultural resources. No separation of grain from cob or stock is made on the farm. Everything is cut off at the ground and brought in. From the industry in that area comes industrial wastes that can also be processed in the refinery. The refinery then has many options to produce food, feed, fiber, or fuel, or use the material as a substrate for fermentation, chemical production, or compost. The market structure determines what processing will be done and the best decisions based on total needs will be made. Since it has been shown that farm land must have some of the crop residue left as nutrient, it is important that part of the products produced include some type of compost. The compost produced could be improved over the original plant residues by enrichment with other needed ingredients.

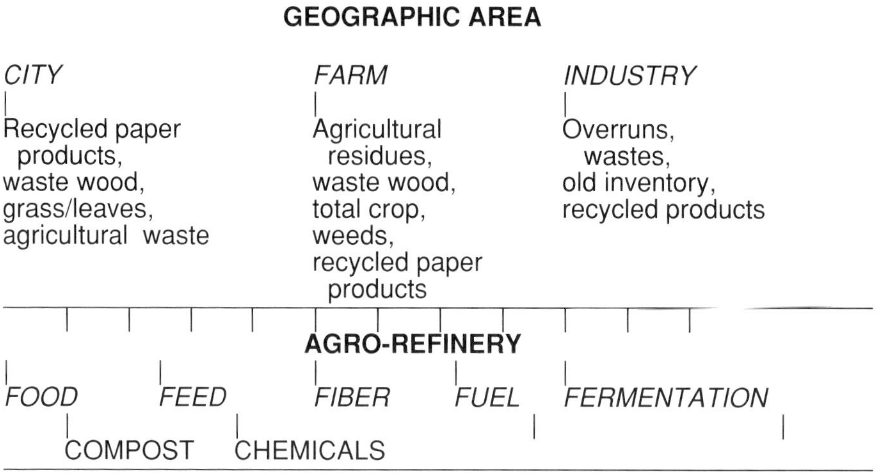

GEOGRAPHIC AREA

CITY	*FARM*	*INDUSTRY*
Recycled paper products, waste wood, grass/leaves, agricultural waste	Agricultural residues, waste wood, total crop, weeds, recycled paper products	Overruns, wastes, old inventory, recycled products

AGRO-REFINERY

FOOD *FEED* *FIBER* *FUEL* *FERMENTATION*

COMPOST CHEMICALS

Figure 7.19 Concept of an agro-refinery.

Our present paradigm in the food production industry is to grow grain for food and use the residue for feed. In many parts of the United States, however, there is excess grain and there are schemes to burn grain for fuel. Plants, such as kenaf, are now grown for fiber on lands that once were used to grow grain. We have, for generations, used herbicides to remove weeds from food and feed crops, but, if we are after fiber, the "weeds" may be the best crop. There are schemes to ferment grains to ethanol for use as an additive to gasoline. There are also schemes to convert agro-based resources to methanol, phenols, and other chemicals.

In our scheme, all producers and consumers work together to manage resource production, utilization, and waste management. It expands the concept of sustainable agriculture to include all needs on a long range basis.

Rather than these product decisions being made in advance and individual manufacturing facilities built to produce only one product, the agro-refinery would be the central processing and decision making hub for the entire agro-based industry. It would be the recycling center for agro-waste resources as well as the processing center for all agricultural resources in the region and would lead to many more options for the use of this valuable resource.

REFERENCES

Rowell, R. M., Opportunities for composite materials from jute and kenaf. *International consultation on jute and the environment*, Food and Agricultural Organization of the United Nations, ESC:JU/IC 93/15, 1, 1993.

Rowell, R. M., Potentials for composites from jute and allied fibers. Proceedings, *International Jute and Allied Fibre Symposium on Biocomposites and Blends*, New Dehli, India, 1, 1994a.

Rowell, R. M., *International task force on diverse jute applications*, Food and Agricultural Organization of the United Nations, ESC: JU/EGM 94/7, 9, 1994b.

Rowell, R. M., Caulfield, D. F., Sanadi, A., O'Dell, J., and Rials, T. G., Thermoplasticization of kenaf and compatibilization with other materials, in *Proceedings, Sixth Annual International Kenaf Conference*, New Orleans, LA, 1, 1994a.

Rowell, R. M., O'Dell, J. L., and Rials, T. G., Chemical modification of agro-fiber for thermoplasticization, in *Proceedings, Pacific Rim Bio-Based Composite Symposium*, Vancouver, B.C., Canada, 144, 1994b.

Rowell, R. M., Spelter, H., Arola, R. A., Davis, P., Friberg, T., Hemingway, R. W., Rials, T., Luneke, D., Narayan, R., Simonsen, J., and White, D., *Forest Prod. J.*, 43(1) 55, 1993.

Sanadi, A. R., Caulfield, D. F., Jacobson, R. E., and Rowell, R. M., Reinforcing polypropylene with agricultural fibers, in *Proceedings, International Jute and Allied Fibre Symposium on Biocomposites and Blends,* New Dehli, India, 163, 1994a.

Sanadi, A. R., Caulfield, D. F., and Rowell, R. M., Reinforcing polypropylene with natural fibers, *Plastic Engineering*, Vol 1 (4), 27, 1994b.

Sanadi, A. R., Caulfield, D. F., Walz, K., Wieloch, L., Jacobson, R. E., and Rowell, R. M., Kenaf fibers – Potentially outstanding reinforcing fillers in thermplastics, In *Proceedings, Sixth Annual International Kenaf Conference*, New Orleans, LA, 155, 1994c.

CHAPTER 8

Processing into Composites

Brent English, Poo Chow, and Dilpreet Singh Bajwa

CONTENTS

1. INTRODUCTION

Globally, many lignocellulosic fiber options exist for the production of composite products. A literature search was conducted at the USDA Forest Service, Forest Products Laboratory to survey the worldwide use of agricultural fibers in composite building products. A total of 1,039 citations were selected from the vast number available. Youngquist et al. (1993a) summarized the work on composite products from agricultural fibers and part of this report was used here.

2. CHARACTERISTICS OF LIGNOCELLULOSIC COMPOSITES

In a broad sense, a composite can be defined as any combination of two or more resources, in any form, and for any use. For the purposes of this discussion, the term "composite" describes two situations. The first is when the lignocellulosic serves as the main ingredient in the composite. The second is when the lignocellulosic serves as a reinforcing filler or aggregate within a matrix material. Whatever scenario is used, the objective of composite development is to produce a product whose performance characteristics combine the beneficial aspects of each constituent component.

2.1 General Manufacturing Issues

Successful manufacture of any product requires control over raw materials. Ideally, raw materials are uniform, consistent, and predictable. Lignocellulosics do not offer these qualities but instead vary widely between species. Plants of the same species also have differences. Plant stems differ from leaves, which differ from seed husks. Within the stem, bast fibers differ greatly from pith fibers; even the harvest time affects fiber quality.

For the purpose of producing a composite product, uniformity, consistency, and predictability are accomplished by reducing separated portions of the plants into small, relatively uniform and consistent particles, flakes, or fibers where effects of differences will average out. Depending on the lignocellulosic material, size reduction is sometimes augmented by chemical treatments designed to weaken the bonds between the components. The degree of size reduction and the shape of the individual lignocellulosic component will depend on the application. Different composites tolerate or demand different sizes and shapes.

Another limiting factor in the use of agro-based resources for large scale industrial purposes is bulk density. Low bulk density can significantly increase transportation costs. For instance, chunked wood (sawed into short lengths and stacked) has a dry basis gross bulk density of 240–320 kg/m^3. In the United States, the economics of processing and transporting wood with such bulk density indicate a practical procurement radius of about 65 km (Vaagen, 1991). In contrast, annual fiber stems of a plant such as kenaf or straw cannot be compacted much beyond 135 kg/m^3, which may limit the feasible supply basin to a range of 25–35 km (Sandwell, 1991).

2.2 General Lignocellulosic Composite Classifications

While there is a broad range of lignocellulosic composites, and many applications for them, for the purposes of this chapter they will be grouped into three general categories. Complete books have been written about each of the categories, and the constraints of this chapter necessitate that our discussion be general and brief. References are provided to lead the reader to more detailed information.

The first category of lignocellulosics that we discuss is that of conventional composites. Conventional composites are already in the marketplace with a high degree of customer acceptance. Here, the lignocellulosic serves as the main ingredient, and a small percentage (generally less than 10%) of a heat-curing adhesive holds the composite together. Particleboards and fiberboards are common examples of this type of lignocellulosic composite.

The second classification is that of inorganic-bonded composites. In these composites, inorganic materials like gypsum or Portland cement hold the composite together. The lignocellulosic might be the main ingredient or serve as an aggregate.

The third composite classification to be discussed here is that of lignocellulosic/thermoplastic composites. In this class, the lignocellulosic can serve as a reinforcing filler in a thermoplastic matrix, or conversely, the thermoplastic may serve as a binder to the lignocellulosic.

Whether used as the main ingredient of the composite or used as a reinforcing filler, much of the raw material factors are the same. Some of these are discussed in the next section, followed by individual sections on conventional panel type composites, inorganic-bonded composites and lignocellulosic/thermoplastic composites.

3. RAW MATERIALS: CHARACTERIZATION, STORAGE, AND PREPARATION

Agro-based lignocellulosics suitable for composites stem from two main sources. The first is agricultural residues. These materials, like rice husks or cereal straws, are the by-products of food or feed crops. While value-added uses are found for portions of some of these residues, most are used for more mundane purposes like animal bedding or fuel. Others are simply left on the field or burnt to reduce mass. The second class is those lignocellulosics grown specifically for their fiber. Two examples are jute and kenaf. These plants also have residues, which are often used for bedding or fuel as well.

Technically speaking, almost any agricultural fiber can be used to manufacture composition panels. However, it becomes more difficult to use certain kinds of fibers when restrictions in quality and economy are imposed. The literature has shown that several kinds of fibers have existed in sufficient quantity, in the right place, at the right price and at the right time to have merited at least occasional commercial use. While others may exist, we choose to discuss bagasse, cereal straw, coconut coir, corn stalks, cotton stalks, jute, kenaf, and rice husks. The remainder of this section briefly addresses the issues of fiber harvesting, storage, and fiber preparation to use these fiber sources in composite production.

3.1 Bagasse

Bagasse is the residue fiber remaining when sugarcane is pressed to extract the sugar. Some bagasse is burned to supply heat to the sugar refining operation, some is returned to the fields, and some finds its way into various board products. Bagasse is composed of fiber and pith. The fiber is thick walled and relatively long (1–4 mm). For use in composites, fibers are obtained mostly from the rind, but there are fibrovascular bundles dispersed throughout the interior of the stalk as well (Hamid et al., 1983). In what could be considered a definitive work, Atchison and Lengel (1985) told of the history and growth of bagasse fiberboard and particleboard at the 19th Washington State Particleboard Symposium. Their paper describes the various success and failure stories of bagasse utilization in composite panel production.

Bagasse is available wherever sugarcane is grown. As such, almost no harvesting problems exist, and large volumes are available at sugar mills. In northern climates, the cane harvest usually lasts about 2.5 months. In warm climates, bagasse may be available for as long as 10 months out of the year. During this time, bagasse supply is relatively constant; the remainder of the year, it must be stored. Special care must be taken during storage to prevent fermentation, because bagasse does have a high sugar content. To reduce the sugar content and increase storage life, bagasse is usually depithed before storage. The pith is an excellent fuel source for the sugar refining operation. Generally, if the bagasse is depithed, dried, and densely baled, it can be stored outside. If handled in a careful manner, bagasse can also be stored wet. In the wet method, large bales of bagasse are specially fabricated and stacked to insure adequate air flow. Heat from fermenting sugars effectively sterilizes the bales. Bagasse can be stored for several years using this method (Chapman, 1956). Other storage options are available, including some that keep the bagasse wet beyond the fiber saturation point.

As previously mentioned, only bagasse fiber is utilized for the production of high-quality composition panels. Various schemes are available to separate the bagasse fiber from the pith, some of which are described in Chapter 6. The fibers after depithing are more accurately described as fiber bundles that can be used "as is" to make particleboard, or they can be refined to produce fibers for fiberboard. Recently, the Tilby process (Sugartree, 1992) has been developed for the separation of the bagasse fiber and pith before the stalk is crushed to extract the juice. This system uses rollers to flatten and then guide the stalk over blades that cleanly separate the two materials. It is claimed that reclamation is increased for both the fiber and the juice.

3.2 Cereal Straw

After bagasse, cereal straw is probably the second most important agricultural fiber for composite panel production. For the purpose of this paper, cereal straw is meant to include straw from wheat, rye, barley, oats, and rice. Straw, like bagasse, is an agricultural residue. Unlike bagasse, large quantities of cereal straw are generally not available at one location. Storage is usually accomplished by baling. The bales must then be transported to a manufacturing facility. Straws have a high ash content, thus tending to fill fireboxes in boilers. Their high inherent silica content results in increased tool wear compared to other lignocellulosic composites. Conversely, the high silica content also tends to make them naturally fire-resistant.

Plants have existed in several countries to make thick (5–15 cm) straw panels with kraft paper faces (UNIDO, 1975). The panels are made by heating the straw to about 200°C, at which point springback properties are virtually nil. The straw is fed through a reciprocating arm extruder and made into a continuous low density (0.25 specific gravity) panel. Kraft paper is then glued to the faces and edges of the panels. These panels can then be cut for prefabrication into housing and other structures. The low density of these panels makes them fairly resilient, and test data show that housing built using these panels is especially earthquake resistant. In the 1980s, such a plant was set up in California to produce straw panels from wheat and rye straw (Galassco, 1992).

Straw can be used to supplement part of the fiber content in wood particleboard. A large particleboard plant in the United States, located in La Grande, Oregon, substituted straw at a rate of 8% and found no major problems except that the sander dust from the faces deposited additional ash in the boiler. This plant then stopped using straw in the face and used it only in the core. At a rate of 10% or less, the effect on tool wear was not significant (Knowles, 1992).

The time of harvest for the straw is important to board quality (Rexen, 1977). The quality of the straw is highest when the grain is at its optimum ripeness for harvesting. Under-ripe straw has not yet yielded its full potential, and over-ripe straw becomes brittle. For particleboards, straw is reduced by hammer milling or knife milling. For the production of fiber-based products, straw can be pulped by using alkali treatments and refining (see Chapter 6). Ryegrass straw particleboard was commercially produced in the United States in Oregon (Loken et al., 1991).

3.3 Coconut Coir

Coconut coir is the long fiber (15–35 cm) from the husk of the mature coconut, and the average husk weighs 400 grams (Singh, 1979). Coir is a fiber source for many cottage industries and it is readily woven into mats and made into ropes and other articles for both domestic use and export.

Coir has been used to produce a variety of composite products including particleboards and fiberboards. When used as a reinforcing fiber in inorganic-bonded composites, coir is very resistant to alkalinity and variations in moisture, when compared to other lignocellulosics (Savastano, 1990).

3.4 Corn Stalks

Based on our literature search, there is currently no commercial utilization of corn stalks or cobs in lignocellulosic composite production. However, a low-density insulation board was produced in Dubuque, Iowa for several decades in the middle of this century. In addition, a three-layer board having a corn cob core and wood veneer face was produced for a short time in Czechoslovakia after World War II (UNIDO, 1975).

Corn stalks, like many agricultural fiber sources, consist of a pithy core with an outer layer of long fibers. Currently in the United States, corn stalks are chopped and used for forage, left on the field, or baled for animal bedding. The cobs are occasionally used for fuel. Research shows that corn stalks and cobs can be made into reasonably good particleboard and fiberboard (Chow, 1974). In the research, corn stalks and cobs were either hammermilled into particles or reduced to fibers in a pressurized refiner.

3.5 Cotton Stalks

Cotton is cultivated primarily for textile fibers, and little use is made of the cotton plant stalk. Stalk harvest yields tend to be low and storage can be a problem. The cotton stalk is plagued with parasites, and stored stalks can serve as a breeding ground for the parasites to winter over for next year's crop. Attempted commercialization of cotton stalk particleboard in Iran was unsuccessful for this reason (Brooks, 1992).

If the parasite issue could be addressed, cotton stalks could be an excellent source of fiber. With respect to structure and dimensions, cotton stalk fiber is similar to common species of hardwood fiber (Mobarak and Nada, 1975). As such, debarked cotton stalks can be used to make high grades of paper. The stalk is about 33% bark and quite fibrous. Newsprint quality paper can be made from whole cotton stalks. For particleboard production, cotton stalks can be hammermilled like other materials. For fiberboards, cotton stalks can be refined with or without chemical treatment, depending on the quality of fiber desired.

3.6 Jute

Jute is an annual plant in the genus *Corchorus*. The major types grown are generically known as white jute and tossa jute. Jute, grown mainly in India and Bangladesh, is harvested at 2 to 3 months of growth, at which time it is 3–5 meters tall. Jute has a pithy core, known as jute stick, and the bast fibers grow lengthwise around this core.

Jute bast fiber is separated from the pith in a process known as retting. Retting is accomplished by placing cut jute stalks in ponds for several weeks. Microbial action in the pond softens the jute fiber and weakens the bonds between the individual fibers and the pith. The fiber strands are then manually stripped from the jute stick and hung on racks to dry. Very long fiber strands can be obtained this way. If treated with various oils or conditioners to increase flexibility, the retted jute fiber strands are suitable for manufacturing into textiles.

Most composites made using jute exploit the long fiber strand length. Commercially, both woven and non-woven jute textiles are resin- or epoxy-impregnated and molded into fairly complex shapes. In addition, jute textiles are used as overlays over other composites. Jute stick is used for fuel, and in poor areas it is stacked on end, tied into bundles, and used as fences and walls.

3.7 Kenaf

Kenaf is a hibiscus, and is similar to jute or hemp in that it has a pithy stem surrounded by fibers. The fibers make up 20–25% of the dry weight of the plant (LaMahieu et al., 1991). Kenaf grows well in warm climates and does not have the narcotic effect found in the non-fibrous parts of the hemp plant. Mature kenaf plants can be 5 m high.

Kenaf is currently generating much interest from government and industry. The U.S. Department of Agriculture is promoting kenaf, and other non-food, non-feed agricultural crops, because these crops are not subject to subsidies (AARC, 1992). As an indication of the interest in kenaf, a recent bibliography devoted solely to kenaf had 241 scientific citations (USDA, 1992). Also, the International Kenaf Association was formed and is devoted to the study and promotion of kenaf.

Historically, kenaf fiber was first used as cordage. Industry is now exploring the use of kenaf in papermaking and nonwoven textiles. Like jute, most kenaf composite products exploit the long aspect ratio of kenaf fibers and fiber bundles. One way to do this is to form the kenaf into a nonwoven textile mat that can be used for erosion control, seedling mulches, or oil spill absorbents. After a resin is added to the kenaf mats, they can be pressed into flat panels or molded into shapes, such as interior car door substrates. In addition, low-density particleboard based on kenaf pith is produced in Spain (Riccio, 1993). Kenaf pith is used in other parts of the world for animal bedding and other absorbent applications.

Environmental concerns prevent the retting of kenaf fiber in the United States; therefore, alternate means of separating the bast from the pith are employed. If dry, separation begins by chopping the kenaf stalk into shorter lengths, which fractures the pith. Standard screening and air separation techniques can then be used to separate the two different materials. Commercially, kenaf bast fiber separated this way can be purchased 98% pith-free. The Tilby process has also been employed experimentally for separating green kenaf stalk. Kenaf is generally stored in a dried and baled state.

3.8 Rice Husks

Rice husks are an agricultural residue that are available in fairly large quantities in one area. Trees "store on the stump" and with rare exception other lignocellulosics do not share this luxury. Rice husks are a notable exception because they are stored on the grain. The grain is stored to be milled year round, making the availability of rice husks reasonably uniform, at least in the United States (Haislip, 1994).

Rice usually comes to the mill at about an 8% moisture content level (Vasishth and Chandramouli, 1974). Rice husks are quite fibrous by nature and little energy input is required to prepare the husks for board manufacture. To make high-quality boards, the inner and outer husks are separated and broken at their "spine." This can be accomplished by hammer milling or refining. Rice husks have a high silica content, and present the same cutting tool problems outlined in Section 3.3.

3.9 Other Fiber Sources

Other important fiber sources include flax shives, bamboo, papyrus, and reed stalks. Many countries like France, Sweden, Belgium, and Germany use flax shives to produce resin-bonded particleboards. In the United States, flax is also used in the manufacture of insulation board by mixing with wood pulp. There are two varieties of flax; one is for fiber and the other is for linseed oil production. Bamboo is an important source of raw material for fiberboards in tropical countries. Most varieties of bamboo are fast-growing and produce strong fibers; particleboards from bamboo have also been made.

Papyrus is used in making hardboard in East Africa. Suitable quality insulation boards and hardboards have also been made in a pilot plant in Sefan, Israel. Particleboards have been produced from reed/typha mixtures that meet or exceed specifications. Insulation boards and plastic-bonded boards have also been prepared from reeds. Other miscellaneous fibers include: banana leaves, grasses, palm, sorghum, etc. While many fibers have been used successfully in the laboratory to produce boards, most of these materials have not been used commercially because of cost or other factors.

4. CONVENTIONAL PANEL-TYPE COMPOSITES

Many conventional lignocellulosic composites are in the marketplace with a high degree of customer acceptance. They are usually available in panel form and are widely used in housing and furniture. Conventional composites typically use a heat curing adhesive to hold the lignocellulosic component together.

Conventional composites fall into two main categories based on the physical configuration of the communited lignocellulosic: fiberboards and particleboards. Within these categories are low, medium, and high-density classifications. Within the fiberboard category, both wet and dry processes exist.

Within limits, the performance of a conventional type composite can be tailored to the end use. Varying the physical configuration of the communited lignocellulosic and adjusting the density of the composites are just two ways to accomplish this. Other ways include varying the resin type and amount, and incorporating additives to increase water resistance or to resist specific environmental factors. On an experimental basis, lignocellulosics have also been chemically modified to change performance.

4.1 Resins and Additives for Conventional Composites

Worldwide, the most common resin for lignocellulosic composites is urea formaldehyde (UF). About 90% of all lignocellulosic composite panel products are bonded with UF (Maloney, 1989). UF is inexpensive, reacts quickly when the composite is hot-pressed, and is easy to use. UF is water-resistant, but not waterproof. As such, its use is limited to interior applications unless special treatments or coatings are applied. A more durable adhesive is phenol formaldehyde (PF). PF is two to three times as expensive as UF, but the increased durability for exterior applications makes it a popular resin. A third common resin, melamine formaldehyde (MF), falls roughly midway between UF and PF on both cost and performance.

Natural options exist that might someday replace or supplement the synthetic resins listed above. Tannins, which are natural phenols, can be modified and reacted with formaldehyde to produce a satisfactory resin. Resins have also been developed by acidifying spent sulfite liquor, generated when wood is pulped for paper. Wet process fiberboards frequently use the lignin inherent in the lignocellulosic as the resin (Suchsland and Woodson, 1986).

Except for two major uncertainties, expectations are that urea–formaldehyde and phenol–formaldehyde systems will continue to be the dominant wood adhesives for lignocellulosic composites. The two uncertainties are the possibility of much more stringent regulation of formaldehyde-containing products and the possibility of limitations or interruptions in the supply of petrochemicals. One result of these uncertainties is that considerable research has been conducted in developing new adhesive systems from renewable resources.

Although research results have indicated that a number of new adhesive systems have promise, their commercial use is currently limited. One example is the use of isocyanate adhesives. The slow adoption of this material is due to the relatively high cost and to toxicity concerns. This material does have some definite advantages, including rapid cure at moderate temperatures, insensitivity of cure to moderately high moisture, good durability, and the absence of formaldehyde emissions.

The most common additive to lignocellulosic composite panels other than resin is wax. Even small amounts, 0.5–1%, act to retard the rate of liquid water pick up. This is important when the composite is exposed to wet environments for short periods of time. Wax addition, however, has little effect on long-term equilibrium moisture content. Flame retardants, biocides, and dimensional stabilizers are also added to panel products. They are discussed in more detail elsewhere in this chapter.

4.2 Particleboards

The wood particleboard industry grew out of a need to dispose of large quantities of sawdust, planer shavings, plywood trim, and other relatively homogeneous waste materials produced by other wood industries. Simply put, particleboard is produced by hammermilling the material into small particles, spray application of adhesive to the particles, and consolidating a loose mat of the particles into a panel product with heat and pressure (Figure 8.1). All particleboards are currently made using a dry process, where air is used to randomize and distribute the particles prior to pressing.

Figure 8.1 Particles are also used to produce composites like particleboards. They are typically produced by hammermilling.

Particleboards are often made in three layers. The faces of the boards are made up of the fines from the communition, while the core is made of the coarser material. Producing a panel this way gives better material utilization and the smooth face presents a better surface for overlaying or veneering. Particleboards are readily made from a variety of agricultural residues. Low density insulating or sound-absorbing particleboards can be made from kenaf core or jute stick. Low, medium, and high density panels can be produced with cereal straw. Rice husks are commercially manufactured into medium and high density products in the Middle East.

All other things being equal, reducing lignocellulosic materials to particles requires less energy than reducing the same material into fibers. Particleboards are generally not as strong as fiberboards, however, because the fibrous nature of lignocellulosics is not exploited as well. Particleboards find use as furniture cores, where they are often overlaid with other materials for decorative purposes. Thick particleboard can be used in flooring systems and as underlayment. Thin panels can be used as paneling substrate. Most particleboard applications are interior, and so they are usually bonded with UF, although some use of PF and MF exists for applications requiring more durability. The various steps involved in particleboard manufacturing are included below.

4.2.1 Particle Preparation

There are two basic particle types: hammermill-type particles and flake-type particles. Hammermilled particles are often roughly granular or cubic in shape, and thus have no significant length-to-width ratio. For non-woody materials, flake-type particles are the most common. Their sizes are usually in the range of 0.2–0.4 mm in thickness, 3.0–30 mm in width, and 10.0– 60.0 mm in length. Particle geometry significantly influences the board properties; the length of flake-type particles is probably most important as it influences maximum strength.

The most common type of machines used to produce flake-type particles are the cylinder type and the rotating disc type. The cylinder type has knives mounted either on the exterior of the cylinder similar to a planer or on the interior of a hollow cylinder. For the rotating disc type, the knives are mounted on the face of the disc at various angles. The knife angle and spacing influence the nature of the flake obtained.

4.2.2 Classification and Conveying of Particles

It is desirable to classify the particles before they proceed to further operations. Very small particles increase furnish surface area and thus increase resin requirements. Oversized particles can adversely affect the quality of final product because of internal flaws in the particles. While some classification is accomplished using air streams, screen classification methods are the most common. In screen classification, the particles are fed over a vibrating flat screen, or series of screens. The screens may be wire cloth, plates with holes or slots, or plates set on edge.

The two basic methods of particle conveying are mechanical and air conveying. The choice of conveying method depends upon the size of particles. In air conveying, care should be taken that the material does not pass through many fans resulting in particle size reduction. In some types of flakes, damp conditions are maintained to reduce break-up during conveying.

4.2.3 Drying

The moisture content of particles is critical during hot pressing operations. Thus, it is essential to carefully select the proper dryers and control equipment. The moisture content of the material depends on whether resin is to be added dry or in the form of a solution or emulsion. The moisture content of materials leaving the dryers is usually in the range of 4–8%. The main methods used in drying particles include rotary, disc, and suspension drying.

A rotary dryer consists of a large horizontal rotating drum that is heated either by steam or direct heat from 100–200°C. The drum is set at a slight angle, and material is fed in on the high end and discharged at the low end. The rotary movement of the drum allows movement of the material from the input to the output end.

A disc drier consists of a large vertical drum. It is equipped with a vertical shaft mounted with several horizontal discs with flaps. The particles move from the upper disc to the lower disc as drying progresses. Air is circulated from the bottom to the top. Drying time is usually from 15–45 min while the temperature is about 100°C.

A suspension drier consists of a vertical tube where the particles are introduced. The particles are kept in suspension by ascending air, resulting in rapid drying. As drying progresses, the particles leave the tube and are carried away by the air stream to be deposited as dried material. The drying temperature varies from 90° to 180°C. High flashpoint drying is similar to suspension drying. It consists of a looped length of ducting approximately 40 cm in diameter. The temperature applied is high, approximately 400°C. It may be necessary to pass the dried particles through a cooling drum to reduce the fire hazard and to bring the particles to the proper temperature for resin addition.

4.2.4 Resins and Wax Addition

Frequently used resins for particleboards include urea formaldehyde, phenol formaldehyde, and to a much lesser extent, melamine formaldehyde, as described in Section 4.1. The type and amount of resin used for particleboards depend on the type of product desired. Based on the weight of dry resin solids and ovendry weight of the particles, resin content is usually in the range of 4–15%, but most likely 6–9%. Resins are usually introduced in water solutions containing about 50– 60% solids. Besides resin, paraffin wax emulsion is added to improve moisture resistance. The amount of wax ranges from 0.3–1% based on the oven-dry weight of the particles.

4.2.5 Mat-Forming

After the particles have been prepared, they must be laid into an even and consistent mat to be pressed into a panel. This can be accomplished in a batch mode or by continuous formation. The batch system employs a caul or tray on which a deckle frame is placed. Mat formation is induced either by the back and forth movement of the tray or the back and forth movement of the hopper feeder. After formation, the mat is usually pre-pressed prior to hot-pressing. Three layer boards can also be produced in this system, in which case, three forming stations are necessary. For three layer boards, the two outer layers consist of particles differing in geometry from those of the core. The resin content of the outer layers is usually higher, about 8–15%, with the core having a resin content of about 4–8%.

In continuous mat forming systems, the particles are distributed in one or several layers on traveling cauls or on a moving belt. Mat thickness is controlled volumetrically. Like batch forming, the formed mats are usually pre-pressed, commonly with a single-opening platen press. Pre-pressing reduces the mat height and helps to consolidate the mat for pressing.

4.2.6 Hot-Pressing

After pre-pressing, the mats are hot-pressed into panels. Hot press temperatures are usually in the range of 100–140°C. Urea-based resins are usually cured between 100 and 130°C. Pressure depends on a number of factors, but is usually in the range of 14 to 35 kg/cm^2 for medium density boards. Upon entering the hot press, the mats usually have moisture content of 10–15% but are reduced to about 5–12 percent during pressing.

Alternatively, some particleboards are made by the extrusion process. In this system, formation and pressing occur in one operation. The particles are forced into a long, heated die (made of two sets of platens) by means of reciprocating pistons. The board is extruded between the platens. The particles are oriented in a plane perpendicular to the plane of the board, resulting in properties which differ from those obtained with flat-pressing.

4.2.7 Board Finishing

After pressing, the board is trimmed to bring the board to the desired length and widths, and to square the edges. Trim losses usually amount to 0.5–8%, depending on the size of the board, the process employed, and the control exercised. Trimmers usually consist of saws with tungsten carbide tips. After trimming, the boards are sanded or planed prior to packaging and shipping. The particleboards may also be veneered or overlaid with other materials to provide better surface and improve strength properties. In such products, further finishings with lacquer or paint coatings may be done, or some fire-resistant chemicals may be applied.

4.3 Fiberboards

Several things differentiate fiberboards from particleboards (Figure 8.2); the most notable of these is the physical configuration of the comminuted material. Because lignocellulosics are fibrous by nature, fiberboards exploit their inherent strength to a higher degree than particleboards. To make fibers for composite production, bonds between the fibers in the plant must be broken. In its simplest form, this is accomplished by attrition milling. Attrition milling is an age-old concept whereby material is fed between two discs, one rotating, one stationary. As the material is forced through the pre-set gap between the discs, it is sheared, cut and abraded into fibers and fiber bundles. Grain has been ground this way for centuries.

Figure 8.2 Fibers can be made from many lignocellulosics and they form the raw materials for many composites, most notably fiberboards. They are typically produced by the refining process.

Attrition milling, or as it is commonly called, refining, can be augmented by water soaking, steam cooking, or chemical treatments. By steaming the lignocellulosic, the lignin bonds between the cellulosic fibers are weakened. As a result, the fibers more readily separate, usually with less damage. Chemical treatments, usually alkali, are also used to weaken the lignin bonds. All of these treatments help increase fiber quality and reduce energy requirements, but may reduce yield as well. Refiners are available with single- or double-rotating discs, as well as steam-pressurized and unpressurized configurations. Fibers can also be produced by steam explosion. In this system, lignocellulosic material is subjected to high pressure steam for a short period of time, usually less than a minute. The pressure is then rapidly dropped. The pressure differential within the lignocellulosic explodes it into fibers and forces the fibers from the pressure vessel.

Fiberboards are normally classified by density and can be made by either dry or wet processes. Dry processes are applicable to boards with high density (hardboards) and medium density (medium density hardboard or MDF). Wet processes are applicable to high density hardboards and low density insulation boards as well. The following subsections briefly describe the manufacturing of high and medium density dry process hardboards, wet process hardboards, and wet process, low density insulation boards.

4.3.1 Dry Process Fiberboards

Dry process fiberboards are made in a similar fashion to particleboards. Resin (UF, PF) and other additives may be applied to the fibers by spraying in short retention blenders, or introduced as the wet fibers from the refiner are fed into a blow line dryer. Alternatively, some fiberboard plants add the resin in the refiner. The adhesive coated fibers are then air-laid into a mat for subsequent pressing much the same as particleboard (Section 4.2.5).

Pressing procedures for dry process fiberboards differ somewhat from particleboards. After the fiber mat is formed (Figure 8.3), it is typically prepressed in a band press. The densified mat is then trimmed by disc cutters and transferred to caul plates for the pressing operation. Dry-formed boards are pressed in multi-opening presses with temperatures of around 190–210°C (Figure 8.4). Continuous-pressing large, high pressure band presses are also gaining in popularity. Board density is a basic property and is an indicator of board quality. Moisture content greatly influences density; thus, the moisture content is constantly monitored by moisture sensors using infrared light.

Figure 8.3 A laboratory produced air-laid mat before pressing. Approximate dimensions are 600 × 600 mm × 150 mm thick. Resin was applied to the fibers before mat production. This mat will be made into a high density fiberboard approximately 3 mm thick.

4.3.2 Wet Process Hardboards

Wet process hardboards differ from dry process fiberboards in several significant ways. First, water is used as the distribution medium for the fibers to be formed into

Figure 8.4 A similar mat to that in Figure 8.3 about to enter the press.

a mat. As such, this technology is really an extension of paper manufacturing technology. Secondly, some wet process boards are made without additional binders. If the lignocellulosic contains sufficient lignin, and if the lignin is retained during the refining operation, the lignin can serve as the binder. Under heat and pressure, the lignin will flow and act as a thermosetting adhesive, enhancing the naturally occurring hydrogen bonds.

Refining is an important step for the development of strength in wet process hardboards. The refining operation must also yield a fiber of high "freeness," that is, it must be easy to remove water from the fibrous mat. The mat is typically formed on a Fourdrinier wire, like paper making, or on cylinder formers.

Wet process hardboard pressing is done in multi-opening presses heated by steam or hot water. The press cycle consists of 3 phases and lasts 6–15 min. The first phase is at high pressure and removes most of the water while bringing the board to the desired thickness. The second phase is mainly for water vapor removal. The final phase is relatively short and results in the final cure. Maximum pressures used are about 5 MPa. Heat is essential during pressing to induce fiber-to-fiber bond. High temperatures of up to 210°C are used to increase production by faster evaporation of the water. Lack of sufficient moisture removal during pressing adversely affects strength and may result in "spring back" or blistering.

4.3.3 Post Treatments of Wet and Dry Process Hardboards

Several treatments exist to increase dimensional stability and mechanical performance of hardboard. They are heat treatment, tempering, and humidification and may be done singularly, or in conjunction with each other. Heat treatment is the exposure of the pressed fiberboards to dry heat that improves the dimensional stability and mechanical properties of the boards. The process also reduces water adsorption and improves the bond between fibers.

Tempering is the heat treatment of pressed boards, preceded by the addition of oil. Tempering improves the board's surface hardness and is often done on S2S (smooth two sides) wet-formed hardboard. It also further improves resistance to abrasion, scratching, scarring, and improves the resistance to water. The most common oils used include linseed oil, tung oil, and tall oil.

Humidification is the addition of water to bring the board moisture content into equilibrium of the air. Initially, a pressed board has almost no moisture content. When it is exposed to air, it expands linearly by taking on 3–7% moisture. The most common humidifiers for this purpose are the continuous or progressive type. Air of high humidity is forced through the stacks where it provides water vapor to the boards. The entire process is controlled by a dry bulb/wet bulb controller. Other methods include spraying water on the back side of board and the application of vacuum to force the moist air through the board.

4.3.4 Insulation Boards

Insulation boards are low density, wet-laid panel products used for insulation, sound deadening, carpet underlayment, and similar applications. In the manufacture of insulation board, the need for refining and screening is a function of the raw material available, the equipment used, and the desired end product. Insulation boards typically do not use a binder, and rely on hydrogen bonds to hold the board together. Sizes are usually added to the furnish at about 1 percent to provide the finished board with a modest degree of water resistance and dimensional stability. Sizes often used include rosin, paraffin, cumarone, resin, asphalt, and asphalt emulsions.

Like wet process hardboard, insulation board manufacture is a modification of paper making. A thick fibrous sheet is made from a low consistency pulp suspension in a process known as wet felting. Felting can be accomplished through use of a deckle box, Fourdrinier, or cylinder screen. A deckle box is a bottomless frame that sets over a screen. A measured amount of stock is put in the box to form one sheet; vacuum is then applied to remove most of the water. Use of Fourdrinier screen for felting is similar to paper making, except that line speeds are reduced to 1.5–15 m/min.

Like the Fourdrinier, the cylinder method uses a screen, except with this system the screen is placed around a cylinder. The cylinder screen rotates in a pulp slurry, picking up fiber through vacuum.

Insulation boards are usually cold pressed to remove most of the free water after forming. From there, the wet mats are dried to the finished moisture content. Dryers may be a continuous tunnel, or a multi-deck arrangement. The board is generally dried in stages; temperatures employed range from 120–190°C. It takes about 2–4 h to bring the moisture content to about 1–3%.

After drying, some of the boards are treated for various applications. Boards may be given tongue and groove or shiplap edges or grooved to produce a plank effect. Some are laminated by means of asphalt to produce roof insulation boards.

In the United States, about one-third of production is treated with a sealer coat to facilitate painting.

4.3.5 Fiberboard Finishing

Trimming: Consists of reducing the products into standard sizes and shapes. Generally, double-saw trimmers are used to saw the boards. Trimmers consist of overhead mounted saws or movable saw drives. The trimmed boards are stacked in piles for future processing.

Sanding: If thickness tolerance is critical, the hardboard is sanded prior to finishing. S1S boards require this process. Sanding reduces thickness variation and improves surface paintability. In sanding, single head, wide belt sanders are used, with abrasive grits varying from 24–36.

Finishing: Finishing involves surface treatments to give the board a good appearance and improve performance. The boards are cleaned using water sprays followed by drying at about 240°C for 30 sec. The board's surfaces are then modified using paper overlay, paints, stains, or prints.

Punching: Punched boards are perforated sheets used as peg board. Most punching machines punch 3 rows of holes at a time while the board advances in position.

Embossing: Embossing consists of pressing the board with a textured form. This process results in a slightly contoured board surface that can enhance the resemblance of the board to sawed wood, weathered wood, brick, and others.

4.4 Special Purpose Conventional Composites

Special purpose composites are produced to obtain desirable properties like water resistance, mechanical strength, acidity control, and decay and insect resistance. Overlays and veneers can also be added for both structural and appearance enhancement, as in Figure 8.5.

4.4.1 Moisture Resistance Conventional Composites

Sizes cover the surface of fibers, reduce their surface energy, and render the fibers relatively hydrophobic. The application of a sizing agent can occur in two ways. In the first, water is used as a medium to assure the thorough mixing of size and fiber. The size is forced to precipitate out of the water and is fixed to the fiber surface. In the second method, the size is applied directly to the fibers. Rosin is a common sizing agent; it is obtained from living pine trees, pine stumps, and as a by-product from kraft pulping of pines. Rosin size is added in amounts of less than 3% solids based on dry fiber weight.

Waxes are high molecular weight hydrocarbons derived from crude oil. Wax sizes are used in dry fiberboard processes, and wax is added in solid form or sometimes together with liquid resin solutions. Wax sizes tend to lower strength properties to a greater extent than rosin.

Asphalt is also used to increase water resistance, especially in low-density wet process insulation boards. Asphalt is a black-brown solid or semi-solid cement

Figure 8.5 A medium density fiberboard with a veneer overlay. The edges can be shaped and finished as required by the end-product.

material which liquifies when heated. The predominant component of asphalt is bituminous. It is precipitated onto the fiber by the addition of alum.

4.4.2 Flame Retardant Conventional Composites

Lignocellulosic products are combustible and develop combustible gases that, at high temperature, can be a fire hazard. Most fire-retardant chemicals are thermally stable inorganic salts like aluminum trihydrate or borate ester. By coating the surface of the lignocellulosic, they inhibit the release of combustible gases.

4.4.3 Preservative Treated Conventional Composites

Wood is highly susceptible to attack by fungi and insects; thus, treatment is essential for maximum durability in adverse conditions. Common preservative treatments include chromated copper arsenate (CCA), creosote, and pentachlorophenol (PCP), which was recently prohibited in the United States because of its toxicity. Generally, application of 0.50–0.75 weight percent of these compounds provides adequate protection.

5. INORGANIC-BONDED COMPOSITES

Inorganic-bonded wood composites have a long and varied history being first manufactured commercially in Austria in 1914. A plethora of building materials can be made using inorganic binders and lignocellulosics, and they run the normal gamut of panel products, siding, roofing tiles and pre-cast building members, (Figure 8.6.)

Figure 8.6 A laboratory produced low-density, cement-bonded composite panel. Full scale panels such as these are used in construction.

There are three main categories of inorganic binders. They are gypsum, Magnesia cement, and Portland cement. Gypsum and magnesia cement are moisture-sensitive, and their use is generally restricted to interior applications. Portland cement-bonded composites are more durable and are used in both interior and exterior applications. Inorganic-bonded composites are made by blending proportionate amounts of the lignocellulosic with inorganic materials in the presence of water and allowing the inorganic material to cure or "set up" to make a rigid composite. All inorganic-bonded composites are very stable, highly insect and vermin resistant, and very fire resistant.

A unique feature of inorganic-bonded composites is their manufacture adaptable to either end of the cost and technology spectra. This is facilitated by the fact that no heat is required to cure the inorganic material. For example, in the Philippines, Portland cement-bonded composites are fabricated using mostly manual labor and are used in low cost housing. In Japan, the fabrication of these composites is automated, and they are used in very expensive modular housing.

The versatility of inorganic composite manufacture makes it ideally suited to a variety of lignocellulosic materials. With a very small capital investment and the most rudimentary of tools, satisfactory inorganic-bonded lignocellulosic composite building materials can be produced on a small scale using largely unskilled labor. If the market for the composite increases, technology can be introduced to increase manufacturing throughput. The labor force can be trained concurrently with the gradual introduction of the more sophisticated technology.

5.1 Gypsum Bonded Composites

Gypsum can either be derived by mining from natural sources or can be obtained as flue gas gypsum. Flue gas gypsum, now being produced in very large quantities in the United States because of Clean Air Act regulations, is the result of the introduction of lime into the combustion process to reduce sulfur dioxide emissions. By 1995, more than 100 power plants throughout the United States will be producing gypsum. Flue gas gypsum can be used in lieu of mined gypsum.

Gypsum panels are frequently used to finish interior wall and ceiling surfaces. In the United States, these products are generically called "dry wall" because they replace wet plaster systems. To increase the bending strength and stiffness of the gypsum panel, it is frequently wrapped in paper, which provides a tension surface. An alternative to wrapping the gypsum with fiber to increase strength and stiffness is to put the fiber within the panel. Several firms in the United States and Europe are doing this with recycled paper fiber. There is no reason that other lignocellulosics cannot be used. Gypsum is widely available and does not have the highly alkaline environment presented by cement. Experimentally, sisal and coir have been successfully used in gypsum panels (Mattone, 1990).

Gypsum panels are normally made from a slurry of gypsum, water, and lignocellulosic fiber. In large scale production, the slurry is extruded onto a belt. The belt carries the slurry through a drying oven to drive off the water and facilitate the cure of the gypsum. The panel is then cut to length, and trimmed if necessary.

5.2 Magnesia Cement-Bonded Composites

Magnesia cement-bonded boards have not seen the high level of commercial activity that cement-bonded or gypsum-bonded panels have, mainly due to price. Magnesia cements do, however, offer several manufacturing advantages over Portland cement. First, the various sugars in lignocellulosics do not seem to have as much effect on the curing and bonding of the binder. Second, Magnesia cements are more tolerant of high water content during production (Pazner and Klemarevski, 1989). This opens up possibilities to use lignocellulosics not amenable to Portland cement composites, without leaching or other modification, and opens up alternative manufacturing processes, and thus, products. In addition, while Magnesia cement-bonded composites are considered water-sensitive, they are much less so than gypsum-bonded composites.

One successful application of Magnesia cements is a low density panel made for interior ceiling and wall applications. In the production of this panel product, wood wool, or excelsior, is laid out in a low density mat. The mat is then sprayed with a aqueous solution of Magnesia cement, then the mat is pressed and cut into panels. It is easy to envision this technique being used for lignocellulosics with long fibers, like jute, hemp, or kenaf.

In Finland, Magnesia cement-bonded particleboard is manufactured using a converted conventional particleboard plant. The Magnesia oxide is applied to the lignocellulosic particles in a batch blender along with water and other chemicals. Depending on application and other factors, the boards may be cold- or hot-pressed (Loiri, 1989).

Other manufacturing processes have been suggested. One possible application may be to spray a slurry of Magnesia cement; water, and lignocellulosic fiber onto existing structures as fireproofing. Extrusion into pipes or other profiles is also possible.

5.3 Portland Cement-Bonded Composites

The most apparent and widely used example of inorganic-bonded composites is cement. Portland cement, when combined with water, immediately reacts in a process called hydration to eventually solidify into a solid stonelike mass. Successfully marketed Portland cement-bonded composites consist of both low density products made with excelsior or coir and high density products made with particles and fibers.

The low density products may be used as interior ceiling and wall panels in commercial buildings. Along with the previously mentioned advantages, they offer sound control and can be quite decorative. In some parts of the world, these panels function as complete wall and roof decking systems. The exterior of the panels is then stuccoed, while the interior is plastered. High density panels can be used as flooring, roof sheathing, fire doors, load bearing walls, and cement forms. Fairly complex molded shapes can be molded or extruded. Thus, decorative roofing tiles or non-pressure pipes can be made.

While the entire sphere of inorganic-bonded lignocellulosic composites is attractive, with cement-bonded ones being especially so, there are limitations and trade-offs when cement is considered. Marked embrittlement of the lignocellulosic component is known to occur and to be caused directly by the alkaline environment provided by the cement matrix. In addition, hemicellulose, starch, sugar, tannins, and lignin, all to a varying degree, affect the cure rate and ultimate strength of the composites. To make strong and durable composites, measures must be taken to ensure long-term stability of the lignocellulosic in the cement matrix.

To overcome these problems, various schemes have been developed. The most common is leaching, whereby the lignocellulosic is soaked in water for a day or two to extract some of the detrimental components. However, in some parts of the world, the water containing the leachate is difficult to dispose. Low water–cement ratios are helpful, as are the use of set accelerators like calcium carbonate. Conversely, low alkali cements have been developed, but they are not readily available in all parts of the world. Two other strategies—natural pozzolans and carbon dioxide treatment are discussed below.

5.3.1 Role of Natural Pozzolans

Pozzolans are defined as siliceous or siliceous and aluminous materials that can react chemically with calcium hydroxide (lime) at normal temperatures, in the presence of water to form cement compounds (ASTM, 1988). Some common pozzolanic materials include volcanic ash, fly ash, rice husk ash, and condensed silica fume. All of these can react with lime at normal temperatures to make a water resistant, natural cement.

As a general statement, when blended with Portland cement, pozzolans increase the strength of the cement, but slow the cure time. More important is that pozzolans decrease Portland cement alkalinity (Swamy, 1990), which indicates that addition of lignocellulosic-based material (rice husk ash) to cement-bonded lignocellulosic composites may be advantageous.

5.3.2 Carbon Dioxide Treatment of Portland Cement-Bonded Composites

In the manufacture of a cement-bonded wood fiber composite, the cement hydration process normally requires from 8 to 24 hours to develop sufficient board strength and cohesiveness to permit the release of consolidation pressure. By exposing the cement to CO_2, the initial hardening stage can be reduced to under 5 minutes. The phenomenon results from the chemical reaction of CO_2 with calcium hydroxide to form calcium carbonate and water.

Reduction of initial cure time of the cement-bonded wood fiber composite is not the only advantage of using CO_2 injection. The inhibiting effects of sugars, tannins, etc. on cement hydration are also greatly reduced which is especially important with the variety of lignocellulosics available. In addition, research has demonstrated that CO_2-treated composites can be twice as stiff and strong as untreated composites (Geimer et al., 1993). Finally, CO_2-treated composites do not experience efflorescence, a condition whereby calcium hydroxide migrates to the surface of the material, so the appearance of the final product is not changed.

6. LIGNOCELLULOSIC/THERMOPLASTIC COMPOSITES

As described elsewhere in this chapter, there is a long history of the use of lignocellulosics with thermosetting polymeric materials, like phenol or urea formaldehyde in the production of composites. The use of lignocellulosics with thermoplastics, however, is a more recent innovation (see Chapter 12). Broadly defined, a thermoplastic is any material that softens when heated and hardens when cooled. Thermoplastics selected for use with lignocellulosics must melt at or below the degradation point of the lignocellulosic component, normally 200–220°C. This group includes polypropylene, polystyrene, vinyls, both low- and high-density polyethylenes, and others.

There are two main strategies for the use of thermoplastics in lignocellulosic composites. In the first, the lignocellulosic component serves as a reinforcing filler in a continuous thermoplastic matrix. In the second, the thermoplastic serves as a binder to the majority lignocellulosic component.

In an ideal system with regular and oriented filler material, the point where the thermoplastic matrix ceases to be continuous may be as low as 20% matrix material.

From a practical view, the point where the thermoplastic phase ceases to be continuous in a lignocellulosic composite is thought to be around 35–40%. This is because of the random and irregular nature of the lignocellulosic component.

The presence or absence of a continuous thermoplastic matrix may also determine the processability of the composite material. As a general statement, if a continuous matrix exists, it may be possible to process the composite using conventional thermoplastic processing equipment; however, if no continuous matrix exists, other processes may be required. For the purposes of discussion, we will present these two scenarios as high thermoplastic content composites and low thermoplastic content composites.

6.1 High Thermoplastic Content Composites

High thermoplastic content composites are those in which the thermoplastic component exists in a continuous matrix and the lignocellulosic component serves as a reinforcing filler (Figure 8.7). The great majority of reinforced thermoplastic composites available commercially use inorganic materials as their reinforcing fillers, e.g., glass, clays, and minerals. These materials are heavy, abrasive to processing equipment, and non-renewable. Lignocellulosics are lighter, much less abrasive, and of course, renewable. As a reinforcement, lignocellulosics stiffen, and somewhat strengthen the thermoplastic when compared to unfilled material. As a filler material, the lignocellulosic can usually be prepared for composite production for one third to one half the cost of the thermoplastic.

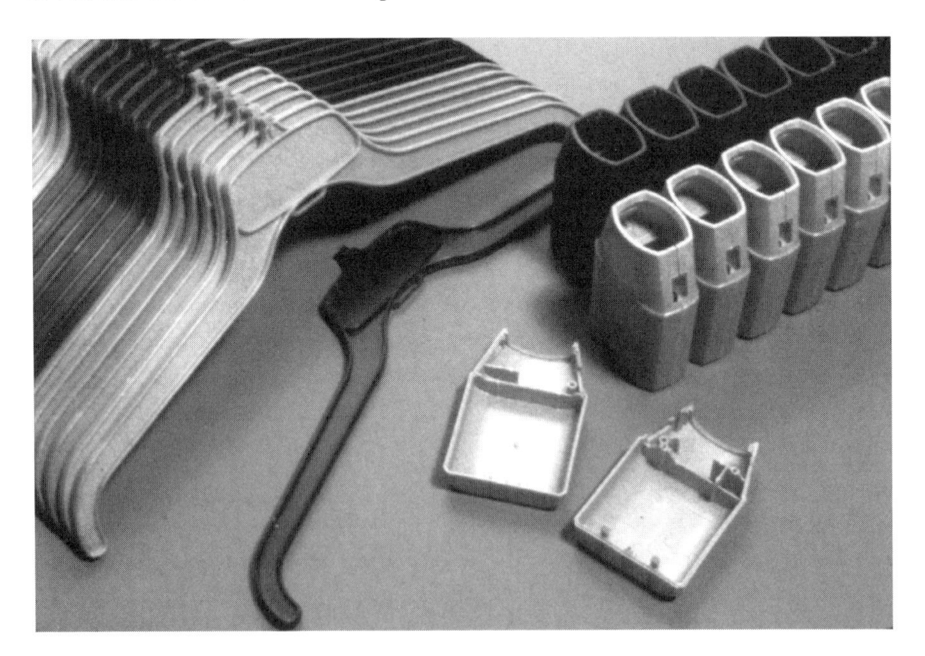

Figure 8.7 Using lignocellulosics as reinforcing fillers in thermoplastics allows them to be molded into a wide variety of shapes and forms.

These composites are not without trade-offs. Impact resistance of the composite decreases significantly when compared to the unfilled thermoplastic, but for most applications, this is not significant. By point of comparison though, the impact resistance is often better than similar composites made with thermosetting resins. The composite is also somewhat more sensitive to moisture than the unfilled material or an inorganic-filled composite. From a practical standpoint, though, the temperature sensitivity of the composite, because of the thermoplastic component, is usually more significant than any change in properties brought about by moisture absorption.

The lignocellulosic may be in fiber or particle form. Particles are easier to prepare and are usually used more as a filler, although they do have a reinforcing effect. Fibers are used more as a reinforcement, although their relatively low cost (when compared to the thermoplastic) still makes them viable fillers. The lignocellulosic component size can also vary widely. One commercial product uses rather large wood filler in the 20 mesh size range. Conversely, the lignocellulosic component might exist as individual fibers or fiber fragments.

6.1.1 Compounding

Perhaps the greatest challenge to continued commercialization of high thermoplastic content lignocellulosic composites is their compounding, or blending. Compounding consists of the feeding and dispersion of the lignocellulosic component throughout the thermoplastic matrix. The ultimate goal of any compounding operation is usually the production of a compounded, pelletized feedstock. The feed pellets are suitable for use in almost any plastic processing operation and this is the normal form in which they are sold. Problems in this area focus on the differences in bulk density of the two components and the degree of shear of the compounding equipment, and thus fiber length retention or loss.

Thermoplastics in pellet form have a bulk density in the range of 500–600 kg/m^3. Lignocellulosics have an uncompacted bulk density of 50–250 kg/m^3. Fibers are at the low end of the lignocellulosic bulk density continuum; particles are at the high end. Dry blending of the thermoplastic and the lignocellulosic generally results in settling of the heavier component. This problem is exasperated if the thermoplastic is in molten form, at which time its bulk density or specific gravity will be closer to 900–1000 kg/m^3.

Bulk density issues are simplified by selecting a lignocellulosic component with a bulk density as close to the thermoplastic component as possible. This usually involves selecting a particle or flour as opposed to a fiber. Wood flour is commercially used in the United States, and feeding problems are minimal. The extension of the wood flour feeding to ground nut shell flours and other particulate materials like finely hammermilled rice hulls would seem reasonable. Particles do not give the level of reinforcement as that obtained from fibers; however, additional measures must be taken for fiber handling.

The feeding of fibers can be made easier if batch-style, kneading-type compounding equipment is selected. Kneading-type compounding equipment is used by the chemical, pharmaceutical, cosmetics, foodstuffs, rubber and plastics industries as well as many other special fields for compounding a diverse range of materials.

The operation of this type of equipment is quite simple. Pre-weighed amounts of various materials are loaded into a heated mixing chamber. A ram feeder facilitates the feeding of low bulk density materials like lignocellulosics. In the chamber are two low-speed, high-torque kneading rotors. The chamber is closed, the rotors turn, and the mixing is conducted. After mixing, the compounded material is discharged from the chamber en masse. The compounded mass is usually fed through a single screw extruder and pelletizing line for production of a pellet. The three compounding variables controllable by this type of equipment are mixing time, batch temperature, and energy consumption.

Batch-style, kneading-type compounding equipment offers several advantages. This equipment can accommodate a wide range of feedstocks and work with extremely high viscosity materials. Special ram feeders can accommodate very low bulk density materials. Relatively low shear forces help retain fiber length, and importantly, excellent quality control is reached because the formulation components of each batch can be weighed individually.

On the down side, batch-style equipment has lower output than continuous compounders at the same power rating and the en masse discharge requires expensive downstream processing equipment for practical material utilization by pelletization. Also, the addition of time sensitive components to the formulation is not feasible due to the batch nature of the compounding.

Instead of discharging en masse, a discharge screw can be added to empty the mixing chamber. This enables the material to be extruded into a sheet or strand that can be fed into a pelletizer for subsequent operations. This simplifies, but does not eliminate, the downstream equipment needed for further processing.

Continuous kneading mixers are also available; they differ substantially from batch-style equipment and can be used to some advantage. In operation, this equipment generally consists of two long, intermeshing rotors in a heated barrel. Control of the formulation depends on accurate feeding of the raw materials. In operation, materials are fed into one end of the barrels either through gravity or ram feeders. The action of the rotors forces the material through the mixing chamber where it is blended and subsequently discharged through a sheet or strand die, usually directly into a pelletizer. As compared to the batch-style equipment, some advantages and disadvantages become apparent.

First, output is increased over batch-style machines of the same power rating. Second, the discharged material can be fed directly into a pelletizer. In addition, time-sensitive components, like thermosetting crosslinking agents, can be introduced at various stages of the compounding operation. Shear is relatively moderate, so fiber length retention is generally good. Disadvantages include a degree of loss of formulation control because the formulation is dependent on the ability to consistently feed the components. These types of compounders also do not function as pumps; therefore, they cannot be used to extrude or mold finished products.

Both batch and continuous kneading compounders need downstream processing equipment to make a product. A more sophisticated and more expensive compounder is a twin screw extruder. Similar in operation to a continuous kneader, twin screw extruders have two screws, or augers running in a common barrel. Materials can be fed into the extruder at various points to accommodate various compounding schemes. Special ram feeders, or crammers, are needed for lignocellulosic materials. Twin screws act as pumps, and are often used in profile extrusion. They can also be used for injection molding.

If fiber length retention is not a concern, other high shear compounding equipment may be used. Two batch-style machines, a thermokinetic mixer known as a K-mixer (Myers and Clemons, 1993) and film densification equipment (English and Schneider, 1994), can be used. Both of these machines use kinetic energy from a high speed rotor enclosed in a chamber to melt the thermoplastic component and blend in the lignocellulosic component.

6.1.2 Profile Extrusion

After the materials have been compounded, they can be made into products. One process to accomplish this is profile extrusion where the composite material is first heated so that the thermoplastic component can flow. It is then continually pumped and forced through a die of a given cross section configuration. The material is supported as it cools, normally in a cold water bath, and then cut to length. As mentioned above, twin screw extruders can be used as compounders and as pumps for profile extrusion. Many products are made this way; common ones include pipe and tubing, furniture edging and moldings, and sheet goods.

6.1.3 Injection Molding

Injection molding differs from profile extrusion in that after the material is heated, it is pumped into a permanent mold, where it takes shape and cools. The mold is then opened, and the finished part discharged. Injection-molded parts range from buttons to computer cases to automotive components.

Currently, the primary application of high thermoplastic content lignocellulosic composites is for interior door panels and trunk liners in automobiles. Some producers of plastic lumber are also adding lignocellulosic fiber to their product to increase stiffness and reduce creep. Additional large-volume, low-to-moderate cost applications are expected in areas such as packaging (trays, cartons, pallets), interior building panels, and door skins.

6.2 Low Thermoplastic Content Composites

Low thermoplastic content composites can be made in a variety of ways. In their simplest form, the thermoplastic component acts much the same way as a thermosetting resin, that is, as a binder to the lignocellulosic component. An alternative way is to use the thermoplastic in the form of a textile fiber. The thermoplastic textile fiber enables a variety of lignocellulosics to be incorporated into a low-density, nonwoven textile-like mat. The mat may be a product in itself, or it may be consolidated into a high density product.

6.2.1 Conventional-Type Low Thermoplastic Content Composites

Experimentally, low thermoplastic content composites have been made that are very similar to conventional lignocellulosic composites in many performance characteristics (Youngquist et al., 1993b). In their simplest form, lignocellulosic particles or fibers can be dry-blended with thermoplastic granules, flakes or fibers and pressed into panel products. Because the thermoplastic component remains molten when hot, different pressing strategies must be used than when using thermosetting binders.

Two options have been developed to accommodate these types of composites. In the first, the material is placed in the hot press at ambient temperature. The press then closes and consolidates the material, and heat is transferred through conduction to melt the thermoplastic component, which flows around the lignocellulosic component. The press is then cooled, "freezing" the thermoplastic so that the composite can be removed from the press.

Alternatively, the material can be first heated in an oven or hot press. The hot material is then transferred to a cool press where it is quickly consolidated and cooled to make a rigid panel. Some commercial non-structural lignocellulosic/thermoplastic composites are made this way.

6.2.2 Nonwoven Textile-Type Composites

In contrast to high-thermoplastic content composites and conventional low-thermoplastic content composites, nonwoven textile type composites typically require long fibrous materials for their manufacture. These fibers might be treated jute or kenaf, but more typically are synthetic thermoplastic materials. Nonwoven processes allow and tolerate a wider range of lignocellulosic materials and synthetic fibers dependent on application. After the fibers are dry blended, they are air-laid into a continuous, loosely consolidated mat. The mat then passes through a secondary operation in which the fibers are mechanically entangled or otherwise bonded together. This low density mat may be a product in itself, or the mat may be shaped and densified in a thermoforming step.

If left as a low density mat and used without significant modification by post processing these mats have a bulk density of 50–250 kg per cubic meter. These products are particularly well-known in the consumer products industry, where nonwoven technology is used to make a variety of absorbent personal care products, wipes, and other disposable items. These products use high quality pulps in conjunction with additives to increase their absorptive properties. Other applications, as described below, can use a much wider variety of lignocellulosics.

One interesting application for low-density, nonwoven mats is for mulch around newly planted seedlings. The mats provide the benefits of natural mulch; in addition, controlled-release fertilizers, repellents, insecticides, and herbicides can be added to the mats as discussed in Chapter 7. Research results on the combination of mulch and pesticides in agronomic crops have been promising (Crutchfield et al., 1985). The addition of such chemicals could be based on silvicultural prescriptions to ensure seedling survival and early development on planting sites where severe nutritional deficiencies, animal damage, insect attack, and weed problems are anticipated.

Low-density nonwoven mats can also be used to replace dirt or sod for grass seeding around new homesites or along highway embankments. The grass seed can be incorporated directly into the mat. These mats promote seed germination and good moisture retention. Low-density mats can also be used for filters. The density of the mats can be varied, depending on the material being filtered and the volume of material that passes through the mat per unit of time.

High density fiber mats can be defined as composites made using the aforementioned nonwoven mat process which are post formed into rigid shapes by heat and pressure. To insure good bonding, the lignocellulosic can be precoated with a thermal setting resin, e.g., phenol–formaldehyde, or it can be blended with synthetic fibers, thermal plastic granules, or any combination of these. These products typically have a specific gravity of 0.60–1.40. After thermoforming, the products possess good temperature resistance. Because longer fibers are used, these products exhibit better mechanical properties than those obtained with high thermoplastic content composites; however, high lignocellulosic content leads to increased moisture sensitivity.

7. CONCLUSIONS

Agricultural fibers have been successfully utilized in a variety of composite panels, most notably conventional composite panels and inorganic-bonded composites. Lignocellulosic/thermoplastic composites are a newer area of lignocellulosic utilization. It is anticipated that interest and commercial development will continue in this area. More than enough agricultural fiber residues are available to support composite manufacturing needs, although the agro-based materials may not have a suitable geographical distribution to provide an economically feasible endeavor.

Lignocellulosics are attractive material sources for composites because they are lightweight, economical, and require low amounts of energy for processing. In addition, their growth, use, and disposal are generally considered environmentally friendly. As renewable materials, they can be used to replace or extend non-renewable materials such as those based on petroleum.

REFERENCES

ASTM, American Society of Testing Materials, Concrete and mineral aggregates, *1988 Annual Book of ASTM Standards*, Sec. 4, Vol. 4.02, 4.03, ASTM, U.S., 1988.

Atchison, J. E. and Lengel, D. E., Rapid growth in the use of bagasse as a raw material for reconstituted panel board, *Proceedings, Nineteenth Particleboard Symposium*, Washington State University, Pullman, WA, 145, 1985.

Brooks, H., Brooks Associates, Ltd., Rochester Hills, MI, Personal communication, 1992.

Chapman, A. W., Purchasing, handling, and storing of bagasse (for insulating board manufacture), *Proceedings, FAO UN Conference on Pulp and Paper Prospects in Latin America*, United Nations, NY, 335, 1956.

Chow, P., Dry formed composite board from selected agricultural residues, *World Consultation on Wood Based Panels. Food and Agriculture Organization of the United Nations*, New Delhi, India, 1974.

Crutchfield, D. A., Wicks, G. A., and Burnside, O. C., Effect of winter wheat (*Triticum aestivum*) straw mulch level on weed control, *Weed Sci.*, 34(1), 110, 1985.

English, B. E. and Schneider, J. S., Paper fiber/low-density polyethylene composites from recycled paper mill waste: Preliminary Results, in *Proceedings, Recycling Conference*, May 15–18, Boston, MA, 1994.

Galassco, R., Mansion Industries, Inc., City of Industry, CA, Personal communication, 1992.

Geimer, R. L., Souza, M. R., Moslemi, A. A., and Simatupang, M. H., Carbon dioxide application for rapid production of cement particleboard, in *Proceedings of Inorganic-bonded Wood and Fiber Composites*, 2nd Intl. Conference, Moscow, ID, Forest Products Research Society, Madison, WI., 31, 1993.

Haislip, J., Dry Creek Trading, Inc., Elk Grove, CA, Personal communication, 1994.

Hamid, S. H., Maadhah, A. G., and Usmani, A.M., Bagasse-based building materials, *Polymer-Plastics Technology and Engineering*, 21(2), 173, 1983.

Knowles, L., A better alternative: Boise gold, *Timberline Tribune*, 1(6), 1992.

LaMahieu, P. J., Oplinger, E. S., and Putman, D. H., *Alternative Field Crops Manual: Kenaf*, University of Wisconsin-Madison Extension Publication, Madison, WI, 1991.

Loiri, V., Plant experience in the manufacture of magnesite and slag particleboards, in *Proceedings, Fiber and Particleboards Bonded with Inorganic Binders*, Forest Products Research Society, Moscow, ID, 1989.

Loken, S., Spurling, W., and Price, C., *Guide to resource efficient building elements*, Center for Resourceful Building Technology, Missoula, MT, 1991.

Maloney, T. M., *Modern Particleboard and Dry-Process Fiberboard Manufacturing*, Miller Freeman Publications, San Francisco, CA, 1989.

Mattone, R., Comparison between gypsum panels reinforced with vegetable fibers: Their behavior in bending and under impact, *Proceedings, 2nd Int. Rilem Symposium: Vegetable Plants and their Fibres as Building Material,* Salvador, Bahia, Brazil, 1990.

Mobarak, F. and Nada, A.M., Fiberboard from exotic raw materials–2. Hardboard from undebarked cotton stalks, *Journal of Applied Chemistry and Biotechnology*, 25(9), 659, 1975.

Myers, G. E. and Clemons, C. M., *Solid waste reduction demonstration grant program, Project No. 91-5*, Sponsored by the Wisconsin Department of Natural Resources, 1993.

Paszner, L. and Klemarevski, A., New developments in wood bonding with magnesium oxyphosphate cement, in *Proceedings, Fiber and Particleboards Bonded with Inorganic Binders*, Forest Products Research Society, Moscow, ID, 1989.

Rexen, F., Straw as an industrial material, in *Proceedings, Sol. Energy Agric., Jt. Conf.*, Intl. Sol. Energy Soc, London, England, 38, 1977.

Riccio, F., Danforth International Trade Associates, Inc., Point Pleasant, NJ, Personal communication, 1992.

Sandwell and Associates, *Kenaf assessment study, Report 261323/2*. Tallahatchie County Board of Supervisors, Charleston, MS, 1991.

Savastano, H., The use of coir fibres as reinforcement to Portland cement mortars, *Proceedings, 2nd Int. Rilem Symposium: Vegetable Plants and Their Fibres as Building Material*, Salvador, Bahia, Brazil, 1990.

Sing, S. M., Physico-chemical properties of agricultural residues and strengths of Portland cement-bound wood products, *Research and Industry*, 24, 1, 1979.

Suchsland, O. and Woodson, G. E., Fiberboard manufacturing practices in the United States, *USDA Forest Service Agricultural Handbook No. 640,* USDA, Washington, DC, 1986.

Sugartree, *Promotional literature*, The Sugar Tree Group, Boulder, CO, 1992.

Swamy, R. N., Vegetable fibre reinforced cement composites–a false dream or a potential reality? in *Proceedings, 2nd Int. Rilem Symposium: Vegetable Plants and Their Fibres as Building Material,* Salvador, Bahia, Brazil, 1990.

UNIDO, Review of agricultural residues utilization for production of panels, *Food and Agricultural Organization of the United Nations. United Nations Industrial Development Organization. FO/WCWBP/75, Document No. 127.* New York, 1975.

USDA, Potential new crop: Kenaf, commercial fiber and pulp source, *QB92-54 Quick Bibliography Series*, United States Department of Agriculture, National Agricultural Library, Washington, D.C., 1992.

Vaagen, W., Vaagen Bros. Lumber, Inc., Colville, WA, Personal communication, 1991.

Vasishth, R. C. and Chandramouli, P., New panel boards from rice husks and other agricultural by-products, *World Consultation on Wood Based Panels,* Food and Agriculture Organization of the United Nations. FO/WCWB/75, Doc. No. 30, New York, 1974.

Youngquist, J. A., English, B. E., Spelter, H., and Chow, P., Agricultural fibers in composition panels, *Proceedings 27th International Particleboard/Composite Materials Symposium*, Washington State University, Pullman WA, 1993(a).

Youngquist, J. A., Myers, G. E., Muehl, J., Krzysik, A., and Clemons, C. C., Composites from recycled wood and plastics, *Final report prepared for U.S. Environmental Protection Agency*, Cincinnati, OH, 1993(b).

CHAPTER 9

Properties of Composite Panels

John A. Youngquist, Andrzej M. Krzysik, Poo Chow, and Roger Meimban

CONTENTS

1. INTRODUCTION

The opportunities offered by lignocellulosic composites—such as optimized performance, minimized weight and volume, cost effectiveness, fatigue and chemical resistance, and resistance to biodegradation—are available to virtually every manufacturer. Researchers in the area of materials technology are showing increased interest in the benefits of composite technology. The objective of composite technology is to produce a product with performance characteristics that combine the beneficial aspects of each constituent. New composites are produced with an aim to either reduce the cost of production or to improve performance, or both.

Standards for composite panels are essential for product acceptance in major markets because they give distributors and wood users some assurance that the products possess minimum specific quality levels. Standards which first were recognized for their value as mass production techniques came into common use and the industrial revolution accelerated. At least three standards organizations have a major influence on the quality of composite panels manufactured for most domestic U.S. markets and many foreign markets (Carll, 1982). The American Society for

Testing and Materials (ASTM) and the American National Standards Institute (ANSI) are the organizations currently most active in developing voluntary standards in the United States; ANSI is the U.S. representative regarding standards with the International Standards Organization (ISO).

Each country generally has developed standards for the production and/or use of various panel products. Because of the complexity of this subject on a worldwide basis, we have chosen to discuss the standards and test methods for composite panels produced in the United States.

By way of a brief historical review, it is interesting to note that the first insulating board plant in the United States that used agricultural byproducts, such as bagasse or sugarcane, wheat straw, and corn stalk, was established at Marrero, Louisiana in 1920 (Youngquist et al., 1993). The first hardboard plant was built in the late 1920s. Particleboard was first developed in Germany during World War II and was introduced into the United States in the early 1950s.

This chapter addresses the physical and mechanical properties of composite board products made from wood- and agro-based lignocellulosic materials. To evaluate the performance of various panel products, we first review the classification of the major types of composite panels and then describe the test methods generally used to determine the behavior of these composites. Then, we discuss property requirements of various panel-type composite products. We conclude with a description of properties of composite board made from various types of lignocellulosic-based particles and fibers.

2. EXISTING WOOD COMPOSITE PANELS

2.1 Background Information

Because wood properties vary among species, between trees of the same species, and between pieces from the same tree, wood in the solid form cannot match wood in the comminuted and reconstituted form in providing materials with a wide range of properties that can be controlled in processing. When processing variables are properly selected, the end result can sometimes surpass nature's best effort. Material science normally deals with the influence of changes on properties at the molecular level. With reconstituted wood materials, the level at which change is produced is the fiber, particle, or flake. Changes in properties are created partially by combining, reorganizing, or stratifying these elements. Although the molecular structure of chemically changing the wood element itself, addition of chemicals to improve the product's resistance to decay and insect attack, and treatments for fire retardancy may also be made. Application of these treatments is relatively easy when wood is in the comminuted form, before it is converted into the final product configuration, whereas it is quite difficult in solid wood because the treating chemicals do not diffuse easily throughout solid wood.

The basic element for reconstituted wood materials may be the fiber, as it is in paper, but it can also be larger wood particles composed of many fibers and varying in size and geometry. These characteristics, along with control of their variations, provide the chief means by which materials can be fabricated with predetermined properties.

2.2 Definition of Wood-Based Composites

Marra (Marra, 1972) discussed a number of wood elements and developed a nonperiodic table of wood elements (Figure 9.1). These elements range from logs to lumber to thin lumber or thick veneer, and down to paper, fibers, wood flour and cellulose.

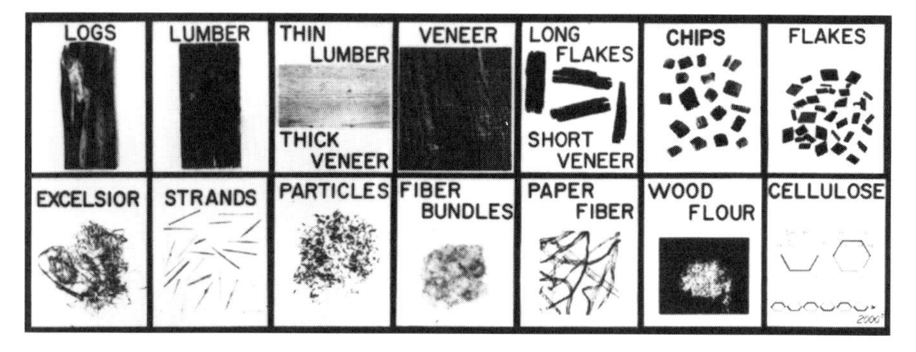

Figure 9.1 Basic wood elements from largest to smallest. Marra, G. G., *Forest Prod. J.*, 22(9), 43, 1972.

This table provides an overall view of the many types of wood components or elements that can be used to produce a wood-based composite product.

Currently, the term *composite* is being used to describe any wood material glued together. This product mix ranges from fiberboards to laminated beams and components. A logical basis (Maloney, 1986) for classifying wood composites has been suggested as follows:

The Family of Composite Materials

Veneer-Based Materials	**Laminates**
• Plywood	• Laminated Beams
• Laminated Veneer Lumber (LVL)	• Overlayed Materials
• Parallel-Laminated Veneer (PLV)	• (panels or shaped materials combined with non-wood materials such as metal, plastic, and fiberglass)
Composite Materials	
• Fiberboard	• **Edge-Glued Material**
• Medium Density Fiberboard (MDF)	• Lumber panels
• Particleboard	•
• Waferboard	•
• Oriented Waferboard (OWB)	• **Components**
• Oriented Strandboard (OSB)	• I-Beams
• Comply®[1] PLY	• T-Beam Panels
	Stress-Skin Panels

[1] A registered trademark of APA-Engineered Wood Assocation, Tacoma, WA.

The above noted composite materials fill a number of non-structural and structural applications in product lines ranging from panels for interior covering purposes to panels for exterior uses to applications in furniture and in support structures in many different types of buildings. This chapter will concentrate on those products that can be made from either wood or other agro-based particle or fiber resources, and which fall into Maloney's subcategory termed composite materials.

Figure 9.2 provides a useful way to further classify composite materials, presents a very good overview of the types of products that are discussed in this chapter, and provides a quick reference to how these composite materials compare to solid wood and plywood from a density and general processing standpoint.

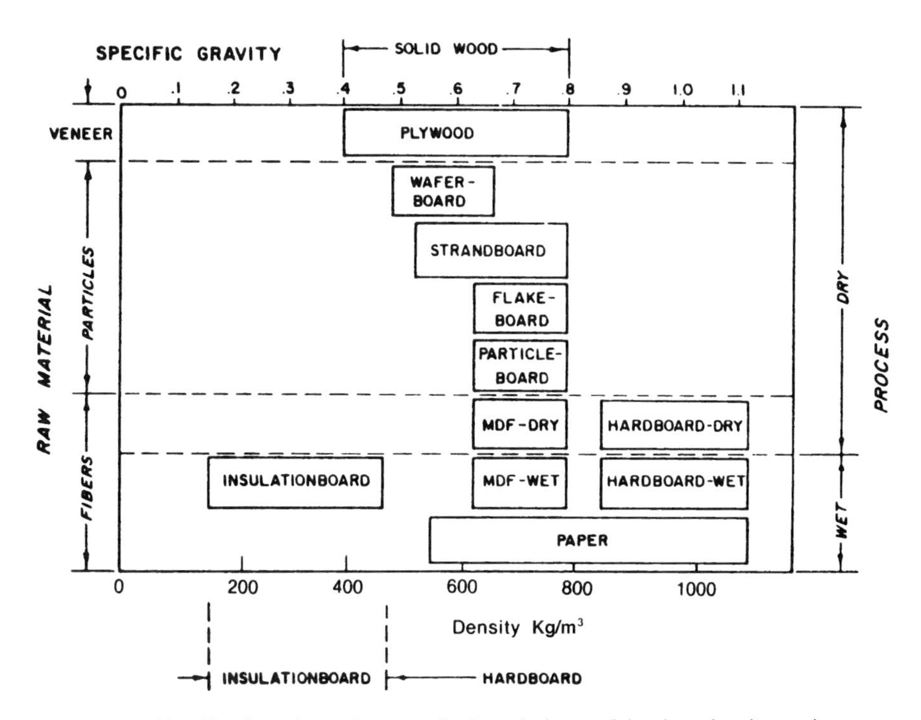

Figure 9.2 Classification of wood composite boards by particle size, density, and process type. Suchsland, D. and Woodson, G. E., Fiberboard manufacturing practices in the United States, *Agric. Handbook*, 640, USDA, Washington, D.C., 1986.

This shows the raw material classifications of fibers, particles, and veneers on the left-hand y-axis and shows the specific gravity on the x-axis. The right-hand y-axis describes in general terms the processing that takes place to produce a particular product and is classified either as wet or dry processed materials. Note that wet-processed materials are those usually dealt with in fiber form and may include up to 1% adhesive and a small added component of wax. The dry process includes roundwood or wood which is a waste product in a lumber or planing operation. This material is fiberized and dried; adhesive is added in a separate operation, and then the particles are hot-pressed into final configuration. Figure 9.3 provides examples of some of the composite materials that are represented in the schematic format in Figure 9.2.

Figure 9.3 Examples of various composite products.

3. CLASSIFICATION OF COMPOSITE PANELS

The major types of composite panels are generally categorized either by the size of the material from which they are made or by a term that describes the broad end-use of the product. This section describes and explains various composite panels which can easily be produced from various lignocellulosic resources.

3.1 Fiberboard

Lignocellulosic materials are first reduced to fibers or fiber bundles and then put back together by special forms of manufacture into fiberboard panels. Fiberboard is broadly classified into three groups: insulating board, medium density fiberboard, and hardboard. Insulating board is referred to as cellulosic fiber insulating board in ASTM C208 (ASTM, 1994d) and as cellulosic fiberboard in ASTM D1554 (ASTM, 1994e) and ANSI standard ANSI/AHA A194.1 (AHA, 1985). The range of uses and specially developed products within these broad classifications require further division of the products, as shown in Table 9.1.

Table 9.1 Classification of Fiberboard Panel Products

Board type	Density	
	(kg/m³)	(lb/ft³)
Insulation board	160–500	10–31.2
Medium density fiberboard	64–800	40–50
Medium density hardboard	500–800	31.2–50
Hardboard	500–1,450	31.2–90
High density hardboard	800–1,280	50–80

3.1.1 Insulating Board

Insulating board is a generic term for a homogeneous panel that is made from interfelted lignocellulosic fibers and that has been consolidated under heat to a density range between 160 and 500 kg/m³. The many different types, names, and uses of these boards are given in Table 9.2..

3.1.2 Medium Density Fiberboard

Medium density fiberboard (MDF) is made from lignocellulosic fibers combined with a synthetic resin. The dry-process technology utilized to manufacture MDF is a combination of that used in the particleboard industry and that used in the hardboard industry. There are three density levels for MDF:

Low <640 kg/m³
Medium 640–800 kg/m³
High >800 kg/m³

3.1.3 Hardboard

Hardboard is a generic term for a homogeneous panel that is made from interfelted lignocellulosic fibers and has been consolidated under heat and pressure to a density of 500 kg/m^3 or more.

3.2 Particleboard

Particleboard panel products typically are made from small lignocellulosic particles and flakes that are bonded together with a synthetic adhesive under heat and pressure. The density levels for particleboard are the same as those for MDF.

3.3 Mineral-Bonded Panels

In mineral-bonded panels, lignocellulosic fibers are mixed with inorganic binders like magnesium oxysulphate, magnesite gypsum, or Portland cement. The panels range in density from 290–1,250 kg/m^3. Agro-fibers can be blended with cement, formed into mats, and pressed to a density of 460–640 kg/m^3 to form a panel product4.

Table 9.2 Types, Grades, and Intended Uses for Cellulosic Fiber Insulating Board[a]

Types	Grades	Name	Intended Use
I	—	Sound-deadening board	In wall assemblies to control sound transmission
II	—	Roof insulating board	
	1	Roof insulating board	Under built-up roof systems
	2	Roof insulating board	Under single-ply roof systems
III	—	Ceiling tiles and panels	
	1	Non-acoustical uses	Decorative wall and ceiling coverings
	2	Acoustical uses	Decorative sound-absorbing wall and ceiling coverings
IV	—	Wall sheathing	
	1	Regular	Wall sheathing in frame construction
	2	Structural	Wall sheathing in frame construction for exterior wall bracing
V	—	Backer board	Behind exterior finish in wall assemblies—no structural requirement
VI	—	Roof deck	Roof decking for flat, pitched, or shed-type, open-beamed, ceiling-roof construction

[a] ASTM C208-94 (ASTM, 1994d).
ASTM D1554-94 (ASTM, 1994e).

STANDARDS AND TEST METHODS FOR COMPOSITE PANELS

Standards for composite panel products are voluntary in the United States. However, certification of conformance with a standard is advantageous to a product manufacturer from the standpoint of marketing and product conformance. Standards also permit ready identification of product quality and suitability and protect producers and distributors from cost-cutting competitors. Because the construction market in the United States is so important for composite panel producers, building code approval is a significant marketing consideration: reference to a standard in the building code requires the use of products manufactured under that standard. In addition, building codes usually demand conformance of composite panel products to a specific standard. Commodity standards, frequently referred to as product standards, can be classified further as manufacturing method standards or laboratory test standards. Panel performance is generally evaluated using dimensional tests, physical property tests, and mechanical property tests.

4.1 Dimensional Tests

Methods of measuring panel dimensional properties and the required accuracy of these measurements for composite panel products are defined in ASTM D1037. Tolerance limits for size, thickness, squareness, and straightness are listed in the standards discussed in Section 4.2 on property requirements of composite panels.

4.2 Physical Property Tests

The physical property tests discussed in this section refer to American Society for Testing and Materials (ASTM) standard D1037 (ASTM, 1994a) unless otherwise noted.

4.2.1 Moisture Content

The average moisture content of a panel at the time of shipment from the manufacturer cannot exceed 10% (based on the ovendry weight) for all grades of particleboard. In the United States, the average moisture content of hardboard should not be less than 2% nor more than 9%. Three specimens should be cut from different locations in the panel and the test results averaged. Generally, a 76 mm wide by 152 mm long specimen of full thickness is used to obtain the dimensions to an accuracy not less than ±0.3% and the weight to an accuracy of not less than ±0.2%. The ovendry weight of the sample is then obtained after drying the sample at 103 ± 2°C until constant weight is reached.

Moisture content is calculated as follows:

$$M = 100\left[(w - f)/f\right]$$

M = moisture content (%)
w = initial weight
f = final weight when ovendry

4.2.2 Density

Density is an important indicator of a composite's performance. It virtually affects all properties of the material. The density of the specimen is determined using the full thickness of the composite. The dimensions are measured to an accuracy of not less than 0.3%, and the weight is measured to an accuracy of not less than ±0.2%. The density of wood-based composites is generally based on the ovendry weight, which is obtained after drying a specimen at 103 ± 2°C until constant weight is reached. The density is calculated as follows:

$$\text{Density (kg/m}^3) = f/Lwt$$

f = ovendry weight (kg)
L = length of sample (m)
w = width of sample (m)
t = thickness of sample (m)

Some industries use specific gravity instead of density when referring to a panel product. Specific gravity is the ratio of the density of a material compared to the density of water.

4.2.3 Water Soak Test

The 24 h water soak test determines the water absorption behavior of the composite and the effects of the absorbed water on composite dimensions. The test specimen is 304 by 304 mm or 152 by 152 mm, with edges smoothly and squarely trimmed. The specimen is conditioned to a constant weight and moisture content in a conditioning chamber maintained at a relative humidity (RH) of 65 ± 1% and a temperature of 20 ± 3°C. The specimen is weighed to an accuracy of not less than ±0.2% and the width, length, and thickness are measured to an accuracy of not less than ±0.3%. The thickness is measured at four points midway along each side 25 mm. After 24 h of submersion in distilled water at 20 ± 1°C, the specimen is weighed after the excess water drains off. The thickness is measured at the same four points and the average is obtained. The following calculations can then be made:

(a) Thickness swelling (%) = $[(T_w - T_i)/T_i] \times 100$

T_w = wet thickness
T_i = initial thickness

(b) Water absorption (%) = $[(W_w - W_i)/W_i] \times 100$

W_w = wet weight
W_i = initial weight

Thickness swelling is critical where the composite board is exposed to water or moisture for extended periods.

4.2.4 Linear Expansion

The linear expansion test measures the dimensional stability of a composite to changes in moisture content (Figure 9.4). The specimen is generally 76 mm wide and 304 mm long; it needs to be at least 152 mm long. The specimen is first conditioned at a RH of 50 ± 2% and a temperature of 20 ± 3°C. The length of each specimen is measured to the nearest 0.02 mm.

The specimen is then conditioned at a RH of 90 ± 5% and a temperature of 20 ± 3°C. Linear expansion is then calculated as follows:

$$\text{Linear expansion (\%)} = \left[\left(L_{90} - L_{50}\right)/L_{50}\right] \times 100$$

L_{90} = length at 90% RH
L_{50} = length at 50% RH

Thickness swelling and water absorption can also be calculated at these different conditions of relative humidity.

Figure 9.4 Dial gauge comparator for determining linear variation with change in moisture content.

4.2.5 Thermal Insulation

Thermal insulation (thermal conductivity) is an important property that relates to heat flow through a composite board. The standard test method is entitled *Standard Test Method for Steady-State Heat Flux Measurements and Thermal Transmission Properties by Means of the Guarded-Hot-Plate Apparatus*, and it is described in the ASTM C177-92 specification (ASTM, 1994b). Specimens approximately 25 mm thick are placed on each side of a hot plate, and the thermal conductivity is measured. Conducting this test is complex because thermal conductivity depends upon environmental and apparatus test conditions, as well as the formulation and density of the composite product.

4.3 Mechanical Property Tests

The mechanical property tests discussed in this section refer to ASTM D1037 standards (ASTM, 1994a) unless otherwise noted. The mechanical properties of composite boards depend on the moisture content at the time of test. Material tested dry is conditioned to a constant weight and moisture content in a climate chamber maintained at 20° ± 3°C and an RH of 65 ± 1%. There are also several tests for measuring the properties of a composite product at various moisture contents and humidity levels. Material tested wet is soaked in 20° ± 3°C water for 24 h prior to mechanical property testing.

One method of obtaining a measure of the inherent ability of a material to withstand severe exposure conditions is to use an accelerated aging test. Using this method, each sample is subjected to six complete cycles of aging:

- Immerse specimen in water at 49 ± 2°C for 1 h
- Spray specimen with steam and water vapor at 93 ± 3°C for 3 h
- Store at –12 ± 3°C for 20 h
- Heat at 99 ± 2°C in dry air for 3 h
- Spray again with steam and water vapor at 93 ± 3°C for 3 h
- Heat in dry air at 99 ± 2°C for 18 h
- After six cycles, further condition the material at 20 ± 3°C and RH 65 ± 1% for at least 48 h before the test.

A test commonly used in Canada and Europe involves submerging the specimen in boiling water at 100°C for 2 h before property testing (CSA, 1978). The following section describes tests for static bending, tensile strength, dent and impact resistance, and fastener holding strength. Regardless of the moisture content of the specimen at the time of test, the following procedures are commonly used to determine the mechanical properties of composite products.

4.3.1 Static Bending

Static bending tests determine the modulus of rupture and modulus of elasticity of composites (Figure 9.5).

4.3.1.1 Modulus of Rupture

Modulus of rupture (MOR) has become a common measurement of composite board bending strength. The MOR is the ultimate bending stress of a material in flexure or bending, and it is frequently used in comparing one material to another.

$$MOR = 3PL/2bd^2$$

MOR = modulus of rupture (for midspan loading), kPa
P = maximum bending load, N
L = length of span, mm, 24× the depth

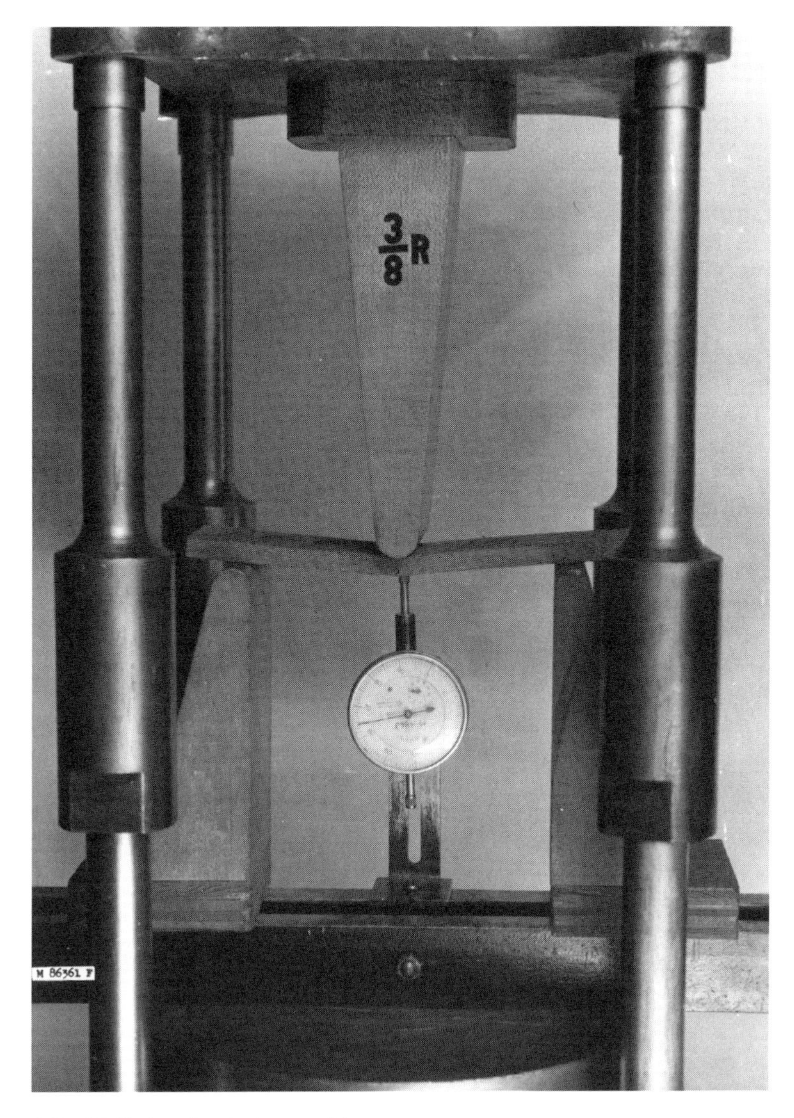

Figure 9.5 Static bending test assembly.

b = width of specimen, mm
d = thickness (depth) of specimen, mm

4.3.1.2 Modulus of Elasticity

Modulus of elasticity (MOE) tests the specimen's ability to resist bending. This property is determined from the slope of the straight-line portion of the load-deflection curve (P_1/Y_1). The MOE is then calculated by the following formula:

$$MOE = \left(P_1 L^3\right) / \left(4bd^3 Y_1\right)$$

MOE = stiffness (apparent modulus of elasticity), kP
P_1 = load at proportional limit, N
L = length of span, mm, 24× the depth
b = width of specimen, mm
d = thickness (depth) of specimen, mm
Y_1 = center deflection at proportional limit, mm

4.3.2 Tensile Strength

Tensile strength is measured perpendicular (internal bond strength) and parallel to the face of the specimen.

4.3.2.1 Perpendicular to Face

Tensile strength perpendicular-to-face is a measure of the resistance of a material to be pulled apart in the direction perpendicular to its surface. A 50 mm square specimen is bonded with an adhesive to steel or aluminum alloy loading blocks of the same dimensions. The internal bond strength is an important property of composite boards; it is calculated as follows:

$$IB = P/bL$$

IB = internal bond, kPa
P= maximum load, N
b= width of specimen, mm
L= length of specimen, mm

4.3.2.2 Parallel to Face

Tensile strength in the parallel-to-face orientation (Figure 9.6) is the resistance of a board material to be pulled apart parallel to its surface. The maximum load at the time of fracture is divided by the cross-sectional area (width × thickness) of the specimen to give maximum strength.

4.3.3 Dent and Impact Resistance

Two tests—face hardness and falling ball impact resistance—are used to measure the resistance of boards to indentation or to the damage that occurs in service when boards are struck by moving objects.

4.3.3.1 Face Hardness

The face hardness (dent resistance) of a composite board specimen is determined by the modified Janka ball test, which records the load required to imbed a steel

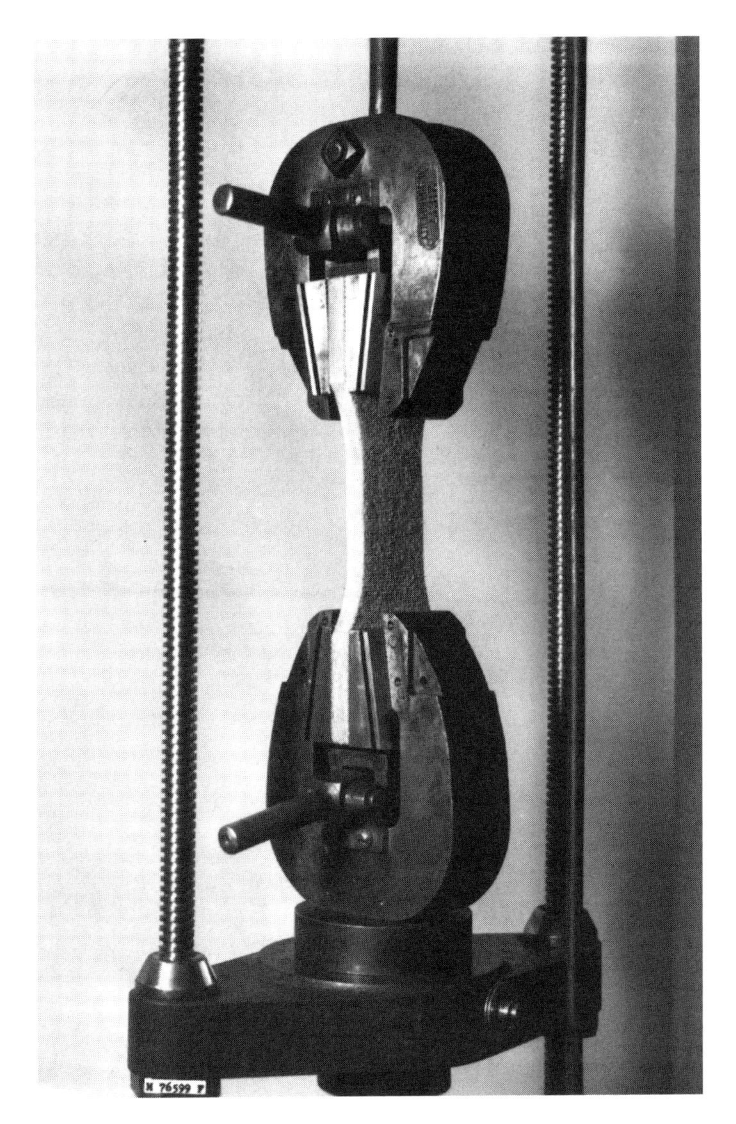

Figure 9.6 Assembly for tension test parallel to surface.

"ball" 11.28 mm in diameter to a depth of one half its diameter. The value obtained using this technique is referred to as the hardness value for the specimen.

4.3.3.2 Falling Ball Impact Resistance

The specimen for the impact resistance test is 304 × 304 mm or 152 × 152 mm, with all four edges smoothly and squarely trimmed. The impact strength is the resistance to fracture when a sudden localized load is applied against the face of a panel held between supports. This value is usually obtained by dropping a 50 mm diameter steel ball from increasing heights at the center of the board until the specimen fails. The height of drop, in millimeters (inches), that produces visible failure on the opposite face of the board is recorded as the index of resistance to impact.

4.3.4 Fastener Holding Strength

The procedures for testing fastener holding strength follow basic ASTM D1037 standards (ASTM, 1994a) as well as individual standards: ANSI A208.1-1993 Particleboard (National Particleboard Association, 1993), ANSI A208.2-1994 Medium Density Fiberboard (National Particleboard Association, 1994), and ANSI/AHA A135.6-1989 Hardboard Siding (AHA, 1989). Fastener resistance is an important property of composite panels used in structural applications such as sheathing. The tests discussed in this section are related to the capability of a composite material to be fastened with either screws or nails.

4.3.4.1 Face Screw Holding

This test measures the withdrawal resistance of screws from the face of the board. The specimen for this test is at least 76 mm wide by 102 mm long. Number 10 Type AB 25 mm sheet-metal screws are threaded into the specimen to a depth of 17 mm. Lead holes are predrilled using a bit 3.2 mm in diameter. If the boards are less than 19 mm thick, the specimen is made from two thicknesses of a sample product, which are laminated together with an adhesive. The screw is withdrawn immediately after it has been imbedded.

4.3.4.2 Edge Screw Holding

This test measures the withdrawal resistance of screws from the edge of the board. The size of the test specimen is the same as that noted for the face screw holding test. The same type of sheet metal screw is threaded into the edge of the board at the mid-thickness 17 mm. A lead hole, the same size as that for the face screw holding test, is predrilled. Boards less than 16 mm thick are not tested. The screw is withdrawn immediately after it has been imbedded.

4.3.4.3 Direct Nail Withdrawal

This test is made using nails that are driven through the specimen from face-to-face to measure the resistance to withdrawal, in a plane normal to the face (Figure 9.7). The specimen is at least 76 mm wide and 152 mm long. Nails 2.8 mm in diameter are used, and the withdrawal tests are made immediately after the nails have been driven into the specimen.

4.3.4.4 Nail-head Pull-through

This test measures the resistance of a composite product to pulling the head of a nail or other fastener through the board. It is designed to simulate the conditions

Figure 9.7 Test assembly for measuring the resistance of nails to direct withdrawal.

encountered with forces that tend to pull paneling or sheathing from a wall or siding. The specimen is 76 mm wide by 152 mm long. A common wire nail, 2.8 mm in diameter, is driven through the board specimen at a right angle to the face, with the nail head flush with the surface of the board.

4.3.4.5 Lateral Nail Resistance

This test is made to measure the resistance of a nail to the lateral movement through a composite board (Figure 9.8). The specimen is 76 mm wide and can be any length. A nail 2.8 mm in diameter is driven at a right angle into the face of the panel, with the nail centered on the width. For hardboard siding, a 3.3 mm diameter nail is used, spaced 9.5 mm from the edge of the specimen.

4.4 Chemical Tests

There are many chemical tests for measuring various properties of composite panels, such as procedures for determining acidity and the amount of extractives. Many of the more recently established chemical tests relate to monitoring and controlling air or water quality during the manufacture of composite products.

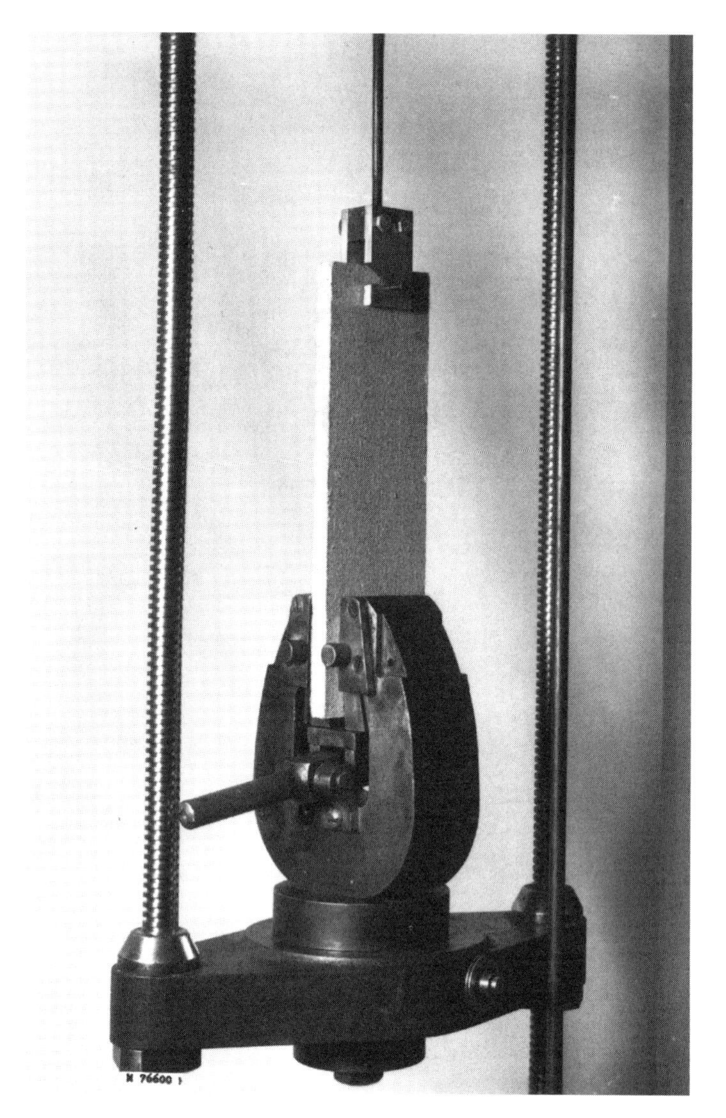

Figure 9.8 Test assembly for measuring the resistance of nails to lateral movement.

Two important chemical properties related to composite panels include formaldehyde content and ash content. Tests for determining these properties are described in the following text.

4.4.1 Formaldehyde

Formaldehyde emissions are generally determined for composite panels bonded with urea formaldehyde. The test method is described in ASTM E1333-90—*Standard Test Method for Determining Formaldehyde Levels From Wood Products Under Defined Test Conditions Using a Large Chamber* (ASTM, 1991). This test method measures the formaldehyde level from wood products under conditions designed to simulate product use in structures such as manufactured homes.

The material is generally tested within 30 days and is sealed before conditioning. Specimens are conditioned on edge with a minimum distance of 153 mm between each panel for 7 days at 24°C and 50% RH. Specimens are then inserted into the test chamber at these same conditions for a minimum of 16 h but no more than 20 h, at which time the chamber air is sampled and analyzed for formaldehyde concentration. The formaldehyde concentration in the air is set by standards for different products and is dependent on application of those products. Chamber test concentrations are useful in comparing the formaldehyde emission performance of products.

4.4.2 Ash Content

The test for ash content covers the determination of ash in wood or wood products. The test method is described in ASTM D1102-84—*Standard Test Method for Ash in Wood* (ASTM, 1994c). The method requires a crucible, a muffle furnace, an analytical balance, and a drying oven. Ash content is expressed as percentage of the residue after dry oxidation at 580°C to 600°C of the total ovendry lignocellulosic materials. The test specimen consists of 2 g of wood, ground to pass through a No. 40 sieve. The empty and covered crucible is ignited over a burner or in the muffle furnace at 600°C. The crucible is then cooled in a desiccator and weighed to the nearest 0.1 mg. Then, 2 g of the specimen is placed in the crucible. The uncovered crucible and the specimen are then weighed and dried in an oven at 100–105°C. After 1 h, the cover is replaced on the crucible. The specimen is then weighed after cooling. This drying operation is repeated until the weight is constant to 0.1 mg. After the weight of the ovendry specimen is recorded, the crucible and contents are placed in a muffle furnace and ignited until all the carbon is eliminated. The crucible is then heated slowly to avoid flaming and loss of the specimen. The final ignition temperature is 580–600°C. The specimen is accurately weighed after cooling. The ash content is calculated as follows:

$$\text{Ash content (\%)} = \left(W_1/W_2\right) \times 100$$

W_1 = weight of ash
W_2 = weight of ovendry sample

5. PROPERTY REQUIREMENTS OF COMPOSITE PANELS

A number of properties could be considered critical to the performance of a specific wood-based composite product, depending upon its end use. Important in some respects to all composites are dimensional stability, strength and stiffness, and fastener-holding properties.

The properties of most composite board products are determined according to ASTM standard (ASTM, 1994a), and to a considerable extent, these properties either suggest or limit the uses. In the following text, fiberboard and particleboard are grouped into various categories suggested by manufacturing process, properties, and use. In general, these products are ones that can easily be made from either wood or other agro-based fibrous materials.

5.1 Fiberboard

Fiberboard includes insulating board, medium density fiberboard, and hardboard.

5.1.1 Fiber Insulating Board

Table 9.3 shows requirements for some physical and mechanical properties of insulating board, published in ASTM C208-94—*Standard Specification for Cellulosic Fiber Insulating Board* (ASTM, 1994d). Physical properties are also included in the ANSI Standard for Cellulosic Fiberboard, ANSI/AHA 194.1 (AHA, 1985). These products are used for such applications as sound deadening, sheathing, shingle backing, and insulation.

5.1.2 Medium Density Fiberboard

Minimum property requirements, as specified by the American National Standard for MDF (ANSI A208–1994) (National Particleboard Association, 1994), are given in Table 9.4. The furniture industry is by far the dominant MDF market. Medium density fiberboard is frequently used in place of solid wood, plywood, and particleboard in many furniture applications. It is also used for doors, moldings, and trim components.

5.1.3 Hardboard

Property requirements for hardboard are presented in Table 9.5 which classifies hardboard by surface finish, thickness, and minimum physical and mechanical properties, for three classifications, as specified by the American National Standard for Basic Hardboard (ANSI/AHA A135.4-1995) (AHA, 1995). The uses for hardboard generally can be subdivided according to uses developed for construction, furniture and furnishings, cabinet and store work, appliances, and automotive and rolling stock. Typical hardboard products are prefinished paneling, house siding, floor underlayment, and concrete form board.

5.2 Particleboard

Tables 9.6 and 9.7 show requirements for grades of particleboard and particleboard flooring products, as specified by the American National Standard for Particleboard (ANSI A208.1-1993) (National Particleboard Association, 1993). Today, approximately 85% of interior-type particleboards are used as core stock for a wide variety of furniture and cabinet applications. Floor underlayment and manufactured

Table 9.3 Physical Property Requirements for Cellulosic Fiber Insulating Board[a]

Physical requirements	Sound-deadening board	Roof insulation board							Ceiling tiles and panels (Grades 1&2)[b]	Wall sheathing			Backer board	Roof deck
		Grade 1				Grade 2				Regular	Structural			
	13 mm	11 mm	13 mm	25 mm	51 mm	13 mm	25 mm	51 mm	13 mm 14 mm 16 mm	13 mm	13 mm	20 mm	11 mm 9 mm	38 mm 51 mm 76 mm
Thermal conductivity (k) (max), W/m·0 at 24 ± 3°C	0.055	0.055	0.055	0.055	0.055	0.058	0.058	0.058	0.055	0.058	0.063	0.058	0.058	0.058
Transverse strength (either direction) (min), N	53.4	31.1	31.1	62.3	124.6	62.3	107	160	44.5	62.3	89.0[c]	111.2	27	—
Tensile strength parallel to surface (min),[d] kPa	1,034	345	345	345	—	1,034	1,034	—	1,034	1,034	1,379	1,034	1,034	—
Tensile strength perpendicular to surface (min), kPa	28.7	23.9	23.9	23.9	23.9	28.7	28.7	28.7	28.7	28.7	38.3	28.7	28.7	28.7
Water absorption by volume (max), %	7	10	10	10	10	7	7	7	—	7	(e)	7	7	10
Linear expansion, 50–90% RH (max), %	0.5	0.5	0.5	0.5	0.5	0.5	0.5	0.5	0.5	0.5	0.6	0.5	0.5	0.5
Flame spread index, finish surface (max)	—	—	—	—	—	—	—	—	200	—	—	—	—	
Vapor permeance, pressure differential, mg/s·m²·kPa	0.287	—	—	—	—	—	—	—	—	0.287	0.287	0.287	(f)	
Modulus of rupture (min), kPa	1,655	9,665	965	552	276	1,896	965	483	—	1,896	2,758[c]	1,379	1,379	(g)
Deflection at specified min load (max), mm	22		32	16	8	19	11	5	—	19	19	14	30	—

Table 9.3 Physical Property Requirements for Cellulosic Fiber Insulating Board[a]

Physical requirements	Sound-deadening board	Roof insulation board							Ceiling tiles and panels (Grades 1&2)[b]	Wall sheathing			Backer board	Roof deck
		Grade 1				Grade 2				Regular	Structural			
	13 mm	11 mm	13 mm	25 mm	51 mm	13 mm	25 mm	51 mm	13 mm 14 mm 16 mm	13 mm	13 mm	20 mm	11 mm 9 mm	38 mm 51 mm 76 mm
Modulus of elasticity (min), mPa	—	—	—	—	—	—	—	—	—	—	—	—	—	276
Deflection span ratio [h]	—	—	—	—	—	—	—	—	—	—	—	—	—	1/240
Moisture content by weight (max), %	10	10	10	10	10	10	10	10	10	10	10	10	10	10

[a] Board dimension is thickness.
[b] Physical properties listed in this column, except flame spread index, apply to the base material before punching, drilling, perforating, or embossing.
[c] The 89-N transverse load and modulus of rupture (MOR) requirements are not required for wall sheathing applications where the manufacturer certifies that the product will meet the specified racking requirements when run in accordance with Methods E72, in conjunction with Appendix D of the HUD Minimum Property Standards, 4900.1 REV-1. In addition, the product is applied vertically and fastened 152 mm apart to intermediate framing and 76 mm apart around the edges of the sheets. Racking strengths of 23.2 kN dry and 17.8 kN wet are required. As an alternative, building code requirements may be specified.
[d] Tensile strength requirements shall be applicable only on thicknesses up to and including 25 mm.
[e] Water absorption for 13 mm structural wall sheathing is determined by the 24-h test in accordance with Test Methods D 1037 using 15% as the maximum. Water absorption for all other products is determined by the 2-h test in accordance with Test Methods C 209.
[f] For roof deck products with a vapor retarder, the maximum should be 0.029. For roof deck products without a vapor retarder, there is no requirement for permeance.
[g] For roof decking, MOR is determined using Methods D 2164. Matched samples are to be tested before and after accelerated aging. Minimum MOR for unaged samples shall be 155 kPa. For aged samples, the minimum shall be no less than 50% of the unaged test result.
[h] Using Methods D 2164.
Source: ASTM ... C208-94 (1994d).

Table 9.4 Medium Density Fiberboard (MDF) Property Requirements[a,b]

Product class[e]	Nominal thickness mm	MOR[c] MPa	MOE[d] MPa	Internal Bond MPa	Screwholding Face N	Screwholding Edge N	Formaldehyde emission[f] (ppm)
			Interior MDF				
HD		34.5	3,450	0.75	1,555	1,335	0.3
MD	≤21	24	2,400	0.60	1,445	1,110	0.3
	>21	24	2,400	0.55	1,335	1,000	0.3
LD		14	1,400	0.30	780	670	0.3
			Exterior MDF				
MD–Exterior glue	≤21	34.5	3,450	0.9	1,445	1,110	0.3
	>21	31	3,100	0.7	1,335	1,000	0.3

[a] MD-Exterior glue panels shall maintain at least 50% of the listed MOR after ASTM D 1037–1991, accelerated aging (Paragraph 3.3.4). Source: ANSI 208.2-1994 MDF (National Particleboard Association, 1994).
[b] Metric property values shall be the primary values used in determining product performance requirements.
[c] MOR = modulus of rupture.
[d] MOE = modulus of elasticity.
[e] HD = density greater than 800 kg/m^3, MD = density between 640–800 kg/m^3 (40–50 lb/ft^3), LD = density less than 640 kg/m^3.
[f] Maximum emission when tested in accordance with ASTM E 1333–1990, Standard test method for determining formaldehyde levels from wood products under defined test conditions using a large chamber (ASTM, 1991).

National Particleboard Association, Medium density fiberboard (MDF), ANSI A208.2–1994, American National Standards Institute, Gaithersburg, MD, 1994.

ASTM, Standard test method for determining formaldehyde levels from wood products under defined test conditions using a large chamber, ASTM E1333–90, American Society for Testing and Materials, Philadelphia, PA, 1991.

home decking represent particleboard construction products. Low-density panels produced in a thickness of 38 mm are used for solid core doors.

6. COMPOSITE PANELS FROM AGRO-BASED FIBERS

Composite panels made from agricultural materials are in the same product category as wood-based composite panels and include low-density insulating board, medium-density fiberboard, hardboard, and particleboard. Composite panel binders may be synthetic thermosetting resins or modified naturally-occurring resins like tannin or lignin, starches, thermoplastics, and inorganics. There seems to be little restriction of what has been tried and what may work.

The following section describes some properties of composite panels made from particles and fibers of bagasse, bamboo, banana stem, coconut and coir, coffee husk, flax, kenaf, reed, rubber tree, rice husk, and miscellaneous fibers. The data were

Table 9.5 Hardboard Physical Properties Requirement

		Water resistance (max avg/panel)			Tensile strength (min avg/panel)	
Class	**Normal thickness (mm)**	**Water absorption based on weight (%)**	**Thick. swelling (%)**	**MOR (min avg/panel) MPa**	**Parallel to surface MPa**	**Perpendicular to surface MPa**
Tempered	2.1	30	25	41.4	20.7	0.9
	2.5	25	20			
	3.2	25	20			
	4.8	25	20			
	6.4	20	15			
	7.9	15	10			
	9.5	10	9			
Standard	2.1	40	30	31	15.2	0.62
	2.5	35	25			
	3.2	35	25			
	4.8	35	25			
	6.4	25	20			
	7.9	20	15			
	9.5	15	10			
Service-tempered	3.2	35	30	31	3.8	0.52
	4.8	30	30			
	6.4	30	25			
	9.5	20	15			

Source: AHA, Basic hardboard, ANSI/AHA A135.4–1995, American National Standards Institute/American Hardboard Association, Palatine, IL, 1995.

obtained from references cited in a literature review conducted by the Forest Products Laboratory and the Department of Forestry at the University of Illinois, Urbana-Champaign (Youngquist et al., 1994).

The research studies included in this review focused on the use of nonwood plant fibers for building materials and panels. The studies covered (1) methods for efficiently producing building materials and panels from nonwood plant fibers; (2) treatment of fibers prior to board production; (3) process variables, such as press time and temperature, press pressure, and type of equipment; (4) mechanical and physical properties of products made from nonwood plant materials; (5) methods used to store nonwood plant materials; (6) use of nonwood plant fibers as stiffening agents in cementitious materials and as refractory fillers; and (7) cost-effectiveness of using nonwood plant materials. More than 30% of the studies addressed the use of bagasse and rice as raw materials in building elements. Other materials widely studied included bamboo (10% of studies), coconut and coir (7%), flax (6%), and straw (6%).

Virtually all studies failed to examine the durability of the product. Of the few studies that did investigate durability, most focused on cement and concrete roofing panels and sheets. This literature review indicates that additional research is needed to obtain information on long-term durability and the influence of weathering on the performance of materials. Moreover, future research needs to focus on comparing the product against product standards, such as American Standard for Testing and

Table 9.6 Requirements for Grades of Particleboard[a,b]

Grade[c]	MOR MPa	MOE MPa	Internal bond MPa	Hardness N	Linear expansion max avg (%)	Screw-holding Face N	Screw-holding Edge N	Formaldehyde maximum emission (ppm)
H-1	16.5	2,400	0.9	2,225	NS[d]	1,800	1,325	0.3
H-2	20.5	2,400	0.9	4,450	NS	1,900	1,550	0.3
H-3	23.5	2,750	1	6,675	NS	2,000	1,550	0.3
M-1	11	1,725	0.4	2,225	0.35	NS	NS	0.3
M-S[e]	12.5	1,900	0.4	2,225	0.35	900	800	0.3
M-2	14.5	2,250	0.45	2,225	0.35	1,000	900	0.3
M-3	16.5	2,750	0.55	2,225	0.35	1,100	1,000	0.3
LD-1	3	550	0.1	NS	0.35	400	NS	0.3
LD-2	5	1,025	0.15	NS	0.35	550	NS	0.3

[a] Particleboard made with phenol–formaldehyde-based resins do not emit significant quantities of formaldehyde. Therefore, such products and other particleboard products made with resin not containing formaldehyde are not subject to formaldehyde emission conformance testing. Source: ANSI A208.1–1993 (National Particleboard Association, 1993). National Particleboard Association, *Particleboard*, ANSI A 208.1–1993, American National Standards Institute, Gaithersburg, MD, 1993.

[b] Panels designated as "exterior glue" must maintain 50% MOR after ASTM D 1037 accelerated aging.

[c] H = density greater than 800 kg/m, M = density between 640–800 kg/m^3, LD = density less than 640 kg/m^3.

[d] NS = Not specified.

[e] Grade M-S refers to medium density, "special" grade. This grade was added to the Standard after grades M-1, M-2, and M-3 had been established. Grade M-S falls between M-1 and M-2 in physical properties.

Table 9.7 Requirements for Grades of Particleboard Flooring Products[a,b]

Grade[c]	MOR MPa	MOE MPa	Internal bond MPa	Hardness N	Linear expansion max avg (%)	Formaldehyde max emission (ppm)
PBU	11	1,725	0.4	2,225	0.35	0.2
D-2	16.5	2,750	0.55	2,225	0.3	0.2
D-3	19.5	3,100	0.55	2,225	0.3	0.2

[a] Particleboard made with phenol–formaldehyde-based resins do not emit significant quantities of formaldehyde. Therefore, such products and other particleboard products made with resin not containing formaldehyde are not subject to formaldehyde emission conformance testing. Source: ANSI A208.1–1993 (National Particleboard Association, 1993). National Particleboard Association, *Particleboard*, ANSI A 208.1–1993, American National Standards Institute, Gaithersburg, MD, 1993.

[b] Grades listed in this table shall also comply with the appropriate requirements listed in Section 3. Panels designated as "exterior glue" must maintain 50% MOR after ASTM D–1037 accelerated aging (Paragraph 3.3.3).

[c] PBU = underlayment; D = Manufactured Home Decking.

Materials (ASTM), German Standard Institute (DIN), International Standard Organization (ISO), American National Standards (ANSI), and the U.S. Department of Commerce standards.

The properties reported here are from different research studies from various parts of the world. The references are cited at the bottom of the tables; the data are limited to those appearing in the reports cited in Youngquist et al. (1994) and necessarily differ from fiber to fiber. No attempt was made to determine the test methods used to obtain these data.

6.1 Bagasse/Guar/Sugarcane

Properties of selected composite boards made from bagasse, guar, and sugarcane are shown in Table 9.8.

Table 9.8 Properties of Selected Composite Boards Made from Bagasse, Guar, and Sugarcane[a]

Material	Ref.	Type of board	Thickness (mm)	Density (kg/m³)	MOE (MPa)	MOR (MPa)	Water absorption (24 h) (%)	Thickness swell (24 h) (%)
Bagasse, guar, sugarcane	1,2,3	Particle board	12–20	520–630	1,400–2,000	16.7–25.5	—	—
		Fiberboard	8–35	300–750	—	—	—	—
		Fiberboard	4.3	—	—	58.5	11.3	—
Bagasse, corn,	4	Particle board	—	720	3,800	16.3	—	—
sunflower, flax	5	Fiberboard	—	810–850	—	22.6–26.5	14–15	8–10
Bagasse + $CaSO_4$ + $5H_2O$	6	Plaster-board (cement)	—	560	—	2.0	—	—
Bagasse + asphalt	7	Composite board	—	900–1,000	—	19.6–24.5	<10	5

[a] MOE is modulus of elasticity; MOR, modulus of rupture.

([1]Anonymous, 1968; [2]Cao et al., 1986; [3]Hesch, 1967; [4]Cherkasov and Lodos, 1969, 1971; [5]Shen, 1984; [6]Saneda et al., 1977; [7]Ni et al., 1961). For additiol information, see Kheleyan et al., 1981; Maldarand Kokta, 1990; Moborek et al., 1982; Naffziger et al., 1962; Rionda, 1969; and Wang, 1975.

Some properties of bagasse composite panels are shown in Figure 9.9. Specifications for these panels are 92% bagasse, 8% urea–formaldehyde, 0.74 specific gravity, and 7.6 mm thickness (Salyer and Usmani, 1982).

6.2 Bamboo

Properties of several types of boards made from bamboo are shown in Table 9.9.

6.3 Banana

Particleboards have been made using banana stalk and wood chips (Youngquist et al., 1994). Urea–formaldehyde resin (10%) was used as a binder and boards with

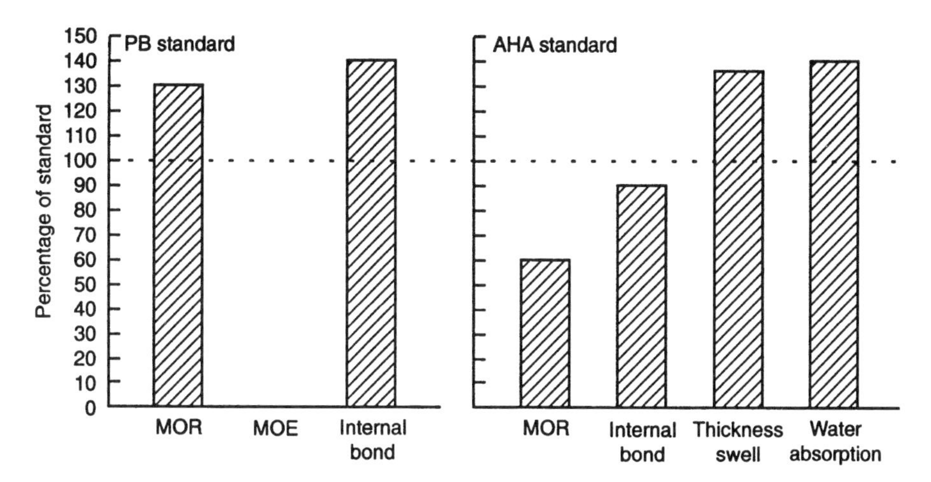

Figure 9.9 Some properties of bagasse composition panels. Thickness swell and water absorption values below dotted line are desirable. (Source: American National Standard Institute [for standards] 1993 [Particleboard Standard} and 1988 [American Hardboard Association]: Salyer, I. O., and Usmani, A. M., *Industrial & Engineering Chemistry, Product Research and Development*, 21(1), 17, 1982.)

Table 9.9 Properties of Selected Composite Boards Made from Bamboo[a]

Material	Ref.	Type of board	MOE (MPa)	MOR (MPa)	Tensile strength (MPa)	Impact strength (kJ/m²)	Compression strength (MPa)	Shear strength (MPa)
Bamboo fiber composite	1	Plastic-bonded	—	151.8	175.27	45.6	—	—
Bamboo + banana + coir + jute	2	Cement-bonded	—	—	0.74	—	—	—
Bamboo + glass	3	Molded plastic	541	17.1	13.1	—	133.8	40.9

[a] MOE is modulus of elasticity; MOR, modulus of rupture.

([1]Jain et al., 1992; [2]Rehsi, 1988; [3]Espleland and Stekhuizen, 1988). For more information, see Chen and Wang, 1981; Chu et al., 1984; and Tsai et al., 1978.

a density of 590–720 kg/m^3 were prepared (Pablo et al., 1975). The strength of the boards increased as the proportion of wood chips increased in the mixture.

6.4 Coconut/Coir

Properties of boards made from coconut or coir (dust, husk, shell, or shell flour) are seen in Table 9.10.

6.5 Coffee Bean

Composite boards made from coffee bean hull and grounds have been reported (Youngquist et al., 1994); however, no strength data was given. The boards were

Table 9.10 Properties of Selected Composite Boards Made from Coconut or Coir (Dust, Husk, Shell, or Shell Flour)[a]

Material	Ref.	Type of board	Density (kg/m³)	MOE (MPa)	MDR (MPa)	Compression strength (MPa)	Water absorption (24 h) (%)	Thickness swell (%)
Coconut /coir	1	Cement board	—	2,500—2,800	9–11	—	14–16	< 1.2
	2	Plastic-bonded	800	—	73.5	84.3	—	—
Coconut fiber polymer	3	Plastic-bonded	1,150 1,180	—	14.2 23.0	— —	— —	— —

[a] MOE is modulus of elasticity; MOR, modulus of rupture.

([1]Aggarwal, 1991; [2]Ohtsuka and Uchihara, 1973a; [3]Ohtsuka and Uchihara, 1973b.)

made using varying amounts of urea-formaldehyde resin, to a thickness of 127 mm and a density of 1,100 kg/m³. Increased resin content resulted in improved board strength and water resistance (Tropical Products Institute, 1963).

6.6 Cotton

Pandey et al. (1979) reported the following properties for particleboard made from cotton (seed hull/husk, stalk):

Density	710 kg/m³
MOR	6.6 MPa
Tensile strength	2.9 MPa
Water absorption (2 h)	38%

Cotton stalk composites have been studied as a substitute for lumber (Zur Burg, 1943).

Some properties of composition panels made from undebarked cotton stalks are shown in Figure 9.10. Specifications for these composition panels are 97% refined undebarked cotton stalk, 3% phenolic resin, 0.82 specific gravity, and 2.8 mm thickness. Thickness swell and water absorption time values are unknown (Fadl et al., 1978).

6.7 Flax, Linseed

Properties of particleboard made from flax and/or linseed shives or straw are shown in Table 9.11.

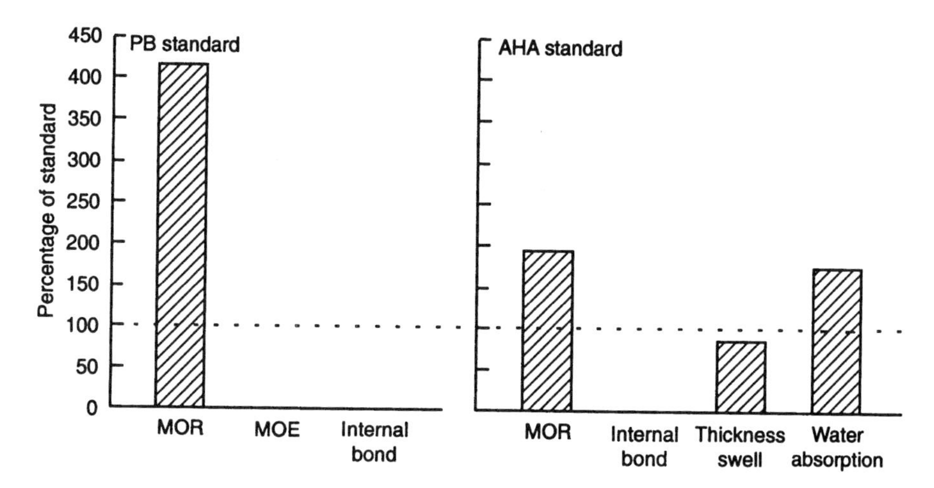

Figure 9.10 Some properties of undebarked cotton stalk composition panels. (Source: American National Standard Institute [for standards] 1993 [Particleboard Standard] and 1988 [American Hardboard Association] data: Fadl, N. A. et al., *Indian Pulp and Paper*, 33(2), 3, 1978.)

Table 9.11 Properties of Selected Particleboard Made from Flax and/or Linseed (Shives, Straw)[a]

Material	Ref.	Type of board	Density (kg/m³)	MOR (MPa)	Tensile strength (MPa)	Compression strength (MPa)	Water absorption (2 h) (%)	Thick ness swell (2 h) (%)	Heat conductivity (Kcal/ m/h/ °C)
Flax, linseed	1	Particle-board	600–650	—	—	—	—	—	—
	2	Fiberboard	—	—	0.7	—	—	6.2	—
	3	Insulating board	180–220	0.59–1.2	—	—	20	—	0.047
Flax + abrolite	4	Plaster-board, cement	790	—	—	5.7	—	—	—

[a] MOR is modulus of rupture.

([1]Eisner and Kolejak, 1958; [2]Verbestel, 1968; [3]Gradovich, 1963; [4]Dvorkin et al., 1987).

6.8 Grass

Narayanamurti and Singh (1963) reported the following properties for particleboard made from grasses:

Density	565–741 kg/m³
Tensile strength	7.2–9.8 MPa

6.9 Kenaf

Bagby and Clark (1976) reported the following properties for hardboard made from kenaf:

Density	780–1,150 kg/m^3
Tensile strength	20.5–76.9 MPa

Some properties for kenaf composition panels are shown in Figure 9.11. Specifications for these panels are 92% depithed kenaf bast fiber, 7% urea–formaldehyde, 1% wax, 0.74 specific gravity, and 12.7 mm thickness. Thickness swell and water absorption values reflect 2 h of immersion (Chow, 1974).

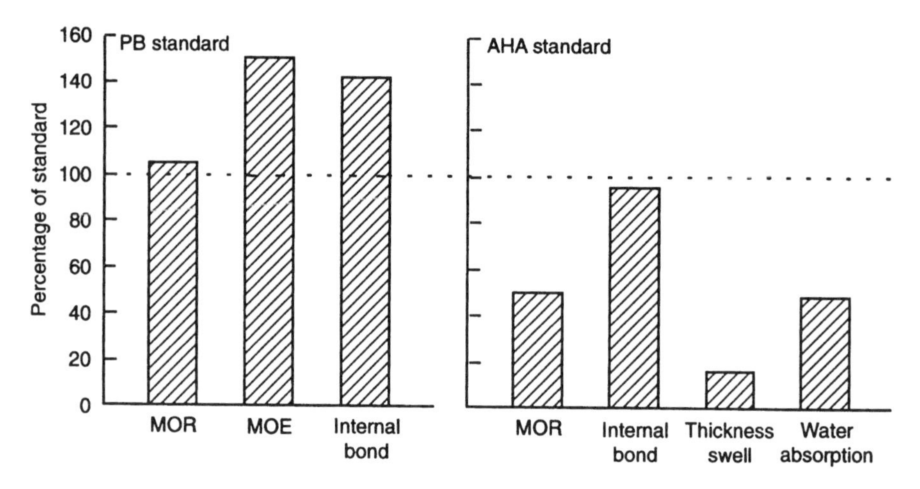

Figure 9.11 Some properties of kenaf composition panels. (Source: American National Standard Institute [for standards] 1993 [Particleboard Standard} and 1988 [American Hardboard Association]; data: Chow, A., *Illinois Research*, 15, 18, 1973.)

6.10 Poppy

Chawla (1978) reported a density of 1000 kg/m^3 for fiberboard made from poppy straw.

6.11 Reed

Al-Sudani et al. (1988) reported thickness of 16 mm and density of 640 kg/m^3 for particleboard made from reed stalks.

6.12 Rice

Properties of boards made from rice husks are shown in Table 9.12. Rice husks or their ash are used in cement block and other cement products. The addition of the hulls increases thermal and acoustic properties (Govindarao, 1980). Some properties of selected rice husk composition panels are presented in Figure 9.12. Specifications for these composition panels are 0.94 specific gravity and 5.1 mm thickness. Husk and resin content are unknown (Govindarao, 1980).

Table 9.12 Properties of Selected Composite Boards Made from Rice Husks[a]

Material	Ref.	Type of board	Density (kg/m³)	MOR (MPa)	Tensile strength (MPa)	Compression strength (MPa)	Water absorption (24 h) (%)	Thickness swell (24 h) (%)	Heat conductivity (Kcal/m/h /°C)
Rice husk	1,2	Particle board	800	13.1	—	—	—	—	—
		Hardboard	1,150	22.8	—	—	13	7.4	—
Rice husk + sawdust	3	Hardboard	1,270	—	10.8–12.3	12.7–14.2	17–18 (7 days)	—	—
Rice husk mineral-bonded[b]	4	Insulating board	200–211	—	—	0.118–0.126	12.6–18	8–10	0.052
Rice husk cement-bonded[c]	5	Plaster board	570	39.2	—	—	—	—	—

[a] MOR is modulus of rupture.
[b] Rice husk bonded to wood with organophosphate binder.
[c] Rice and cement/gypsum and bentonite binder.

([1]Anonymous, 1975; [2]Kluge et al., 1978; [3]Sarkaria and Iyengar, 1968; [4]Gamza and Korol'kov, 1983; [5]Anonymous, 1903).For more information, see Vasishth, 1974.

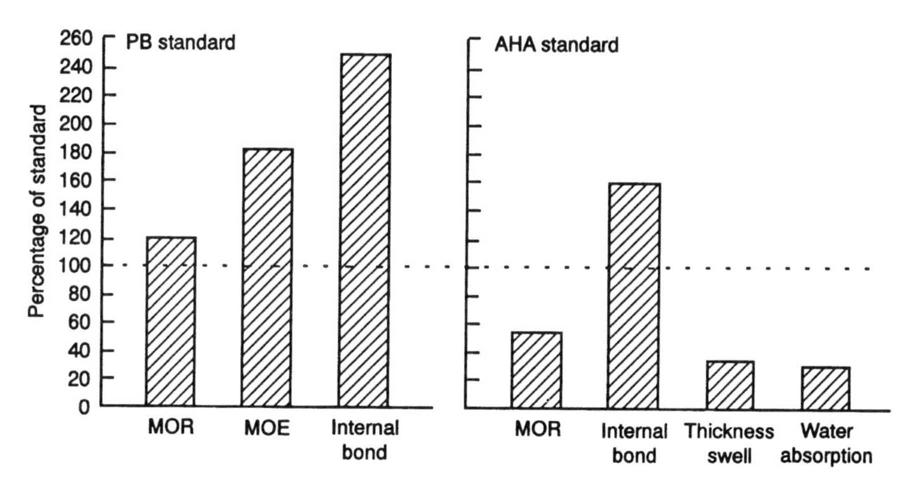

Figure 9.12 Some properties of rice husk composition panels. (Source: American National Standard Institute [for standards] 1993 [Particleboard Standard] and 1988 [American Hardboard Association]; data: Govindarao, V. M. H., *Journal Sci. Ind. Res.*, 39(9), 495, 1980.)

6.13 Rubber

Shimomura et al. (1991) reported a density of 1160 kg/m³ and bending strength of 63.2 MPa for board made from rubber fiber.

6.14 Straw and Other Fibers

Properties of selected composite boards made from various agro-based fibers are shown in Table 9.13.

Table 9.13 Properties of Selected Composite Boards Made from Various Agro-Based Fibers[a]

Material	Ref.	Type of board	Density (kg/m³)	MOE (MPa)	MOR (MPa)	Compression strength (MPa)	Thickness swell (%)	Noise reduction coefficient
Foliage	1	Particle board	790	1,590	6.6	—	10	—
Guar + sorghum stalks	2	Cement board	530–700	—	—	—	—	—
Straw + reed stems	3	Insulating board	188–214	—	10.8–11.0	13.7–14.5	—	—
Straw + polymers	4	Plastic board	520	—	47.6	—	—	—
Straw + polymers								
Corn maize	5	Particle board	—	—	—	—	—	0.39–0.54

[a] MOE is modulus of elasticity; MOR, modulus of rupture.

([1]Chow, 1973; [2]Gabir et al., 1988; [3]Khaleyan et al., 1981; [4]Ohtsuka, 1973; [5]Sampathrajan et al., 1991).

Some properties of straw composition panels are shown in Figure 9.13. Specifications for these panels are 97% pulped rice straw, 3% urea–formaldehyde resin, 0.98 specific gravity, and 2.0 mm (0.08 in.) thickness. Thickness swell and water absorption time values are unknown (Fadl et al., 1984).

7. CONCLUSIONS

Current production trends clearly indicate the increased use of both structural and nonstructural wood-based composite panels for many applications. Each material on the market was developed to provide the strength and other properties needed for a specific end-use. New uses are being developed continuously for these materials. The expanded use of composite materials has resulted in a significant reduction of production costs and greatly improved utilization of the fiber-based resource.

Strength and other property values presented in this report are basic values useful for comparative purposes and for developing improved products. Actual design values are generally available from code authorities, industry associations, and product manufacturers.

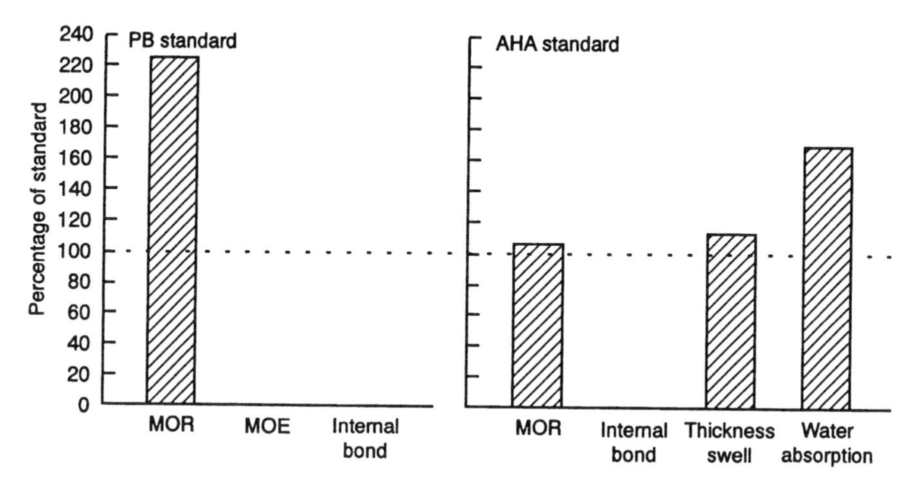

Figure 9.13 Some properties of straw composition boards. (Source: American National Standard Institute [for standards] 1993 [Particleboard Standard] and 1988 [American Hardboard Association]; Fadl et al., data: Fadl, N. A. et al., *Res. Ind.*, 29(4), 288, 1984.)

Economic considerations favor the selection of materials already in production. For specialized uses, new products with special strength and physical property values can be developed. The wood- or agro-based panels now on the market are the product of research in all phases of material selection, material preparation, and development of improved adhesives and manufacturing methods. Key areas in which further research could yield substantial benefits include improving strength and stiffness properties, determining optimum species or plant mixtures, improving durability and weatherability, reducing thickness swell, and improving methods for producing fire- and decay-resistant panels.

REFERENCES

Aggarwal, L.K., Development of coir fiber reinforced composite panels, *Research and Industry*, 36, 273, 1991.

AHA, *Cellulosic fiberboard*, ANSI/AHA A194.1-1985, American National Standards Institute/American Hardboard Association, Palatine, IL, 1985.

AHA, *Basic hardboard*, ANSI/AHA A135.4-1995, American National Standards Institute/American Hardboard Association, Palatine, IL, 1995.

AHA, *Hardboard siding*, ANSI/AHA A135.6-1989, American National Standards Institute/American Hardboard Association, Palatine, IL, 1989.

Al-Sudani, O. A., Daoud, D. S., and Michael, S., Properties of particleboard from reed-type mixtures, *Journal of Petroleum Research*, 7, 197, 1988.

Anonymous, Bagasse panels bonded with urea formaldehyde resin, Assignee: Societe Anon. Verkor. Patent, P.N.: GB 1127700, I.D.: 680918, 1968.

Anonymous, Resin-coated rice hulls and the production of composite articles therefrom, Assignee: Cor. Tech. Research Ltd. Patent, P.N.: GB 1403154, I.D.: 750813, 1975.

Anonymous, Lightweight building materials with high strength and heat resistance, Assignee: Nichias Corp. Patent, P.N.: JP 58140361, I.D.: 830820. [Japanese], 1983.

ASTM, Standard test method for determining formaldehyde levels from wood products under defined test conditions using a large chamber, ASTM E1333-90, American Society for Testing and Materials, Philadelphia, PA, 1991.

ASTM, Standard test methods for evaluating the properties of wood-based fiber and particle panel materials, ASTM D1037-94, American Society for Testing and Materials, Philadelphia, PA, 1994a.

ASTM, Standard test methods for steady-state heat flux measurements and thermal transmission properties by means of the guarded-hot-plate apparatus, ASTM C177-85 (Reapproved 1993), American Society for Testing and Materials, Philadelphia, PA, 1994b.

ASTM, Standard test method for ash in wood, ASTM D1102-84 (Reapproved 1990), American Society for Testing and Materials, Philadelphia, PA, 1994c.

ASTM, Standard specification for cellulosic fiber insulating board, ASTM C208-94, American Society for Testing and Materials, Philadelphia, PA, 1994d.

ASTM, Standard terminology relating to wood-base fiber and particle panel materials, ASTM D1554-94, American Society for Testing and Materials, Philadelphia, PA, 1994e.

Bagby, M. O. and Clark, T. F., Kenaf for hardboards, *TAPPI Rep. 67*, 9, TAPPI Press, Atlanta, GA, 1976.

Cao, Z., Xing, Z., Guan, Z., Wu, Q., and Bai, G., Dry method of manufacturing bagasse fibreboards, Patent, P.N.: CN 109469/85, I.D.: 860723. [Chinese], 1986.

Carll, C. G., Dickerhoof, H. E., and Youngquist, J. A., U.S. wood-based panel industry: standards for panel products, *Forest Products Journal*, 32, 12, 1982.

Chawla, J. S., Poppy straw—a new source for fibre boards, *Indian Pulp and Paper*, 32, 3, 1978.

Chen, T. Y., and Wang, Y. S. A study of structural particleboard made from bamboo waste, *Quarterly Journal of Chinese Forestry*, 14, 39 [Chinese; English summary], 1981.

Cherkasov, M., and Lodos, J., Use of furfural–urea resin for production of particle board from bagasse, *Sobre Derivados de la Cana de Azucar,* 3, 3 [Spanish], 1969.

Cherkasov, M., and Lodos, J., Use of furfural–urea resins in the manufacture of bagasse boards, *Proceedings of Annual Memorial Conference of the Association Tec. Azucar Cuba*, 38, 689 [Spanish], 1971.

Chow, P., New uses found for discarded Christmas trees, *Illinois Research*, 15, 18, 1973.

Chow, P., Dry formed composite board from selected agricultural residues, *World Consultation on Wood Based Panels,* Food and Agriculture Organization of the United Nations, New Delhi, India, 1974.

Chu, B. L., Chen, T. Y., and Yen, T. Influence of the form of bamboo and particleboard on their bending strength and thermal conductivity, *Forest Products Industries*, 3, 291 [Chinese; English summary], 1984.

CSA, Standard test methods for mat-formed wood particleboards and waferboard, CSA CAN3-0188.0-M78, Canadian Standards Association, Ontario, Canada, 1978.

Dvorkin, L. I., Mironenko, A. V., Shestakov, V. L., and Kovtun, A. M., Composites from phosphogypsum, *Stroitel'nye Materialy Konstruktsii*, 4, 15 [Russian], 1987.

Eisner, K., and Kolejak, M., Building board from flax and hemp fibres, *Drevo*, 13, 356 [Polish], 1958.

Espleland, A. J., and Stekhuizen, T. R., Building material composition production—comprises styrenated unsaturated polyester, hydrated alumina, sand or perlite and bamboo filled with glass fibres, Assignee: (ESPE/) Espeland, A.J. Patent, P.N.: US 4730012, I.D.: 880308, 1988.

Fadl, N. A., Sefain, M. Z., Magdi, Z., and Rakha, M., Hardening of cotton stalks hardboard, *Indian Pulp and Paper,* 33(2), 3, 1978.

Fadl, N. A., Nada, A. M. A., and Rkaha, M., Effect of defibration degree and hardening on the properties of rice-straw hardboard, *Res. Ind.*, 29(4), 288, 1984.

Gabir, K., Khristova, P., and Yossifov, N., Composite boards from guar and sorghum (dura) stalks, *Biological Wastes*, 31, 311, 1988.

Gamza, L. B., and Korol'kov, A. P., Fibrous thermally insulating plates based on organophosphates, *Proizvod. Primen. Fosfatnykh Mater. Stroit.*, 7 [Russian], 1983.

Govindarao, V. M. H., Utilization of rice husk–a preliminary analysis, *Journal Sci. Ind. Res.*, 39(9), 495, 1980.

Gradovich, V. A., Prospects for the expansion of the manufacture of constructional boards from flax waste, *Proizv. Stroit. Izdelii iz Plastmass*, Sb. 73 [Russian], 1963.

Hesch, R., Particle board from sugarcane—a fully integrated production plant, *Board Manufacture*, 10, 39, 1967.

Jain, S., Kumar, R., and Jindal, U.C., Mechanical behaviour of bamboo and bamboo composite, *Journal of Materials Science*, 27, 4598, 1992.

Khaleyan, V. P., Bagdasaryan, A. B., and Khaleyan, G. V., Composition for moulding structural panels containing phenol–formaldehyde resin, foamer, urotropin, expanded perlite, ground straw, and reed stems, Assignee: (ERSH=) Erev Shinanyut Wks, (ACSM=) Armn Cons Mater Ind. Patent, P.N.: SU 885206, I.D.: 811130. [Russian], 1981.

Kluge, Z. E., Tsekulina, L. V., and Savel'eva, T. G., Manufacture of hardboards from rice straw, *Tekhnol. Modif. Drev.*, 55 [Russian], 1978.

Maldas, D., and Kokta, B. V., Studies on the preparation and properties of particleboard made from bagasse and PVC. I. Effects of operating conditions, *Journal of Vinyl Technology*, 12, 13, 1990.

Maloney, T. M., Terminology and products definitions—a suggested approach to uniformity worldwide, *18th International Union of Forest Research Organization World Congress*, 1986 September, Ljubljana, Yugoslavia, Yugoslavia IUFRO World Congress Organizing Committee, 1986.

Marra, G., G., The future of wood as an engineered material, *Forest Prod. J.*, 22(9), 43, 1972.

Mobarak, F., Fahmy, Y. A., and Augustin, H., Binderless lignocellulose composite from bagasse and mechanism of self-bonding, *Holzforschung*, 36, 131, 1982.

Naffziger, T. R., Horeiter, B. T., and Rist, C. E., Upgrading insulating board and molded pulp products by minor additions of dialdehyde starch, *TAPPI*, 45, 745, 1962.

Narayanamurti, D., and Singh, K., Boards from *Phragmites karka*, *Indian Pulp and Paper*, 17, 437, 1963.

National Particleboard Association, *Particleboard*, ANSI A208.1-1993, American National Standards Institute, Gaithersburg, MD, 1993.

National Particleboard Association, *Medium density fiberboard (MDF)*, ANSI A208.2-1994, American National Standards Institute. Gaithersburg, MD, 1994.

Ni, C., Yang, C. T., and Shen, T. K., Asphalt-impregnated bagasse board, *Taiwan Sugar Experiment Station Rep. 23*, 125, 1961.

Ohtsuka, M., and Uchihara, S., Boards from resin-impregnated coconut husk, Assignee: Otsuka Chemical Co. Ltd., Patent, P.N.: JP 48022179, I.D.: 730320. [Japanese], 1973a.

Ohtsuka, M., and Uchihara, S., High-strength inorganic hydraulic material–coconut fiber–polymer composite, Assignee: Ohtsuka Chemical Drugs Co. Ltd. Patent, P.N.: JP 51003354, I.D.: 760202. [Japanese], 1973b.

Ohtsuka, M., Composite with natural high polymers and polymethylmethacrylate, Assignee: Otsuka Chemical Drugs Co. Ltd. Patent, P.N.: JP 48015986, I.D.: 730228 [Japanese], 1973.

Pablo, A. A., Ella, A. B., Perez, E. B., and Casal, E. U., The manufacture of particleboard using mixtures of banana stalk (saba: *MU.S. compreso*, Blanco) and Kaatoan bangkal (*Anthocephalus chinensis*, Rich. ex. Walp.) wood particles, *Forpride Digest*, 4, 36, 1975.

Pandey, S. N., Metha, S. A. K., and Tamhankar, S. H. V., Particle boards from cotton plant stalks, Assignee: Indian Council of Agricultural Research, Patent. P.N.: IN 145886, I.D.: 790113, 1979.

Rehsi, S. S., Use of natural fibre concrete in India, In: Swamy, R.N., Ed., *Natural fibre reinforced cement and concrete*, Blackie and Son Ltd., Glasgow, Scotland, 243, 1988.

Rionda, J.A., Composition based on bagasse and heat-hardenable resin, Assignee: Esso Research and Engineering Co. Patent, P.N.: FR 1565996, I.D.690502. [French], 1969.

Salyer, I. O., and Usmani, A. M., Utilization of bagasse in new composite building materials, *Industrial & Engineering Chemistry, Product Research and Development*, 21(1), 17, 1982

Sampathrajan, A., Vijayaraghavan, N. C., and Swaminathan, K. R., Acoustic aspects of farm residue-based particle boards, *Bioresource Techonology*, 35, 67, 1991.

Saneda, Y., Yamada, H., and Takizawa,T., Bagasse based building boards, Assignee: Shadan Hojin Shinzairyo Kenkyu Kaihatsu Center. Patent, P.N.: JP 52093432, I.D.: 770805. [Japanese], 1977.

Sarkaria, T. C., and Iyengar, M. S., Hardboard and tiles from agricultural residues, Assignee: Council of Scientific and Industrial Research, India. Patent, P.N.: 101714, I.D.: 680615, 1968.

Shen, K.C., Composite products from lignocellulosic materials. Patent, P.N.: SA 8308301, I.D.: 840725, 1984.

Shimomura, T., Tanaka, Y., and Inagaki, K., Spinning polyolefin–plant fiber mixtures for reinforcing fibers for building materials, Assignee: Ube Nitto Kasei Co. Ltd. Patent, P.N.: JP 03224713, I.D.: 911003. [Japanese], 1991.

Suchsland, O., and Woodson, G. E., Fiberboard manufacturing practices in the United States, *Agric. Handbook*, 640, U. S. Department of Agriculture, Washington, D.C., 1986.

Tropical Products Institute, *Manufacture of particle board from coffee husks,* Tropical Products Institute (TPI), London, England, Rep. 18/63, 4 p., 1963.

Tsai, C. M., Lo, M. P., and Poon, M. K., Study on the manufacture of uni-layer particleboards made from bamboo and wood particles, *Bulletin of the Experimental Forest of National Taiwan University*, 121, 41 [Chinese], 1978.

Vasishth, R. C., Resin coated rice hulls and compositions containing them, Assignee: Cor. Tech. Research Ltd. Patent, P.N.: US 3850677, I.D.: 741126, 1974.

Verbestel, J. B., Treatment of flax particles for use in fiberboard manufacture. Patent, P.N.: BE 702968, I.D.: 680201. [Dutch], 1968.

Wang, U., Scale-up production of bagasse plastic combinations using gamma radiation, *Ho Tzu K'o Hsueh*, 12, 12, 1975.

Youngquist, J. A., English, B. E., Spelter, H., and Chow, P., Agricultural fibers in composite panels, *Proceedings of the Washington State University Particleboard Conference*, 27, 133, 1993.

Youngquist, J. A., English, B. E., Scharmer, R. C., Chow, P., and Shook, S. R., Literature review on use of non-wood plants fibers for building materials and panels. *General Technical Report FPL-GTR-80,* USDA Forest Service, Forest Products Laboratory, Madison, WI, 1994.

Zur Burg, F. W., Cotton stalks for synthetic lumber, *Paper Industry*, 25, 12, 1943.

CHAPTER 10

Packaging and Lightweight Structural Composites

Theodore L. Laufenberg

CONTENTS

1. INTRODUCTION

Maybe the world's first package was a dry gourd used to carry water. This package was later supplemented by skin bags, earthen jars, alabaster vases, wooden casks, and crates. Now in the modern age, we have paper tubes and cartons. This listing of "natural" materials excludes the steel, plastic, aluminum, and other manmade materials so common in packaging today. Building upon the information in Chapter 6 on pulp and paper, most of this chapter centers upon the corrugated paperboard container as the predominant packaging material from agro-based fibers. The chapter also provides background on pulp molding technology that yields lightweight structural composites for packaging, construction, and industrial applications.

2. PACKAGING: A SHORT HISTORY OF THE PAPER-BASED* PACKAGING INDUSTRY

We seldom realize how recently the cracker barrel and the candy pail were usual equipment in our food stores, sharing perilously close intimacy with the kerosene can. Not until the turn of the century came the development of the packaging systems that relied only upon fibrous paperstock to store, protect, and move our precious materials and goods in neat, economical, and sanitary packages. An invention by Robert Gair in 1879 set the stage for economical mass production of folding cartons and subsequently for the corrugated fiberboard container.

The development of the corrugated container industry did not occur within a single company; rather it grew from the developments for a number of purposes. Examples are carpet linings, bottle wrappings, padding for hat sweat-bands, interior packing, quick setup cartons, etc. Many patents have been issued on both the articles and the machines for fabricating them from paperboard in England, France, and the United States.

Corrugated materials, as distinguished from boxes, were first patented in 1856 in England by Edward C. Healey and Edward E. Allen in the form of fluted material. As a packing material, it was first patented in the U.S. in 1871 by Albert L. Jones for use in protecting vials and bottles. First, corrugated materials were largely used as cushioning and wrappers for bottled goods such as patent medicines and beverages. They then were used with a paper overwrap for light parcel post and express

* Excerpts from: Wilbur F. Howell, A History of the Currugated Shipping Container Industry in the United States. Published by Samuel M. Langston Co., Camden, NJ, 1940.

boxes around 1895. The first corrugated box was used to ship lamp chimneys; then fruit jars and cereals were given the official exceptions to the freight packaging specifications in 1905. Meanwhile, solid fiber boxes were first used for cereals in 1904 with the development of equipment for pasting, scoring, and slotting fiberboard. The Official Classification Committee issued their first formal rule providing general specifications for both corrugated and solid fiber boxes. Rule 2-B was from the Official Classification, along with Rule 14 and Rule 4-C from the Western and Southern Classifications, respectively. These three specifications were merged in 1920 to become Rule 41 (Dolan, 1991), which is still in use today.

A large influence in the growth of fiber-based packaging was the Pridham Decision which ended shippers' discrimination against fiber containers. In 1914, the U.S. Interstate Commerce Commission ruled that all containers should enjoy the same rates for shipment regardless of material type. Besides the boxes for canned goods, dry goods, drugs, and countless other items, other heavy packaged items have made the transition from wood packaging to corrugated fiber. Perhaps the most dramatic development was termed V-boxes in World War II, where under wet and humid conditions, fiber boxes out-performed other types of containers. Continued development has led to use corrugated fiberboard nearly exclusively in commerce offruit, vegetables, frozen meats, and other perishables.

3. CORRUGATED FIBERBOARD SHIPPING CONTAINERS PROCESSING AND DESIGN

Corrugated board is a composite structure made up of at least one corrugated paperboard layer and at least one flat layer. The corrugated layer is formed by running the paperboard through rolls that create a sinusoidal-like wave geometry across the width of the paperboard sheet. These layers are bonded at the tips of the corrugations to the liner. The corrugated or "fluted" layer is called the "medium," due to central location in the structure, and the flat layers are called "liners." According to the way it is constructed, the corrugated board may be a single liner and a medium (single-faced board) which is commonly used for pads, partitions and cushioning wrap for odd-shaped items.

Board made with a liner on each side of a medium is called "single-wall." This design makes up 90% of all corrugated fiberboard packaging. A board with three liners bonded amidst two mediums is called a "double-wall" board and, yes, there is a "triple-wall" board made of four liners and three mediums. These constructions are used to fabricate shipping containers and associated packaging in the form of cushioning pads, partitions, inserts, corner posts, tubes and nearly any imaginable item, e.g., low-cost furniture. The double- and triple-wall constructions are predominantly for packaging large, cumbersome items such as appliances.

Designing a shipping container for maximum service requires consideration of appearance, weight, price, ease of handling, packing, and protection of contents. Each item is important when selecting the best paperboard, flute size, corrugated board construction, and container design. The objectives can be simply stated as identifying the minimum strength requirements of the container to protect its contents from damage at the lowest possible cost. A person will not be able to fully design a container from this presentation, but they will know the broad problems of designing containers for maximum service. (Koning, 1995).

3.1 Corrugation Geometry

Corrugated medium is further defined by the height and number of corrugations per unit length. As shown in Table 10.1, these factors also dictate the amount of paperboard required to produce a unit length of corrugated medium ("take-up factor"). The "A" flute produces a thicker-walled board with higher top-to-bottom compression strength, and the "B" flute yields a board with better flat crush. Generally, the "A" and "C" are used where top-to-bottom stacking strength and cushioning are required. "B" is used for cushioning when stacking strength is less critical. "E" flute has high flat crush, but poor cushioning capability.

Table 10.1 Corrugated Board Flute Specifications and Relative Properties

				Relative Estimates	
Flute	Average flute height (mm)	Average number of flutes/meter	Medium take-up factor	Top-to-bottom compression	Flat crush
A	4.70	110	1.54	100%	100%
B	2.46	154	1.33	75%	150%
C	3.61	129	1.45	85%	125%
E	1.14	295	1.26	60%	350%

Wright, P. G. et al., *Corrugated Fibreboard Boxes*, APM Packaging, Melbourne, VIC, Australia, 289 p., 1988.

3.2 Corrugating Medium

In corrugated board, the thickness of the board gives the high stiffness by separating the flat liners. However, the thickness of the medium itself also contributes to the corrugated board stiffness. The container is usually made with the flutes running vertically to increase stiffness and stacking strength. Flat crush resistance is also a primary concern of the medium's performance. Paperboard for use as a medium may be from 0.20–0.40 mm in thickness. A processing consideration for a medium is that it be capable of being corrugated without cracking and the surface adheres to the adhesive system in use.

3.3 Linerboards

Forming the inner and outer facings of the corrugated "sandwich," the liners are most influential on the structural properties of the corrugated board. The liners will be chosen for their thickness, compression strength, or surface finish (if printing or laminating is needed). Typical thicknesses of liners are 0.25–0.65 mm.

3.4 Corrugated Container Manufacture

There are three components brought together in the first step of the continuous process known as corrugating: medium, adhesive, and liner. The medium is hot-conditioned prior to being run through the corrugated roll (matched corrugated surfaces mating or "fingerless"). The corrugating roll and an adhesive roll are in contact to apply some controlled amount of adhesive. A liner is brought in contact with the adhesive-topped flute of the medium. When the "single-faced" board is bonded, it continues onto a short residence-time conveyor prior to being fed into another adhesive roll and the "double-back" liner applied. Applying the correct pressures, temperatures, adhesives and paperboards to this process is a matter of art combined with science and will not be discussed with here.

3.5 Adhesives

Starch-based adhesives are by far the most common in use by the corrugating and folding box board industries. Cornstarch is most commonly used; however, potato and tapioca are also used. Different formulations are used for each step in the corrugating process and for effecting joints or closures. Application rates are quite variable depending upon the concentration of solids, the flute profile, and the water resistance expectations for the combined board.

Silicate adhesives, once quite common, have fallen out of use due to concerns of product weight, abrasion, wear, alkalinity, and silicosis. However, it is a low cost adhesive that is seemingly unaffected by seasonal fluctuations. In addition it is tolerant of long storage times, high moisture boards, and produces a near-permanent bond. Application rates are typically 3–6 times higher than starch adhesives.

Water-resistant adhesives are typically made using starch adhesives as the base. Some quantity of dry or liquid resin such as urea–formaldehyde and a catalyst (e.g., aluminum sulfate) added to the starch base provide a water-resistant bond. Polyvinyl alcohol is also used in the folding box industry and infrequently in the corrugating industry; however, it is a cold-set adhesive. Melamines have also found some measure of use in the container industry.

3.6 Creasing, Slotting, and Die-Cutting of Box Blanks

Many steps are involved to produce a container from the corrugated paperboard sheet. The progression of steps may be quite different in each installation. Individual machines can perform each of the operations for special containers. In most modern operations, the steps are all done at the die-cutting station, which is either a flatbed or rotary operation. Creases, or scores, are deformations pressed into the corrugated board to cause it to fold in a straight line where desired. Slots are cut into the flat sheet to create individual panels for folding and to create clearance for one panel relative to another. Die-cutting can be used to simply create the perimeter of the box blank, but it is more typically used to create slots, creases, and the perimeter cutting of corrugated board.

3.7 Common Box Styles

Due to the simplicity of cutting flat sheets of corrugated board, the box designer has unlimited flexibility in choosing a pattern for fitting a specific packaging need. The criteria continue to be cost and performance first. To avoid large waste, the design that has proven to be most popular is the regular slotted container (RSC). RSCs require a corrugated board width equal to the container height and width and a small allowance for creases. Since RSCs have no flap overlap, the closures are critical to box performance. For heavy or perishable items that require heavier sidewalls, the full telescope container has full depth sides on the top and on the bottom to provide full overlap of the singlewall sides.

4. SPECIFICATIONS, TESTING, AND MATERIAL PERFORMANCE

In the early stages of the corrugated fiberboard container industry development, the controlling specifications were those of the railroads, known as "Rule 41" in the United States. These still form the cornerstone of the standard grades and classifications. However, most users of containers have developed their own requirements to assure performance of the container for their specific perception of packaging needs. In this environment, the role of testing plays an important role in the marketplace to assure quality, uniformity, and standard practice within the industry.

4.1 Container Specifications

The Uniform Freight Classification of the railroads contains a set of basic standards with which all containers must comply to be accepted for shipment by rail carriers (Rule 41). Their purpose is to form a basis for deciding whether damage to products in shipping containers has been due to excessively severe handling by railroads or to inadequate design and materials of the shipping container. Each boxmaker prints a certificate on his containers showing that they comply with the rule. If the carriers find damaged products in boxes that do not comply with the rule, the boxmaker is liable for the damage. The required test criteria are either:

a. Minimum basis weight of the unfluted paperboard (liners) used in the package construction (expressed as weight per unit area) and minimum burst test strength.
b. Minimum edge crush strength of the corrugated board.

Other Rule 41 criteria in boxes are limitations on the sum of the box dimensions, the maximum weight of the box and its contents, and a minimum basis weight and thickness of the medium. The environmental conditions under which the testing is performed are also specified along with sampling rates and acceptance/rejection criteria.

Trucking companies use a set of rules known as the National Motor Freight Classification (Item 222, 1991) which is nearly identical to the specifications in Rule 41 of the Uniform Freight Classification. Air and sea cargo lines do not have detailed specifications, unless the item to be shipped is listed as a hazardous material and is subject to international control or treaty that the carrier's country has signed. The U. S. Government also specifies containers according to the requirements of Rule 41, but uses special grades in the federal specification for military or overseas use that may require waterproof adhesives, treated paperboards, or other testing requirements to assure performance in hostile environments.

Individual users' specifications typically rely upon Rule 41 to assure a basic standard for their containers. They can then add specifications tailored for their needs such as enhanced flat crush, compression strength, coefficient of friction, vibrational response, etc. (Ostrem and Godshall, 1979). An obvious result of added specifications is added cost for assuring performance under a quality control system. The specifier will balance that added cost against the savings from reduced product damage and increased satisfaction of the receiver.

4.2 Testing Procedures and Conditioning

There are many organizations around the world that are involved in writing standard testing procedures and specifications for paperboard packaging and containers. The two most active in the U. S. are the Technical Association of the Pulp and Paper Industry (TAPPI) and the American Society for Testing and Materials (ASTM). Industry groups such as the Fibre Box Association (FBA) are active in developing and publishing guides to support these standards. Close liaison between these groups provides the container industry with a good network of standards and supporting implementation information.

4.3 Test Conditions

Nearly all TAPPI and ASTM test specifications require the same conditions for testing room environments: 20°C and 50% relative humidity (RH). Prior to placing samples in this controlled atmosphere, they are preconditioned in a warm 45°C and dry (25–30% RH) atmosphere. This use of preconditioning allows the fiber-based paper materials to reach an equilibrium moisture content in the 50% RH test environment from the dry conditions of the preconditioning chamber. Paper absorbs and retains moisture differently depending upon its history of moisture exposure. By preconditioning the paperboard, corrugated board, or containers a more reproducible test result is obtained.

4.4 Paperboard Tests

Basis Weight: Weight of the paperboard per unit area (e.g., grams per square meter).

Burst Test: A strength measure of the liner, medium, or corrugated board to resist a membrane pressure applied over a small area of the fiber sheet.

Caliper: Caliper is the thickness of the liner, medium, or corrugated board, expressed in millimeters or inches and measured with a micrometer.

Cushioning: A test for cushioning can be conducted over a wide array of loads and speeds but the usual report consists of deceleration plots for impacts of different speeds and of different weights.

Edge Crush Test: A rectangular area of corrugated board is cut precisely for edge loading to determine the edge-loaded strength of the section in the direction of the flutes.

Flat Crush of Corrugated Board: A circular specimen of a standard area is subjected to a compression from liner to liner to determine the fluted medium's ability to resist collapse.

Flat Crush of Corrugated Medium: The test uses a strip of medium that is corrugated in the lab and subsequently placed in compression to obtain a prediction of flat crush of the corrugated board.

Four-Point Flexure Test: This test uses a strip of corrugated board cut either along or across the flutes and measures the change in specimen radius between the two central loading points (the outer two being the supports) to calculate the flexural stiffness of the board.

Friction Coefficient: This test measures the "angle of slide" or the horizontal force needed to move a controlled weight over the surface of the paperboard.

Liner Crush: A long strip of liner is held rigidly upright and subjected to a compression force to measure the liner's ability to carry compression loads.

Porosity: This is a test for measuring the resistance to airflow through paperboard and has an influence on the corrugating and handling speeds.

Puncture: As a test for performance of corrugated board, the puncture test measures the ability of a board to resist a tetrahedral solid's puncture through the face of the corrugated board.

Short Span Compression: A narrow strip of paperboard is subjected to compression loads over a very short (>1 mm) gage length.

Tear: This is a measure of the force required to continue a tear in paperboard once an artificial "cut" has been made in the edge of the fiber sheet.

Tension: Regarded as a simple indicator of the structural performance of a paperboard, this indicates the force required to pull a narrow strip to failure.

Water Absorbtivity: This is a measure of the quantity of water absorbed by a paperboard in a specified time under standard conditions and relates to the rate at which a water-based adhesive may be sorbed during the corrugating process.

There is a wide array of other tests that are routinely used for specifying performance of paperboards for specific applications. The advent of the container as a marketing tool has required tests that allow evaluation of a paperboard as a printing substrate in addition to its ability to carry loads. We will not attempt to summarize these tests here.

4.5 Corrugated Fiberboard Container Tests

After fabrication, sealing, and usually, filling, container compression strength is measured on a flat platen compression tester. Measurements of compression strength and the container's load–deflection behavior may be completed for top-to-bottom, end-to-end, and side-to-side strength. Load is applied at a constant rate until the container fails catastrophically. Each of these loading directions may play a role in a container's shipping environment. Top-to-bottom is the most obvious condition when the container is in a stack or part of a palletized system of containers with pallets loaded on top. End-to-end and side-to-side compression may also be loading conditions when containers are palletized or stacked and require the use of clamping trucks or other clamping pressures to secure the load during handling or while in transit.

4.6 Other Container Handling Tests

4.6.1 Effects of Environmental Conditions

When conditions prevailing in storage or in transit may be much more severe than the standard conditions of testing (e.g., fresh or frozen food packaging), the compression testing may be carried out under these more severe conditions. Typically, when a container design is tested under these conditions, designers will be able to extrapolate the results to apply to containers of the same general design. However, when the design changes and a different mode of container failure may be controlling, these tests no longer have relevance.

4.6.2 Stack Life Testing

When goods are to be stored for extended periods under uncontrolled environmental conditions, tests may be needed for determining the time that a container can carry a load in compression. The loss of strength with time is accompanied by creep (deflection) of the container, resulting in stack tilting, which aggravates the situation and hastens the stack's failure. The tests may utilize machines to apply a constant load to determine the rate at which the container shortens (creep rate).

4.6.3 Vibration

These tests are intended to simulate the effects of vibration on the contents when the contents may be susceptible to damage (e.g., electronic devices or fresh produce). The interaction of the container, internal packing, and the product yields a unique set of circumstances in terms of natural frequencies. Thus, the latest test techniques call for application of a vibration spectrum from actual measurements of handling, road, or rail vibration conditions. When tested with expected vibrational spectra, the designer can assess factors of safety by stipulating the number of "transit lifetimes" that a package and its contents should endure.

4.6.4 Inclined Impact

Designed to simulate a massive low speed impact, this test uses a wheeled test platform to which the containers are mounted. The platform is released down an inclined ramp to meet with a solid bulkhead. Due to the variability of the test platforms, bulkheads, and other factors, this test is primarily used to compare one packaging situation to another on the same inclined impact device.

4.6.5 Revolving Drum Test

As the name implies, a six-sided drum with baffles mounted inside is rotated, which allows the container inside the drum to roll or tumble in an uncontrolled manner. This test provides some insight on the integrity of the package joints and closures, and the ability of the package to resist shifting and penetration of the contents through the sidewalls.

5. CUSHIONING MATERIALS FROM AGRO-BASED RESOURCES

A small but identifiable quantity of fibrous material is used in packaging of delicate items to create an extremely soft cushioning material. Prime examples of products that may require this level of softness are vacuum tubes, delicate electronics, crystal glassware, and fine china. Traditional materials used for this purpose are excelsior (wood wool), coir (coconut husk fiber), and shredded wastepaper. For the bulk of packaged items, the corrugated board provides adequate levels of cushioning, especially in the A and C flute configurations. Singlewall is also still in use for cushioning items which can be wrapped in a tubular fashion.

6. PARTITIONS IN PACKAGING

When packaging multiple individual items, the use of partitions may be desired to prevent item-to-item scuffing, rubbing, and impact. Partitions may be made from a wide array of sheet materials but solid fiberboard is the most common. The partitions usually impart no stacking strength and, thus, have little structural requirements other than cushioning or integrity to keep items apart. Corrugated and singlewall are also used when added cushioning is desired.

7. MOLDED PULP PACKAGING

A growing segment of the packaging industry relies upon the use of molded fiber products in packaging. Examples include: spacers to isolate products from container sidewalls; trays for perishable or fragile items such as eggs, apples, and frozen meats; and damage protection for self-packaged large items such as furniture and appliances. The processes for molding are described in the section on Structural Molded Pulp Products.

8. STRUCTURAL MOLDED PULP PRODUCTS

The development of new process technologies to produce products from cellulose pulps has been an active area of research at the USDA Forest Service, Forest Products Laboratory (FPL). A decade ago, Setterholm (1985) introduced the unique method of forming a three-dimensional, wafflelike structure from molded wood pulp. He called the board "Space-board" because of the presence of open cells or "space" between the ribs of the "board" (Figure 10.1). At the time, Setterholm envisioned producing a Spaceboard panel that would have strength characteristics similar to that of corrugated boxboard but could be produced in a one-step forming process. Additionally, the process could accommodate under-utilized fiber sources such as mixed hardwoods and recycled papers. These two goals set the stage for several breakthroughs in molded pulp processing technology at FPL. Subsequent process improvements by other FPL researchers were developed to optimize the formation and densification of Spaceboard. With these improvements, it became possible to produce Spaceboard in a variety of sizes, ranging from thin boxboard (Hunt and Gunderson, 1988) to thick sheathing panels called Spaceboard II (Scott and Laufenberg, 1994).

Figure 10.1 A wide variety of products can be made using the "Spaceboard" technology.

9. WET PULP MOLDING PROCESS

With water as the forming medium, two basic mechanisms determine bond strength development: fiber flexibility (conformation) and hydrogen bonding. When the board is uniformly densified at elevated temperatures, the conformable fibers are pressed into intimate contact with each other. As the water vaporizes, strong hydrogen bonds are produced, resulting in densities of approximately 1.0 g/cc.

Wet pulp molding requires two basic steps: (1) forming a dense network of wet fiber onto a configured surface, and (2) drying the dense network (Figure 10.2). In the vast majority of molded pulp products, the forming is done onto a hard drainable surface. That surface may be a multilayer of stainless steel backing materials with the pulp contact surface of fine mesh or perforated plate to retain the pulp fiber. Deposition of the fiber onto that surface may be through a number of approaches: (1) gravity flow of the pulp suspension through the porous surface, (2) dipping of the mold into the pulp suspension while withdrawing water through the mold surface thus depositing the fiber onto the mold surface, or (3) pumping the pulp into a closed mold space and withdrawing water from the mold faces. Forming may be accomplished with pulp consistencies (dry weight/wet weight) of 0.5–5 percent. For applications that require highly uniform formations of the wet pulp mat, the consistency needs to be quite low.

Figure 10.2 The "Spaceboard" process uses air or water to disperse fibers over deformable molds to make a honeycomb type panel with unique structural properties.

Drying of the wet pulp mat may be accomplished in a wide variety of ways. The major variable that differentiates molded pulp from structural molded pulp products is the extent of densification and pressure applied when the pulp is drying. Minimal density (and strength) of the resulting pulp product will occur if no pressure is applied. This type of molded pulp product is typically formed on the wet mold, vacuum dewatered, then deposited onto a dryer conveyor system for the extended time required to dry the product (0.5–4 h). Products of this type are used as spacers, corner blocking, cushioning materials, or trays for perishable or sensitive items.

REFERENCES

ASTM, *Annual Book of ASTM Standards*, Philadelphia, Vol. 15.09, Paper; Packaging; Flexible Barrier Materials; Business Copy Products, PA, 1970.

Dolan, J. J., Uniform Freight Classification Rule 41, National Railroad Freight Committee, Chicago, IL, 1991.

Hunt, J. F., and Gunderson, D. E., FPL Spaceboard Development, in: *TAPPI Proceedings of the 198th Corrugated Containers Conference*, TAPPI Press: Atlanta, GA, 11, 1988.

Koning, John, Corrugated Crossroads, TAPPI Press, Atlanta, GA, 341p., 1995.

National Motor Freight Classification Item 222, American Trucking Associations, Inc., Alexandria, VA, 1991.

Ostrem, F. E., and Godshall, W. D., An assessment of the common carrier shipping environment, *General Technical Report, FPL 22*, Forest Products Laboratory, Madison, WI, 1979.

Scott, C. T., and Laufenberg, T. L., Spaceboard II panels: Preliminary evaluation of mechanical properties, in PTEC 94 Timber Shaping the Future, *Proceedings of the Pacific Timber Engineering Conference*; 1994 July 11–15, Gold Coast, Australia. Fortitude Valley, Queensland, Australia: TRADA: 632, Vol. 2, 1994.

Setterholm, V. C., FPL spaceboard new structural sandwich concept, *TAPPI* J. 68(6), 40, 1985.

Wright, P. G., McKinlay, P. R., and Shaw, E. Y. N., *Corrugated Fibreboard Boxes*, APM Packaging, Melbourne, VIC, Australia, 289 p., 1988.

CHAPTER 11

Chemical Modification of Agro-Resources for Property Enhancement

Roger M. Rowell

CONTENTS

1. INTRODUCTION

When considering lignocellulosics as possible engineering materials, there are several very basic concepts that must be considered. First, lignocellulosics are hygroscopic resources that were designed to perform, in nature, in a wet environment. Second, nature is programmed to recycle lignocellulosics in a timely way through biological, thermal, aqueous, photochemical, chemical, and mechanical degradations. In simple terms, nature builds a lignocellulosic from carbon dioxide and water and has all the tools to recycle it back to the starting chemicals. We harvest a green lignocellulosic (for example, a tree) and convert it into dry products, and nature, with its arsenal of degrading reactions, starts to reclaim it at its first opportunity (Figure 11.1).

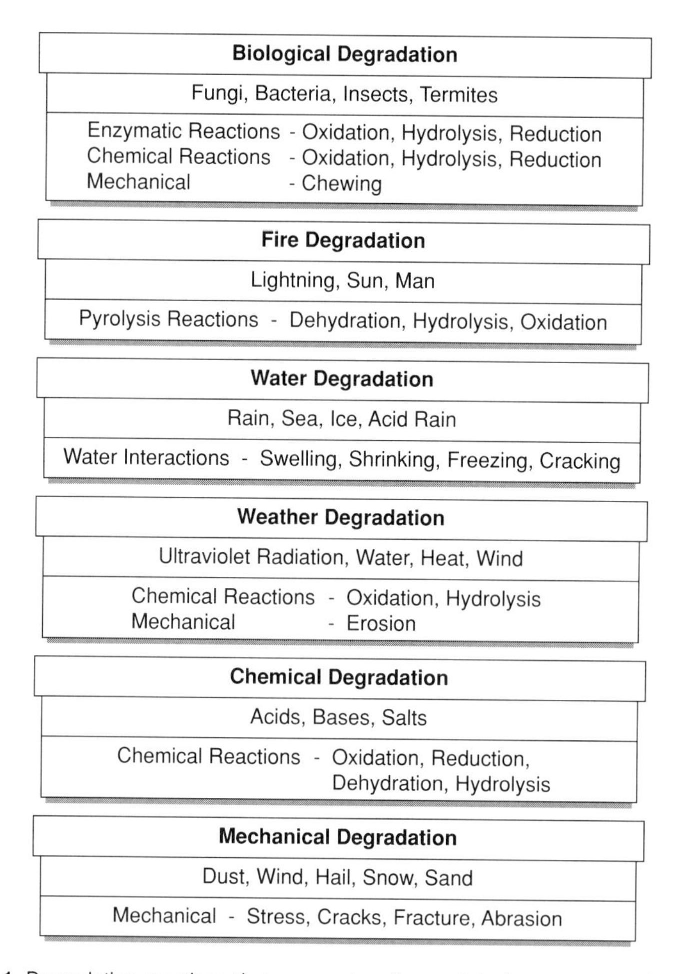

Figure 11.1 Degradation reactions that occur when lignocellulosics are exposed to nature.

In order to produce lignocellulosic-based engineering materials with a long service life, it is necessary to interfere with the natural degradation processes for as long as possible. This can be done in several ways. Traditional methods for decay resistance and fire retardancy, for example, are based on treating the product with toxic or corrosive chemicals which are effective in providing decay and fire resistance but can result in environmental concerns. There is another approach, which is based on the premise that the properties of any resource are a result of the chemistry of components of that resource. In the case of lignocellulosics, cell wall polymers, extractives, and inorganics are the components that, if modified, would change the properties of the resource.

In order to make property changes, one must first understand the chemistry of the components and the contribution each plays in the properties of the resource. Following this understanding, you must then devise a way to modify what needs to be changed to get the desired change in property.

Properties of lignocellulosics, such as dimensional instability, flammability, biodegradability, and degradation caused by acids, bases, and ultraviolet radiation, are all a result of chemical degradation reactions which can be prevented or, at least, slowed down if the cell wall chemistry is altered (Rowell, 1975; Rowell and Youngs, 1981; Rowell, 1983; Rowell and Konkol, 1987; Rowell et al., 1988a; Hon, 1992; Rowell, 1992; Kumar, 1994; Banks and Lawther, 1994).

2. FEATURES OF LIGNOCELLULOSICS

Lignocellulosics are three-dimensional, polymeric composites made up primarily of cellulose, hemicelluloses, and lignin. Other chapters in this book give information of the chemical composition and fiber properties of many different types of natural fibers. Without bias to any given industry (i.e., wood, kenaf, jute, bamboo, etc.), fiber selection for a given product could be made on the basis of chemical composition, fiber dimension, availability, ease of handling, cost, and other factors.

While all types of lignocellulosic fibers differ in chemical composition, within certain limits, all lignocellulosics have very similar properties. That is, they all swell and shrink as the moisture content of the cell wall changes; they burn, they decay, and they are degraded by acids, bases, and ultraviolet radiation. As a general class, all lignocellulosics have similar mechanisms of environmental degradation; it thus might be expected that all types of natural fibers would respond to the same types of chemical treatments to overcome these degradation reactions.

To improve the resistance to the degradation forces acting on lignocellulosics, it is first important to understand the mechanisms of degradation, which components in the cell wall are responsible for these effects, and what can be done to slow down or stop the degradation forces.

3. DEGRADATION OF LIGNOCELLULOSICS

Figure 11.2 shows the cell wall polymers involved in each fiber property as we understand it today (Rowell, 1990). Lignocellulosics change dimensions with changing moisture content because the cell wall polymers contain hydroxyl and other oxygen-containing groups that attract moisture through hydrogen bonding (Stamm, 1964; Rowell and Banks, 1985). The hemicelluloses are mainly responsible for moisture sorption, but the accessible cellulose, noncrystalline cellulose, lignin, and surface of crystalline cellulose also play major roles. Moisture swells the cell wall, and the fiber expands until the cell wall is saturated with water (fiber saturation point, FSP). Beyond this saturation point, moisture exists as free water in the void structure and does not contribute to further expansion. This process is reversible, and the fiber shrinks as it loses moisture below the FSP.

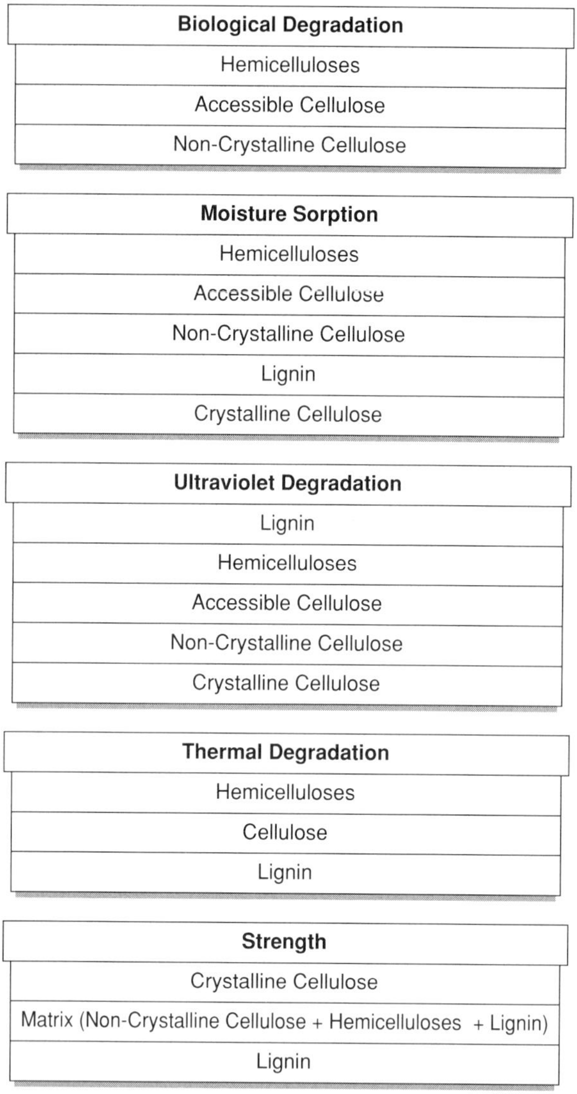

Biological Degradation
Hemicelluloses
Accessible Cellulose
Non-Crystalline Cellulose

Moisture Sorption
Hemicelluloses
Accessible Cellulose
Non-Crystalline Cellulose
Lignin
Crystalline Cellulose

Ultraviolet Degradation
Lignin
Hemicelluloses
Accessible Cellulose
Non-Crystalline Cellulose
Crystalline Cellulose

Thermal Degradation
Hemicelluloses
Cellulose
Lignin

Strength
Crystalline Cellulose
Matrix (Non-Crystalline Cellulose + Hemicelluloses + Lignin)
Lignin

Figure 11.2 Cell wall polymers responsible for the properties of lignocellulosics in order of importance.

Lignocellulosics are degraded biologically because organisms recognize the carbohydrate polymers (mainly the hemicelluloses) in the cell wall and have very specific enzyme systems capable of hydrolyzing these polymers into digestible units.

Biodegradation of the high molecular weight cellulose weakens the fiber cell wall because crystalline cellulose is primarily responsible for the strength of the cell wall (Rowell et al., 1988b). Strength is lost as the cellulose polymer undergoes degradation through oxidation, hydrolysis, and dehydration reactions. The same types of reactions take place in the presence of acids and bases.

Lignocellulosics exposed outdoors undergo photochemical degradation caused by ultraviolet radiation. This degradation takes place primarily in the lignin component, which is responsible for the characteristic color changes (Rowell, 1984). The lignin acts as an adhesive in the cell walls, holding the cellulose fibers together. The surface becomes richer in cellulose content as the lignin degrades. In comparison to lignin, cellulose is much less susceptible to ultraviolet light degradation. After the lignin has been degraded, the poorly-bonded, carbohydrate-rich fibers erode easily from the surface, which exposes new lignin to further degradative reactions. In time, this "weathering" process causes the surface of the composite to become rough and can account for a significant loss in surface fibers.

Lignocellulosics burn because the cell wall polymers undergo pyrolysis reactions with increasing temperature to give off volatile, flammable gasses. The hemicellulose and cellulose polymers are degraded by heat much before the lignin (Rowell, 1984). The lignin component contributes to char formation, and the charred layer helps insulate the composite from further thermal degradation.

4. CHEMICAL MODIFICATION SYSTEMS

For this discussion, chemical modification will be defined as a chemical reaction between some reactive part of a lignocellulosic and a simple single chemical reagent, with or without catalyst, to form a covalent bond between the two. This excludes all simple chemical impregnation treatments which do not form covalent bonds, monomer impregnation treatments that polymerize *in situ* but do not bond with the cell wall, polymer inclusions, coatings, heat treatments, etc.

There are several approaches to chemically modifying the lignocellulosic cell wall polymers. The most abundant single site for reactivity in these polymers is the hydroxyl group, and most reaction schemes have been based on the reaction of hydroxyl groups. Sites of unsaturation in the lignin structure can also be used as a point of reactivity as well as free radical additions and grafting. However, the most studied class of chemical reactions are those involving hydroxyl substitutions.

In modifying a lignocellulosic for property improvement, there are several basic principles that must be considered in selecting a reagent and a reaction system (Rowell, 1975). Of the thousands of chemicals available, either commercially or by synthetic means, most can be eliminated because they fail to meet the requirements or properties listed below.

If hydroxyl reactivity is selected as the preferred modification site, the chemical must contain functional groups which will react with the hydroxyl groups of the lignocellulosic components. This may seem obvious but there are several failed reaction systems in the literature using a chemical that could not react with a hydroxyl group.

The overall toxicity of the chemicals must be carefully considered. The chemicals must not be toxic or carcinogenic to humans in the finished product, and should be as nontoxic as possible in the treating stage. The chemical should be as noncorrosive as possible to eliminate the need for special stainless steel or glass-lined treating equipment.

In considering the ease with which excess reagents can be removed after treatment, a liquid treating chemical with a low boiling point is advantageous. Likewise, if the boiling point of a liquid reagent is too high, it will be very difficult to remove the chemical after treatment. It is generally true that the lowest member of a homologous series is the most reactive and will have the lowest boiling point. The boiling point range for liquids to be considered is 90–150°C. It is also possible to treat fibers with a gas system; however, there may be processing challenges in handling a pressurized gas in a continuous reactor.

Accessibility of the reagent to the reactive chemical sites is a major consideration. To increase accessibility to the reaction site, the chemical must swell the lignocellulosic structure. If the reagents do not swell the structure, then another chemical or co-solvent can be added to meet this requirement. Accessibility to the reactive site is a major consideration in a gas system unless there is a condensation step in the procedure.

Almost all chemical reactions require a catalyst. With lignocellulosics as the reacting substrate, strong acid or base catalysts cannot be used as they cause extensive degradation. The most favorable catalyst from the standpoint of lignocellulosic degradation is a weakly alkaline one. The alkaline medium is also favored as in many cases these chemicals swell the cell wall matrix structure and give better penetration. The properties of the catalyst parallel those of reagents, i.e., low boiling point liquid, nontoxic, effective at low temperatures, etc. In most cases, the organic tertiary amines or weak organic acids are best suited.

The experimental reaction conditions which must be met in order for a given reaction to go is another important consideration. The temperature required for complete reaction must be low enough so there is little or no fiber degradation, i.e., less than 150°C. The reaction must also have a relatively fast rate of reaction with the cell wall components. It is important to get as fast a reaction as possible at the lowest temperature without lignocellulosic degradation.

The moisture present in the lignocellulosic is another consideration in the reaction conditions. It is costly to dry lignocellulosics to less than 1 percent moisture, but it must be remembered that the –OH group in water is more reactive than the –OH group available in the lignocellulosic components, i.e., hydrolysis is faster than substitution. The most favorable condition is a reaction which requires a trace of moisture and the rate of hydrolysis is relatively slow.

Another consideration in this area is to keep the reaction system as simple as possible. Multicomponent systems will require complex separation after reaction for chemical recovery. The optimum would be a reactive chemical that swells the lignocellulosic structure and acts as the solvent as well.

If possible, byproducts should be avoided during the reaction since they may need to be removed. If there is not a 100% reagent skeleton add-on, then the chemical cost is higher and will require recovery of the byproduct for economic and environmental reasons.

The chemical bond formed between the reagent and the lignocellulosic components is of major importance. For permanence, this bond should have great stability to withstand weathering. In order of stability, the types of covalent chemical bonds that may be formed are: ethers > acetals > esters. The ether bond is the most desirable covalent carbon-oxygen bond that can be formed. These bonds are more stable than the glycosidic bonds between sugar units in the lignocellulosic polysaccharides so the polymers would degrade before the grafted ether. It may be desired, however, to have the bonded chemical released by hydrolysis or enzyme action in the final product so that an unstable bond may be desirable from the modification.

The hydrophobic nature of the reagent needs to be considered. The chemical added to the lignocellulosic should not increase the hydrophilic nature of the lignocellulosic components unless that is a desired property.

If the hydrophilicity is increased, the susceptibility to micro-organism attack increases. The more hydrophobic the component can be made, the better the moisture exclusion properties of the substituted lignocellulosic will be.

Single site substitution versus polymer formation is another consideration. For the most part, a single reagent molecule that reacts with a single hydroxyl group is the most desirable. Crosslinking can occur when the reagent contains more than one reactive group or results in a group which can further react with a hydroxyl group. Crosslinking can cause the lignocellulosic to become more brittle. Polymer formation within the cell wall after initial reaction with the hydroxyl groups of the lignocellulosic components gives, through bulking action, dimensional stabilization. The disadvantage of polymer formation is that a higher level of chemical add-on is required for biological resistance than is required in the single site reactions.

The treated lignocellulosic must still possess the desirable properties of lignocellulosics. That is, the fiber strength should not be reduced; no change in color; good electrical insulation properties retained; final product not dangerous to handle; no lingering chemical smells; and still gluable and finishable, unless one or more of these properties are the object of change in the product.

A final consideration is, of course, the cost of chemicals and processing. In laboratory scale experimental reactions, the high cost of chemicals is not a major factor. For commercialization of a process, however, the chemical and processing costs are very important factors. Laboratory scale research is generally done using small batch processing; however, rapid, continuous processes should always be studied for scale-up. Economy of scale can make an expensive laboratory process economical.

In summary, the chemicals to be laboratory tested must be capable of reacting with lignocellulosic hydroxyls under neutral, mildly alkaline or acid conditions at temperatures below 150°C. The chemical system should be simple and capable of swelling the structure to facilitate penetration. The complete molecule should react quickly with lignocellulosic components yielding stable chemical bonds, and the treated lignocellulosic must still possess the desirable properties of untreated lignocellulosics.

5. CHEMICAL MODIFICATION FOR PROPERTY ENHANCEMENT

As was previously stated, because the properties of lignocellulosics result from the chemistry of the cell wall components, the basic properties of a fiber can be changed by modifying the basic chemistry of the cell wall polymers. Many chemical reaction systems have been published for the modification of agro-fiber. These chemicals include anhydrides such as: phthalic, succinic, maleic, propionic and butyric anhydride, acid chlorides, ketene carboxylic acids, many different types of isocyanates, formaldehyde, acetaldehyde, ifunctional aldehydes, chloral, phthaldehydic acid, dimethyl sulfate, alkyl chlorides, β-propiolactone, acrylonitrile, epoxides (such as ethylene, propylene, and butylene oxide), and difunctional epoxides (Rowell, 1983, 1991).

By far, the most research has been done on the reaction of acetic anhydride with cell wall polymer hydroxyl groups to give an acetylated fiber. Many different types of lignocellulosic fibers have been acetylated using a variety of procedures including wood (Rowell, 1983; Rowell et al., 1986), bamboo (Rowell and Norimoto, 1987, 1988), bagasse (Rowell and Keany, 1991), jute (Callow, 1951; Andersson and Tillman, 1989; Rowell et al., 1991), kenaf (Rowell, 1993; Rowell and Harrison, 1993), pennywort, and water hyacinth (Rowell and Rowell, 1989). Without a strong catalyst, acetylation using acetic anhydride alone levels off at approximately 20 weight percent gain (WPG) for softwoods, hardwoods, grasses, and water plants. While acetylation is not the only chemical modification procedure that has been shown to improve properties of lignocellulosics, it has been studied the most and will be used as an example in many cases.

6. PROPERTIES OF CHEMICALLY MODIFIED FIBER

6.1 Moisture Sorption

By replacing some of the hydroxyl groups on the cell wall polymers with bonded chemical groups, the hygroscopicity of the lignocellulosic material is reduced. Table 11.1 shows the equilibrium moisture content (EMC) of several types of lignocellulosic fibers that have been reacted with several types of chemicals. Table 11.1 shows the EMC of pine wood fibers that have been reacted with different chemicals. In all cases, the EMC has been reduced as a result of modification. Both reactions with acetic anhydride and formaldehyde give the best results in lowering the EMC of the treated fiber. Table 11.2 shows the results of acetylating several different types of fibers on the EMC of the modified fiber. In all cases, as the level of acetyl weight gain increases, the EMC of the resulting fiber goes down. All types of fiber show the same level in the reduction in EMC as a function of level of acetyl weight gain (Rowell et al., 1986).

If the reductions in EMC at 65% RH of acetylated fiber referenced to unacetylated fiber are plotted as a function of the bonded acetyl content, a straight line plot results (Rowell and Rowell, 1989). Even though the points shown in Figure 11.3 represent many different types of lignocellulosic resources, they all fit a common

Table 11.1 Equilibrium Moisture Content (EMC) of Control and Chemically Modified Pine Fiber

		Equilibrium Moisture Content at 27°C		
Chemical	**Weight Gain, %**	**30%RH**	**65%RH**	**90%RH**
Control	0	5.8	12	21.7
Acetic Anhydride	20.4	2.4	4.3	8.4
Formaldehyde	3.9	3	4.2	6.2
Propylene Oxide	21.9	3.9	6.1	13.1
Butylene Oxide	18.7	3.5	5.7	10.7

Table 11.2 Equilibrium Moisture Content (EMC) of Fiberboards Made from Control and Acetylated Fiber

		EquilibriumMoisture Content at 27°C		
Fiber	**Weight Gain, %**	**30%RH**	**65%RH**	**90%RH**
Bagasse	0	4.4	8.8	15.8
	9.4	2	5.3	9.5
	13	1.7	4.4	7.7
	17.6	1.4	3.4	5.8
Kenaf	0	4.8	10.5	24.3
	18.4	2.6	5.8	11.3
Bamboo	0	4.5	8.9	14.7
	10.8	3.1	5.3	9.4
	17	2	3.7	6.8
Jute	0	5.8	9.3	18.3
	16.2	2	4.1	7.8
Pine	0	5.6	12.1	22.6
	6.3	4.5	10.2	19.5
	13.8	2.7	6.8	13.2
	18.2	2.1	5.1	9.9
Aspen	0	4.9	11.1	21.5
	8.7	3.1	7.7	14.9
	13	2	5.9	11.8
	17.6	1.6	4.8	9.4

curve. A maximum reduction in EMC is achieved at about 20% bonded acetyl. Extrapolation of the plot to 100% reduction in EMC would occur at about 30% bonded acetyl. This represents a value not too different from the fiber saturation point for water in these fibers. Because the acetate group is larger than the water molecule, not all hygroscopic hydrogen-bonding sites are covered, so it would be expected that the acetyl saturation point would be lower than that of water. This is finding would indicate that it does not matter which type of lignocellulosic resource is acetylated for composites.

The fact that EMC reduction as a function of acetyl content is the same for many different lignocellulosic resources indicates that reducing moisture sorption and, therefore, achieving cell wall stability are controlled by a common factor. The lignin, hemicellulose, and cellulose contents of all the materials plotted in Figure 11.3 are different. Earlier results showed that the bonded acetate was mainly in the lignin and hemicelluloses (Rowell, 1982) and that isolated wood cellulose does not react with uncatalyzed acetic anhydride (Rowell et al., 1994b).

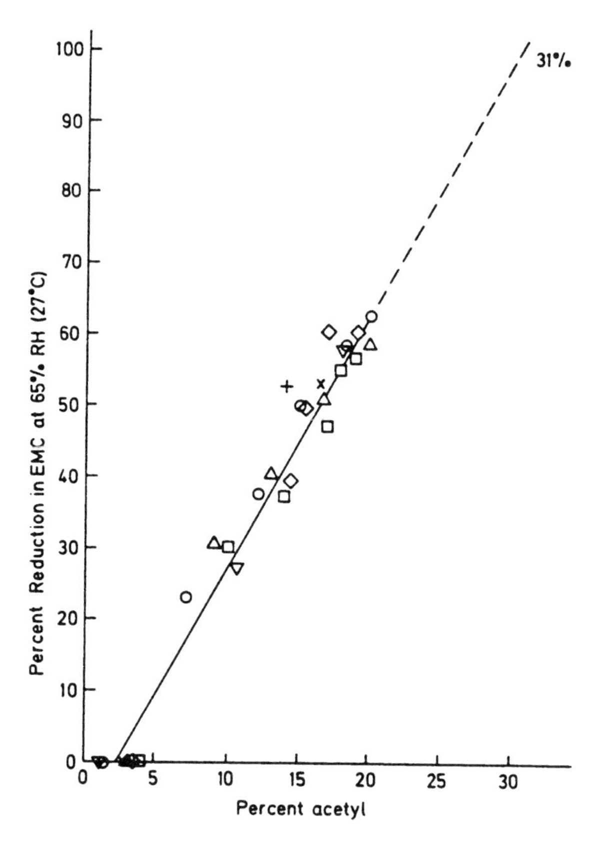

Figure 11.3 Reduction in equilibrium moisture content (EMC) as a function of bonded acetyl content for various acetylated lignocellulosic materials. O, southern pine; □ aspen; Δ bamboo; ◇ bagasse; X, jute; +, pennywort; ∇ water hyacinth.

Because these materials vary widely in their lignin, hemicellulose, and cellulose content, with acetate found mainly in the lignin and hemicellulose polymer, and because isolated cellulose does not acetylate by this procedure—acetylation may be controlling the moisture sensitivity through the lignin and hemicellulose polymers in the cell wall, but not reducing the sorption of moisture in the cellulose polymer.

6.2 Pyrolysis Properties

Chemical modification of agro-based fibers has some effect on the pyrolysis properties of lignocellulosics. In thermogravimetric analysis, control and chemically-modified pine fibers pyrolyze at about the same temperature and rate (Table 11.3, Rowell et al., 1984). Fibers reacted with propylene or butylene oxides have a slightly higher temperature of maximum weight loss. Fibers that were reacted with acetic anhydride or methyl isocyanate showed two peaks in the maximum weight loss data, while both propylene and butylene oxides resulted in only one peak. Since the smaller, lower temperature peak represents the hemicellulose fraction in the fiber, the epoxide-modified hemicelluloses seem to pyrolyze in the temperature range of

Table 11.3 Pyrolysis Properties of Control and Chemically Modified Pine Fiber

Chemical	Weight Percent Gain, %	Temperature of Maximum Weight Loss, °C	Heat of Combustion, kcal/g	Rate of Oxygen Consumption, MM/g sec
None	0	335/375	2.9	0.06/0.13
Acetic Anhydride	21.1	338/375	3.1	0.08/0.14
Methyl Isocyanate	24	315/375	2.6	0.07/0.12
Propylene Oxide	32	380	4.3	0.23
Butylene Oxide	22	385	4.1	0.24

the cellulose fraction. The heat of combustion and rate of oxygen consumption are higher for the epoxide modified fiber as compared to the control, acetic anhydride and methyl isocyanate modified fibers. This data would indicate that reacting fibers with acetic anhydride or methyl isocyanate adds approximately the same carbon, hydrogen, and oxygen content as do the cell wall polymers. Reactive fire retardants could be bonded to the cell wall hydroxyl groups in reactions similar to this approach. The effect would be an improvement in dimensional stability, biological resistance, and fire retardancy.

7. PROPERTIES OF COMPOSITES MADE WITH CHEMICALLY MODIFIED FIBER

7.1 Dimensional Stability

Changes in dimensions, especially in thickness and in linear expansion, are a great problem in lignocellulosic composites because they not only undergo normal swelling (reversible swelling) but also undergo swelling caused by the release of residual compressive stresses imparted to the board during the composite pressing process (irreversible swelling). Water sorption causes both reversible and irreversible swelling, with some of the reversible shrinkage occurring when the board dries. Dimensional instability of lignocellulosic composites has been the major reason for their restricted use.

The EMC values of different types of control and acetylated fibers are given in Table 11.4. The EMC and thickness swelling at three relative humidities for boards made from these fibers are shown in Table 11.4. Comparing the data in Tables 11.1 (fibe alone) and 11.4 (fiberboard) , it can be seen that the EMC for boards is slightly higher than for the fiber alone. The adhesive is more hydrophilic than the acetylated fiber.

Thickness swelling at the three levels of relative humidity is greatly reduced as a result of acetylation. Linear expansion is also greatly reduced as a result of acetylation (Krzysik et al., 1992, 1993). The rate and extent of thickness swelling in liquid water of fiberboards made from control and acetylated fiber are shown in Table 11.5. Both the rate and extent of swelling are greatly reduced as a result of acetylation. At the end of 5 days of water soaking, control boards swelled from

Table 11.4 Equilibrium Moisture Content (EMC) and Thickness Swelling (TS) of Fiberboards Made from Control and Acetylated Fiber

		EMC and TS at 27°C					
Fiber	Weight Percent Gain, %	30% RH EMC, %	TS %	65% RH EMC, %	TS %	90% RH EMC, %	TS %
Pine	0.0	4.5	3.6	9.4	6.6	19.7	29.2
	21.6	1.8	0.4	4.1	1.1	8.3	2.9
Bagasse	0.0	3.8	—	7.6	—	17.1	—
	17.6	1.8	—	4.0	—	7.9	—
Kenaf	0.0	4.8	3.0	10.5	9.6	26.7	33.0
	18.4	2.6	0.8	5.8	2.4	19.3	10.0
Bamboo	0.0	3.2	—	6.6	—	12.3	—
	18.0	1.6	—	4.1	—	7.9	—
Hemlock	0.0	3.3	1.0	7.2	3.1	19.8	11.2
	22.5	1.6	0.2	3.9	1.7	9.3	3.1

18–45%, whereas boards made from acetylated fiber swelled from 3–10%. Drying all boards after the water soaking test shows the amount of irreversible swelling that resulted from water swelling. Control boards show a greater degree of irreversible swelling as compared to boards made from acetylated fiber.

The results of both water vapor and liquid water tests show that acetylation of lignocellulosic fibers greatly improves dimensional stability of composites made from these resources.

7.2 Biological Resistance

Particleboards and flakeboards made from acetylated flakes have been tested for resistance to several different types of organisms. In a 2-week termite test using *Reticulitermes flavipes* (subterranean termites), boards acetylated at 16–17 WPG were very resistant to attack, but not completely (Table 11.6, Rowell et al., 1979, 1988a). This may be attributed to the severity of the test. However, since termites can live on acetic acid and decompose cellulose to mainly acetic acid, perhaps it is not surprising that acetylated wood is not completely resistant to termite attack.

Chemically-modified composites have been tested with decay fungi in several ways. Control and chemically-modified particleboards were exposed to a 12-week soil block test using the brown rot fungus *Gloeophyllum trabeum* and the white rot fungus *Trametes versicolor* (Table 11.7). All boards were made using a phenolic resin (Nilsson et al., 1988; Rowell et al., 1988a). All of the bonded chemicals at a WPG over about 20 show good resistance to brown- and white-rot fungi, except propylene oxide in the brown-rot test. Propylene oxide is not effective in preventing attack by brown-rot fungi even though the same number of hydroxyl groups should be modified as were modified by reaction with butylene oxide, methyl isocyanate, acetic anhydride, β-propiolactone, or acrylonitrile (Rowell et al., 1988b). This exception of propylene oxide to the protection rule is perhaps the key to understanding the mechanism of the resistance to attack by fungi by chemical modification. As was seen in Table 11.4, the EMC of propylene oxide-modified fiber is higher than

Table 11.5 Rate and Extent of Thickness Swelling in Liquid Water of Fiberboards Made from Control and Acetylated Fiber and a Phenolic Resin[1]

	Thickness Swelling at Hours															
	Minutes			Hours						Days						
Fiber	15	30	45	1	2	3	4	5	6	1	2	3	4	5	Oven drying, %	Weight loss after test,%
Kenaf Control	15.5	17.1	21.1	22.6	24.7	26.8	31.1	32.6	34	37.7	41.5	42.6	43.5	44.5	19	2
18.4 WPG	6.7	6.8	6.8	7	7	7	8	8.1	8.3	8.5	8.5	8.7	8.8	9	0.7	2.8
Bagasse Control	19.2	20.2	21	21.6	22	22.7	23	23.6	24	25	25.2	25.3	25.4	25.5	16	1.2
17.6 WPG	1.8	2	2.2	2.3	2.7	2.9	3.3	3.5	3.8	5	5	5.1	5.2	5.2	1.5	1.4
Bamboo Control	4	7.3	8.4	10.2	12.6	13.8	14	14.8	15	16.1	16.5	17.9	18.1	18.2	8.3	—
18.0 WPG	1.5	1.7	1.9	2.3	2.3	2.3	2.4	2.4	2.4	2.5	3.1	3.2	3.2	3.3	2.2	—
Hemlock Control	11.2	11.8	12.3	12.5	14.1	15.2	16.2	16.8	17	17.3	17.5	17.8	17.9	18.1	7.8	2.9
22.5 WPG	2.6	3.3	3.7	3.8	3.9	4	4	4.1	4.2	5.2	5.6	5.8	6	6.6	1.7	1.9
Pine Control	25.7	29.8	30.7	31.6	32.9	33.5	33.8	33.9	34	35	35.6	35.9	36	36.2	24.9	0.5
21.6 WPG	0.6	0.9	1.1	1.2	1.6	1.9	2.1	2.2	2.5	3.7	4	4.2	4.3	4.5	2.6	1.1

[1] Resin content of boards: Kenaf – 8%, Bagasse – 5%, Bamboo – 6%, Hemlock – 8%, Pine – 8%.

Table 11.6 Weight Loss in Chemically Modified Southern Pine After 2 Weeks' Exposure to *Reticulitermes flavipes*

Chemical	Weight Gain, %	Wood Weight Loss, %
Control	0	31
Propylene oxide	9	21
	17	14
	34	6
Butylene oxide	27	4
	34	3
Acetic anhydride	10.4	9
	17.8	6
	21.6	5

any other modified fiber and this may be the reason for the lower biological resistance.

Table 11.7 Biological Resistance of Chemically Modified Pine Against Brown- and White-Rot Fungi

		Weight Loss After 12 Weeks	
Chemical	Weight Gain, %	Brown-rot Fungus %	White-rot Fungus %
None	0	68	7
Acetic anhydride	17	<2	<2
Propylene oxide	25	<15	<2
Butylene oxide	22	<3	<1
Methylisocyanate	20	<3	<1
Formaldehyde	5	<3	<1
β-Propiolactone	25	<2	<2
Acrylonitrile	25	<2	<2

The mechanism of brown-rot fungi attack on lignocellulosics is thought to be as given in Figure 11.4 (Nilsson, 1986). The first biological attack on a lignocellulosic is an enzymatic reaction that results in a metal/peroxide chemical oxidation system. This oxidation system breaks down the large polymers into smaller pieces which results in an early and rapid strength loss as the degree of polymerization of the cellulose molecule is reduced. During this reaction phase, a second enzymatic system starts working in which carbohydrates and lignin are broken down. It is in this phase that weight loss occurs.

This mechanism is consistent with the data that strength losses occur long before weight losses in brown-rot fungi attacked wood (Rowell et al., 1988b). In this mechanism, the key to brown-rot fungi resistance lies in the protection of the hemicellulose polymers. If that single component is protected, attack cannot proceed.

Weight loss resulting from fungal attack is the method most used to determine the effectiveness of a preservative treatment to protect wood composites from decaying. In some cases, especially for brown-rot fungal attack, strength loss may be a more important measure of attack since large strength losses are known to occur in solid wood at very low wood weight loss (Couling, 1961). A dynamic bending–creep

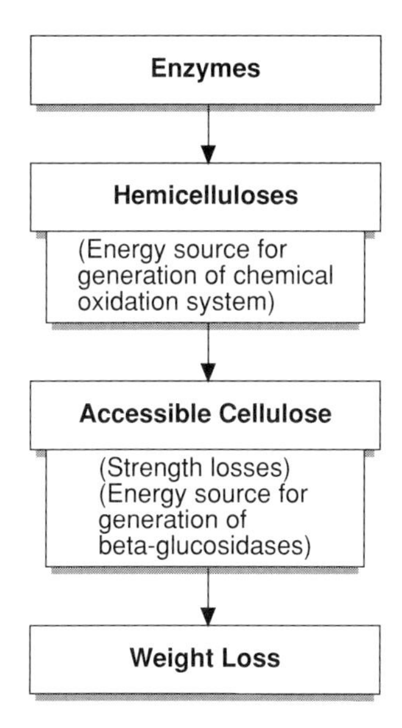

Figure 11.4 Mechanism of brown-rot fungus attack on lignocellulosics.

test has been developed to determine strength losses when wood composites are exposed to a brown- or white-rot fungus (Imamura and Nishimoto, 1985).

Using this bending-–reep test on aspen flakeboards, control boards made with phenol-formaldehyde adhesive failed in an average of 71 days using the brown-rot fungus *T. palustris* and 212 days using the white-rot fungus *T. versicolor* (Rowell et al., 1988b). At failure, weight losses averaged 7.8% for *T. palustris* and 31.6% for *T. versicolor*. Isocyanate-bonded control flakeboards failed in an average of 20 days with *T. palustris* and 118 days with *T. versicolor*, with an average weight loss at failure of 5.5% and 34.4%, respectively (Rowell et al., 1988b). Very little or no weight loss occurred with both fungi in flakeboards made using either phenol–formaldehyde or isocyanate adhesive with acetylated flakes. None of these specimens failed during the 300-day test period.

Mycelium fully covered the surfaces of isocyanate-bonded control flakeboards within 1 week, but mycelial development was significantly slower in phenol–formaldehyde-bonded control flakeboards. Both isocyanate– and phenol–formaldehyde-bonded acetylated flakeboards showed surface mycelium colonization during the test time, but the fungus did not attack the acetylated flakes, so little strength was lost.

In similar bending-creep tests, both control and acetylated pine particleboards made using melamine–urea–formaldehyde adhesive failed because *T. palustris* attacked the adhesive in the glueline (Imamura et al., 1988). Mycelium invaded the inner part of all boards, colonizing in both wood and glueline in control boards, but only in the glueline in acetylated boards. These results show that the glueline is also important in protecting composites from biological attack.

After a 16-week exposure to *T. palustris*, the internal bond strength of control aspen flakeboards made with phenol–formaldehyde adhesive was reduced over 90%

and that of flakeboards made with isocyanate adhesive was reduced 85% (Imamura et al., 1987). After 6 months of exposure in moist unsterile soil, the same control flakeboards made with phenol–formaldehyde adhesive lost 65% of their internal bond strength and those made with isocyanate adhesive lost 64% internal bond strength. Failure was due mainly to great strength reductions in the wood caused by fungal attack. Acetylated aspen flakeboards lost much less internal bond strength during the 16-week exposure to *T. palustris* or 6-month soil burial. The isocyanate adhesive was somewhat more resistant to fungal attack than the phenol–formaldehyde adhesive. In the case of acetylated composites, loss in internal bond strength was mainly due to fungal attack in the adhesive and moisture, which caused a small amount of swelling in the boards.

Another test for biological resistance that has been done on acetylated composites is with brown-, white-, and soft-rot fungi and tunneling bacteria in a fungal cellar (Table 11.8). Control blocks were destroyed in less than 6 months while flakeboards made from acetylated furnish above 16 WPG showed no attack after 1 year (Nilsson et al., 1988; Rowell et al., 1988a). This data shows that no attack occurs until swelling of the wood occurs (Rowell and Ellis, 1984; Rowell et al., 1988a). This is more evidence that the moisture content of the cell wall is critical before attack can take place.

Table 11.8 Fungal Cellar Tests of Aspen Flakeboards Made from Control and Acetylated Flakes[1,2]

	Rating at Intervals (Months)[3]							
Weight Percent Gain, %	**2**	**3**	**4**	**5**	**6**	**12**	**24**	**36**
0	S/2	S/3	S/3	S/3	S/4	—	—	—
7.3	S/0	S/1	S/1	S/2	S/3	S/4	—	—
11.5	0	0	S/0	S/1	S/2	S/3	S/4	—
13.6	0	0	0	0	S/0	S/1	S/2	S/3
16.3	0	0	0	0	0	0	0	0
17.9	0	0	0	0	0	0	0	0

[1] Nonsterile soil containing brown-, white-, and soft-rot fungi and tunneling bacteria.
[2] Flakeboards bonded with 5% phenol–formaldehyde adhesive.
[3] Rating system: 0 = no attack; 1 = slight attack; 2 = moderate attack; 3 = heavy attack; 4 = destroyed; S = swollen.

Table 11.9 shows the data for chemically-modified pine flakeboards in a marine environment (Johnson and Rowell, 1988). As with the termite test, all types of chemical modifications of wood help resist attack by marine organisms. Control flakeboards were destroyed in 6 months to 1 year, mainly because of attack by *Limnoria tripunctata*, while chemically-modified flakeboards show little or no attack after 8 to 10 years.

All laboratory tests for biological resistance conducted to this point show that acetylation is an effective means of reducing or eliminating attack by soft-, white-, and brown-rot fungi, tunneling bacteria, marine organisms, and subterranean termites.

Table 11.9 Ratings of Chemically Modified Southern Pine Exposed to a Marine Environment[1]

Chemical	Weight Percent gain, %	Years of Exposure, years	Mean rating due to attack by *Limnoriid* and *Teredinid* Borers	*Shaeroma terebrans*
Control	0	1	2–4	3.4
Propylene oxide	26	3–11.5	10	3.8
Butylene oxide	28	3–8.5	9.9	8
Butylisocyanate	29	6.5	10	3.4
Acetic anhydride	22	3	8	8.8

[1] Rating system: 10 = no attack; 9 = slight attack; heavy attack; 7 = some attack; 4 = heavy attack; 0 = destroyed.

7.3 Ultraviolet Resistance

Acetylation has also been shown to improve ultraviolet resistance of aspen fiberboards (Feist et al., 1991a). Table 11.10 shows the weight loss, erosion rate, and depth of penetration resulting from 700 h of accelerated weathering. Control specimens erode at about 0.12 μm/h or about 0.02%/h. Acetylation reduces surface erosion by 50%. The depth of the effects of weathering is about 200 μm into the fiber surface for the unmodified boards and about half that of the acetylated boards. Table 11.11 shows the acetyl and lignin content of the outer 0.5 mm surface and of the remaining specimen after the surface had been removed, before and after accelerated weathering. The acetyl content is reduced in the surface after weathering,

Table 11.10 Weight Loss and Erosion of Aspen Fiberboards Made from Control and Acetylated Fiber After 700 Hours of Accelerated Weathering

Specimen	Weight loss, %/h	Erosion, μm/h	Reduction in Erosion, %	Depth of Penetration of Weathering, μm
Control	0.019	0.121	—	199–210
Acetylated	0.01	0.059	51	85–105

Table 11.11 Acetyl and Lignin Analysis Before and After 700 Hours of Accelerated Weathering of Aspen Fiberboards Made from Control and Acetylated Fiber

Specimen	Before weathering: Surface, %	Before weathering: Remainder, %	After weathering: Surface, %	After weathering: Remainder, %
	Acetyl		Acetyl	
Control	4.5	4.5	1.9	3.9
Acetylated	17.5	18.5	12.8	18.3
	Lignin		Lignin	
Control	19.8	20.5	1.9	17.9
Acetylated	18.5	19.2	5.5	18.1

which shows that the acetyl blocking group is removed during weathering. UV radiation does not remove all of the blocking acetyl group, so some stabilizing effect to photochemical degradation still is in effect. The loss of acetate is confined to the outer 0.5 mm since the remaining wood has the same acetyl content before and after accelerated weathering. The lignin content is also greatly reduced in the surface as a result of weathering which is the main cell wall polymer degraded by UV radiation. Cellulose and the hemicelluloses are much more stable to photochemical degradation.

In outdoor tests, flakeboards made from acetylated pine flakes maintain a light yellow color after one year while control boards turn dark orange to light gray during this time (Feist et al., 1991b).

7.4 Strength Properties

The modulus of rupture (MOR), modulus of elasticity (MOE), and bending and tensile strength (TS) parallel to the board surface are shown in Table 11.12 for fiberboards made from control and acetylated pine kenaf and hemlock fiber. Acetylation results in a small decrease in MOR but about equal values in MOE and TS. All strength values given in Table 11.12 are above the minimum standard as given by the American Hardboard Association (ANSI, 1982). It has been shown that there is very little effect on strength properties of thin flakes as a result of acetylation (Rowell and Banks, 1987). The small decrease in some strength properties resulting from acetylation may be attributed to the hydrophobic nature of the acetylated furnish which may not allow the water soluble phenolic or isocyanate resins to penetrate into the flake. The adhesives used in these tests have also been developed for unmodified lignocellulosics. Different types of adhesives may be needed in chemically-modified boards (Vick and Rowell, 1990).

Table 11.12 Modulus of Rupture (MOR), Modulus of Elasticity (MOE), and Tensile Strength (TS) Parallel to the Board Surface of Fiberboards Made from Control or Acetylated Fiber and Phenolic Resin[1]

Board	MOR, MPa	MOE, MPa	TS, MPa
Pine Control	37.1	3.7	19
21.6 WPG	27.9	3.3	13.6
Kenaf Control	47.1	4.6	31
18.4 WPG	38.6	5.1	27.1
Hemlock Control	66	6	33.9
22.5 WPG	51.1	5	32.1
ANSI Standard	31	—	10.3

[1] Resin content of boards: Pine 8%, Kenaf 8%, Hemlock 8%.

It should also be pointed out that strength properties of lignocellulosics are very dependent on the moisture content of the cell wall. Fiber stress at proportional limit, work to proportional limit, and maximum crushing strength are the mechanical properties most affected by changing moisture content by only +/– 1% below FSP (Rowell, 1984; USDA, 1987). Since the EMC and FSP are much lower for chemically-modified fiber than for control fiber, strength properties will be different due to this fact alone.

8. CHEMICAL MODIFICATION FOR THERMOPLASTICIZATION

There have been many research projects over the years studying ways to thermoform lignocellulosics. Most of the efforts have concentrated on film formation and thermoplastic composites. There are presently several research teams in the world continuing this effort. The approach most often used involves the chemical modification of cellulose, lignin, and the hemicelluloses to decrystallize/modify the cellulose and to thermoplasticize the lignin and hemicellulose matrix in order to mold the entire lignocellulosic resource into films or thermoplastic composites.

Shiraishi and his co-workers in Japan have formed thermoplastic films through chemical reactions of wood meal using trifluoroacetic acid (TFAA) or chloride modification (Shiraishi, 1991). Thermoplasticity of esterified wood was found to depend on the acyl group, the method of preparation, and the degree of substitution. They found that as the size of the aliphatic group is increased, the melting temperature of the modified wood at 3 kg/cm^2 is decreased.

Matsuda and his co-workers in Japan (1985) also extensively investigated the esterification of wood in order to make a totally thermoplastic material. They esterified wood without a solvent by simply heating wood meal with succinic anhydride in a mixer for 3 h at temperatures greater than 60°C. The wood meal was readily molded at 180°C under a pressure of 570 kg/cm^2 for 10 min.

Hon and his co-workers in the U.S. (1989) also produced a thermomoldable product by benzylation of wood powder. The degree of substitution was varied by changing the reaction alkalinity, temperature, and time. Sodium hydroxide concentrations greater than 25% were necessary to obtain a high weight gain, presumably because of the need to swell the lignocellulosic substrate. Different species showed variation in reaction rates. The thermoplasticized woods exhibited good melting properties and were readily moldable into bulk materials or extruded into films and sheets. A wide range of glass transition temperatures, from 66–280°C, was observed for the benzylated wood and was largely dependent on weight gain.

Ohkoshi and his co-workers in Japan (1992) have worked on thermoplasticization of wood by allylation. They reported that thermoplasticity increased as the degree of allylation increased. They also found that cellulose was decrystallized during the allylation of wood but this had little effect in lowering the softening temperature of the cellulose. The major contribution to thermoplasticizing wood was modification of lignin. Modification of the xylan had little effect on thermoplasticity.

It is also possible to only modify the matrix of agro-based fiber allowing thermoplastic flow, but keeping the cellulose backbone as a reinforcing filler. This type of composite should have reduced heat-induced deformation (creep) which restricts thermoplastic-based composites from structural uses (Rowell and Clemons, 1992). If a non-decrystallizing reaction system is used, it is possible to chemically modify

Figure 11.5 Scanning electron micrographs of pressed kenaf fiber: A, Control (50X), B, succinic anhydride (SA) reacted (50 WPG, 50X).

the lignin and hemicellulose but not the cellulose. This selective reactivity has been shown to occur if uncatalyzed anhydrides are reacted with wood fiber (Rowell et al., 1994b).

Matsuda (1987) found that the moldability of various esterified woods decreased in the following anhydride order: succinic anhydride > maleic anhydride > pthalic anhydride. Hon and Xing (1992) reported that maleic anhydride reacted faster than succinic anhydride; however, Rowell and Clemons (1992) found the opposite.

Kenaf fiber has been reacted with succinic anhydride to give high weight gains of esterification of the cell wall polymers either by solution or solid state chemistry. The esterified fiber shows a reduced transition temperature from about 170°C down to about 135°C regardless of the weight gain obtained (Rowell et al., 1994a).

Figure 11.5 shows the scanning electron micrograph of hot pressed control and esterified kenaf fiber. The WPG is about 50. The control fiber (A) shows little tendency to thermally flow under the pressure of the hot press, whereas the esterified fiber (B) shows thermal flow at this temperature.

9. FUTURE OF LIGNOCELLULOSIC COMPOSITES

Fiber technology, high performance adhesives, and fiber modification can be used to manufacture structural lignocellulosic composites with uniform densities, durability in adverse environments, and high strength. Fiber modification can also be used to improve properties in composites made of both natural and synthetic resources used for geotextiles, filters, sorbents, packaging, and non-structural composites.

Products having complex shapes can be produced using flexible chemically-modified fiber mats, which can be made by nonwoven needling or thermoplastic fiber–melt matrix technologies. Within certain limits, the mats can be pressed into any desired shape, size, thickness, and density. With fiber mat technology, a complex product can be made directly from a lignocellulosic fiber blend. In general, the present technology requires the formation of flat sheets prior to the shaping of complex parts.

All of this technology can be applied to recycled lignocellulosic fiber as well as virgin fiber, which can be derived from many sources. Agricultural residues, all types of waste paper, yard waste, industrial fiber residues, residential fiber waste, and many other forms of waste lignocellulosic fiber can also be used to make property-enhanced composites (Figure 11.6).

Figure 11.6 Composite products made with improved properties through chemical modificaton.

REFERENCES

Andersson, M. and Tillman, A.-M., Acetylation of jute. Effects on strength, rot resistance and hydrophobicity, *J. Applied Polymer Sci.*, 37, 3437, 1989.

ANSI, American National Standard, *Basic Hardboard.* ANSI/AHA 135.4 (reaffirmed Jan, 1988), American Hardboard Association, Palatine, IL, 1982.

Banks, W. B. and Lawther, J. M., Derivation of wood in composites, *Cellulosic Polymers, Bends and Composites*, Gilbert, R. G. Ed., Hanser Publishers, New York, NY, 1994, 131.

Callow, H. J., Acetylation of cellulose and lignin in jute fiber, *Journal of the Indian Chemical Society*, 43, 605, 1951.

Cowling, E. B., Comparative biochemistry of the decay of Sweetgum sapwood by white-rot and brown-rot fungus, *U.S. Department of Agriculture, Forest Serv. Technol. Bull.*, No. 1258, 50, 1961.

Feist, W. C., Rowell, R. M., and Ellis, W. D., Moisture sorption and accelerated weathering of acetylated and/or methyl methacrylate treated aspen, *Wood and Fiber Sci.*, 23(1), 128, 1991a.

Feist, W. C., Rowell, R. M., and Youngquist, J. A., Weathering and finish performance of acetylated aspen fiberboard, *Wood and Fiber Sci.*, 23(2), 260, 1991b.

Hon, D. N.-S. and Ou, N.-H., Thermoplasticization of wood. I. Benzylation of wood, *J. Applied Polymer Science* (Part A: Polymer Chemistry), 27, 2457, 1982.

Hon, D. N.-S. and Xing, L. M., Thermoplasticization of wood, *Viscoelasticity of Biomaterials*, Glasser, W. G., Ed., Am. Chem. Soc., Washington, D.C., 1992, 118.

Hon, D. N.-S., Chemical modification of lignocellulosic materials: old chemistry, new approaches, *Polymer News*, 17, 102, 1992.

Imamura, Y. and Nishimoto, K., Bending creep test of wood-based materials under fungal attack, *J. Soc. Materials Sci.*, 34(38), 985, 1985.

Imamura, Y., Nishimoto, K., and Rowell, R. M., Internal bond strength of acetylated flakeboard exposed to decay hazard, *Mokuzai Gakkaishi*, 33(12), 986, 1987.

Imamura, Y., Rowell, R. M., Simonson, R., and Tillman, A.-M., Bending-creep tests on acetylated pine and birch particleboards during white- and brown-rot fungal attack, *Paperi ja Puu*, 9, 816, 1988.

Johnson, B. R., and Rowell, R. M., Resistance of chemically-modified wood to marine borers, *Material und Organismen*, 23(2), 147, 1988.

Krzysik, A. M., Youngquist, J. A., Muehl, J. M., Rowell, R. M., Chow, P., and Shook, S. R., Dry-process hardboards from recycled newsprint paper fibers, *Materials Interactions Relevant to Recycling of Wood-Based Materials*, Rowell, R. M., Laufenberg, T. L., and Rowell, J. K., Eds., Materials Research Society, Pittsburgh, PA, 1992, 266, 73.

Krzysik, A. M., Youngquist, J. A., Rowell, R. M., Muehl, J. M., Chow, P., and Shook. S. R., Feasibility of using recycled newspaper as a fiber source for dry-process hardboards, *For. Prod. J.*, 43(7/8), 53, 1993.

Kumar, S., Chemical modification of wood, *Wood and Fiber Sci.*, 26(2), 270, 1994.

Matsuda, H., Preparation and utilization of esterified woods bearing carboxyl groups, *Wood Sci. Technol.*, 21, 75, 1994.

Matsuda, H. and Ueda, M., Preparation and utilization of esterified woods bearing carboxyl groups. IV. Plasticization of esterified woods, *Mokuzai Gakkaishi*, 31(3), 215, 1985.

Nilsson, T., Personal communication, Upsala, Sweden, 1986.

Nilsson, T., Rowell, R. M., Simonson, R., and Tillman, A.-M., Fungal resistance of pine particleboards made from various types of acetylated chips, *Holzforschung*, 42(2), 123, 1988.

Ohkoshi, M., Hayashi, N., and Ishihara, M., Bonding of wood by thermoplasticizing the surfaces III. Mechanism of thermoplasticization of wood by allylation, *Mokuzai Gakkaishi*, 38(9), 854, 1992.

Rowell, R. M., Chemical modification of wood: advantages and disadvantages, *Proceedings, Am. Wood Preservers' Assoc.*, 71, 41, 1975.

Rowell, R. M., Distribution of reacted chemicals in southern pine modified with acetic anhydride, *Wood Sci.*, 15(2), 172, 1982.

Rowell, R. M., Chemical modification of wood: a review, *Commonwealth Forestry Bureau*, Oxford, England, 6(12), 363, 1983.

Rowell, R. M., *The Chemistry of Solid Wood*, Advances in Chemistry Series No. 207, American Chemical Society, Washington, DC, 1984.

Rowell, R. M., Chemical modification of wood: Its application to composite wood products, *Proceedings, Composite Products Symposium, Rotorua, New Zealand, November, 1988, FRI Bulletin*, No. 153, 57, 1990.

Rowell, R. M., Chemical modification of wood, *Handbook on Wood and Cellulosic Materials*, Hon, D. N.-S. and Shiraishi, N., Eds., Marcel Dekker, Inc., New York, 1991, 703.

Rowell, R. M., Property enhancement of wood composites, *Composites Applications: The Role of Matrix, Fber, and Interface*, Vigo, T. L. and Kinzig, B. J., eds., VCH Publishers, Inc., New York, 1992, 365.

Rowell, R. M., Opportunities for composite materials from jute and kenaf, *International Consultation of Jute and the Environment*, Food and Agricultural Organization of the United Nations, ESC:JU/IC 93/15, 1, 1993.

Rowell, R. M. and Banks, W. B., Water repellency and dimensional stability of wood, *USDA Forest Service, General Technical Report FPL 50*, Forest Products Laboratory, Madison, WI, 1985.

Rowell, R. M. and Banks, W. B., Tensile strength and work to failure of acetylated pine and lime flakes, *British Polymer J.*, 19, 479, 1987.

Rowell, R. M., Caulfield, D. F., Sanadi, A., O'Dell, J., and Rials, T. G., Thermoplasticization of kenaf and compatibilization with other materials, *Proceedings, Sixth Annual International Kenaf Conference*, 1994a.

Rowell, R. M. and Clemons, C. M., Chemical and modification of wood fiber for thermoplasticity, compatibilization with plastics and dimensional stability, *Proceedings, International Particleboard/Composite Materials Symposium*, Maloney, T. M. Ed., Pullman, WA, 251, 1992.

Rowell, R. M. and Ellis, W. D., Reaction of epoxides with wood, *USDA Forest Service Research Paper*, FPL 451, Forest Products Laboratory, Madison, WI, 1984.

Rowell, R. M., Esenther, G. R., Youngquist, J. A., Nicholas, D. D., Nilsson, T., Imamura, Y., Kerner-Gang, W., Trong, L., and Deon, G., Wood modification in the protection of wood composites, *Proceedings: IUFRO Wood Protection Subject Group*, Honey Harbor, Ontario, Canada. Canadian Forestry Service, 238, 1988a.

Rowell, R. M. and Harrison, S. E., Property enhanced kenaf fiber composites, *Proceedings, Fifth Annual International Kenaf Conference*, Bhangoo, M. S., Ed., California State University Press, Fresno, CA, 129, 1993.

Rowell, R. M., Hart, S. V., and Esenther, G. R., Resistance of alkylene-oxide treatments on dimensional stability of wood, *Wood Sci.*, 11(4), 271, 1979.

Rowell, R. M. and Keany, F., Fiberboards made from acetylated bagasse fiber, *Wood and Fiber Sci.*, 23(1), 15, 1991.

Rowell, R. M. and Konkol, P., Treatments that enhance physical properties of wood, *USDA, Forest Service, Gen. Technical Report FPL-GTR-55*, Forest Products Laboratory Madison, WI, 1987.

Rowell, R. M. and Norimoto, M., Acetylation of bamboo fiber, *J. Jap. Wood Res. Soc.*, 33(11), 907, 1987.

Rowell, R. M. and Norimoto, M., Dimensional stability of bamboo particleboards made from acetylated particles, *Mokuzai Gakkaishi*, 34(7), 627, 1988.

Rowell, R. M and Rowell, J. S., Moisture sorption of various types of acetylated lignocellulosic fibers, *Cellulose and Wood*, Schuerch, C., Ed., John Wiley and Sons, New York, NY, 1989, 343.

Rowell, R. M., Simonson, R., Hess, S., Plackett, D. V., Cronshaw, D., and Dunningham, E., Acetyl distribution in acetylated whole wood and reactivity of isolated wood cell wall components to acetic anhydride, *Wood and Fiber Sci.*, 26(1), 11, 1994b.

Rowell, R. M., Simonson, R., and Tillman, A. -M., A process for improving dimensional stability and biological resistance of lignocellulosic materials, *European Patent 0213252*, 1991.

Rowell, R. M., Susott, R. A., De Groot, W. G., and Shafizadeh, F., Bonding fire retardants to wood. Part I, *Wood and Fiber Sci.*, 16(2), 214, 1984.

Rowell, R. M., Tillman, A.- M., and Simonson, R., A simplified procedure for the acetylation of hardwood and softwood flakes for flakeboard production, *J. Wood Chem. and Tech.*, 6(3), 427, 1986.

Rowell, R. M., Youngquist, J. A., and Imamura, Y., Strength tests on acetylated flakeboards exposed to a brown rot fungus, *Wood and Fiber Sci.*, 20(2), 266, 1988b.

Rowell, R. M. and Youngs, R. L., Dimensional stabilization of wood in use, *USDA Forest Serv. Res. Note. FPL-0243*, Forest Products Laboratory, Madison, WI, 1981.

Shiraishi, N., Wood plasticization, *Wood and Cellulose Chemistry*, Hon, D. -N. S., and Shiraishi, N., Eds., Marcel Dekker, Inc., New York, 861, 1991.

Stamm, A. J., *Wood and Cellulose Science*, The Ronald Press Co., New York, 1964.

United States Department of Agriculture, Forest Service, *Wood Handbook,* USDA Agri. Handbook 72, Washington, D.C., 1987.

Vick, C. B., and Rowell, R. M., Adhesive bonding of acetylated wood, *Internat. J. Adhesion and Adhesives,* 10(4), 263, 1990.

CHAPTER 12

Agro-Fiber Thermoplastic Composites

Anand R. Sanadi, Daniel F. Caulfield, and Rodney E. Jacobson

CONTENTS

1. INTRODUCTION

Several billion pounds of fillers and reinforcements are used annually in the plastics industry. The use of these additives in plastics is likely to grow with the introduction of improved compounding technology and new coupling agents that permit the use of high filler/reinforcement content (Katz and Milewski, 1987). As suggested by Katz and Milewski, fillings up to 75 parts per hundred (pph) could be common in the future: this could have a tremendous impact in lowering the usage of petroleum based plastics. It would be particularly beneficial, both in terms of the environment and also in socio-economic terms, if a significant amount of the fillers were obtained from a renewable agricultural source. Ideally, of course, an agro-/bio-based renewable polymer reinforced with agro-based fibers would make the most environmental sense.

1.1 Advantages of Using Agro-Fibers in Plastics

The primary advantages of using annual growth lignocellulosic fibers as fillers/reinforcements in plastics are listed below:

Property Advantages
Low densities
Non abrasive
High filling levels possible resulting in high stiffness properties
High specific properties
Easily recyclable
Unlike brittle fibers, the fibers will not fracture when processing over sharp curvatures.

Environmental and Socio-Economic Advantages
Biodegradable
Wide variety of fibers available throughout the world
Generates rural jobs
Non-food agricultural/farm based economy
Low energy consumption
Low cost
Low energy utilization

Material cost savings due to the incorporation of the relatively low cost agro-fibers and the higher filling levels possible, coupled with the advantage of being non-abrasive to the mixing and molding equipment, are benefits that are not likely to be ignored by the plastics industry for use in the automotive, building, appliance, and other applications.

Prior work on lignocellulosic fibers in thermoplastics has concentrated on wood-based flour or fibers, and significant advances have been made by a number of researchers (Woodhams et al., 1984; Klason and Kubat, 1986a, b; Myers et al., 1992; Kokta et al., 1989; Yam et al., 1990; Bataille et al., 1989, Sanadi et al., 1994a). A recent study on the use of annual growth lignocellulosic fibers indicates that these fibers have the potential of being used as reinforcing fillers in thermoplastics (Sanadi et al., 1994b). The use of annual growth agricultural crop fibers such as kenaf has resulted in significant property advantages as compared to typical wood-based fillers/fibers such as wood flour, wood fibers, and recycled newspaper. Composites containing compatibilized polypropylene (PP) and kenaf have mechanical properties comparable with those of commercial PP composites (Sanadi et al., 1994b).

1.2 Limitations

The primary drawback of the use of agro-fibers is the lower processing temperature permissible due to the possibility of lignocellulosic degradation and/or the possibility of volatile emissions that could affect composite properties. The processing temperatures are thus limited to about 200°C, although it is possible to use higher temperatures for short periods. This limits the type of thermoplastics that can be used with agro-fibers to commodity thermoplastics such as polyethylene (PE), PP, polyvinyl chloride (PVC), and polystyrene (PS). However, it is important to note that these lower-priced plastics constitute about 70% of the total thermoplastic consumed by the plastics industry, and subsequently the use of fillers/reinforcement presently used in these plastics far outweighs the use in other more expensive plastics.

The second drawback is the high moisture absorption of the natural fibers. Moisture absorption can result in swelling of the fibers, and concerns about the dimensional stability of the agro-fiber composites cannot be ignored. The absorption of moisture by the fibers is minimized in the composite due to encapsulation by the polymer. It is difficult to entirely eliminate the absorption of moisture without using expensive surface barriers on the composite surface. If necessary, the moisture absorption of the fibers can be dramatically reduced through the acetylation of some of the hydroxyl groups present in the fiber, but with some increase in the cost of the fiber (Rowell, Tillman, and Simonson, 1986) (see Chapter 11). Good fiber-matrix bonding can also decrease the rate and amount of water absorbed by the composite. Research on this area is presently underway at the University of Wisconsin and the USDA Forest Service, Forest Products Laboratory, Madison, Wisconsin.

It is important to keep these limitations in perspective when developing end-use applications. We believe that by understanding the limitations and benefits of these composites, these renewable fibers are not likely to be ignored by the plastics/composites industry for use in the automotive, building, appliance, and other applications.

2. PROCESSING CONSIDERATIONS AND TECHNIQUES

Separation of the fibers from the original plant source is an important step to ensure the high quality of fibers. Details of the different processes used for processing to composites were given in Chapter 8. The limiting processing temperatures when using lignocellulosic materials with thermoplastics are important in determining processing techniques. High processing temperatures (>200°C) that reduce melt viscosity and facilitate good mixing cannot be used (except for short periods), and other routes are needed to facilitate mixing of the fibers and matrix in agro-fiber thermoplastics.

An excellent review by Milewski (1992) on short fiber composite technology covers a variety of reasons that result in problems associated with composite properties falling short of their true reinforcing potential. The major factors that govern the properties of short fiber composites are fiber dispersion, fiber length distribution, fiber orientation, and fiber-matrix adhesion. Mixing the polar and hydrophilic fibers with non-polar and hydrophobic matrix can result in difficulties associated with the dispersion of fibers in the matrix. Clumping and agglomeration must be avoided to produce efficient composites. The efficiency of the composite also depends on the amount of stress transferred from the matrix to the fibers. This can be maximized by improving the interaction and adhesion between the two phases, and also by maximizing the length of the fibers retained in the final composite (Biggs et al., 1988). Using long filaments during the compounding stage can result in higher fiber length distribution. However, long fibers sometimes increase the amount of clumping resulting in areas concentrated with fibers and areas with excessive matrix; this ultimately reduces the composite efficiency. Uniform fiber dispersion cannot be compromised, and a careful selection of processing techniques, initial fiber lengths, process conditions, and processing aids are needed to obtain efficient composites. Several types of compounding equipment, both batch and continuous equipment, have been used for blending lignocellulosic fibers and plastics.

The ultimate fiber lengths present in the composite depend on the type of compounding and molding equipment used. Several factors contribute to the fiber attrition such as the shearing forces generated in the compounding equipment, residence time, temperature, and viscosity of blends. An excellent study on the effect of processing and mastication of several types of short fibers in thermoplastics was conducted by Czarnecki and White (1980). They concluded that the extent of breakage was most severe and rapid for glass fibers, less extensive for kevlar (aramid) fibers, and the least for cellulose fibers. The level of fiber attrition depends on the type of compounding and molding equipment used, level of loading, temperature, and viscosity of the blend (Czarnecki and White, 1980).

The properties of the agro-based thermoplastic composites are thus very process-dependent. Yam et al. (1990) at Michigan State University studied the effect of twin screw blending of wood fibers and high density polyethylene (HDPE) and concluded that the level of fiber attrition depended on the screw configuration and the processing temperature. Average fiber lengths decreased from about 1.26 mm prior to compounding to about 0.49 mm after extrusion. Modification of the screw configuration reduced fiber attrition to an average length of about 0.78 mm. Fiber weight percent up to 60% was reported to have been mixed. The tensile strength of the pure HDPE was higher than that of the wood fiber-HDPE, irrespective of the level of fiber filling. This was explained to be because of a lack of dispersion with fibers clumping in bundles and poor fiber-matrix bonding. Use of stearic acid in HDPE/wood fibers improved fiber dispersion and improved wetting between the fiber and matrix (Woodhams, 1984) and resulted in significant improvement in mechanical properties. Work by Raj and Kokta (1989) indicates the importance of using surface modifiers to improve fiber dispersion in cellulose fibers/PP composites. Use of a small amount of stearic acid during the blending of cellulose fibers in polypropylene decreased both the size and number of fiber aggregates formed during blending in an internal mixer (Brabender roll mill).

Another technique that is gaining acceptance is the high intensity compounding using a turbine mixer (thermokinetic mixer). Woodhams et al. (1990) and Myers et al. (1992) found the technique effective in dispersing lignocellulosic fibers in thermoplastics. Addition of dispersion aids/coupling agents further improved the efficiency of mixing. The high shearing action developed in the mixer decreased the lengths of fibers in the final composite. However, the improved fiber dispersion resulted in improved composite properties. A recent study evaluating fiber lengths of a jute-PP composite blended in a thermokinetic mixer, and then injection molded, was conducted by dissolving the PP in the composite in xylene and then using image analysis to measure the length of the fibers (Karmakar, 1994). The ultimate fibers lengths varied from about 0.10 mm to about 0.72 mm with an average of 0.34 ± 0.13 mm. Recent work using a thermokinetic mixer to blend kenaf in PP (Sanadi, Caulfield, unpublished results) has confirmed the usefulness of the compounding technique in effectively dispersing natural fibers in the thermoplastic matrix. An added advantage is that no pre-drying of the fibers is needed prior to the blending stage in the mixer.

3. SURFACE ENERGIES AND ADHESION

When a fiber composite fails by an interfacial or adhesive type failure, it is presumed that part of the failure arises from the lack of sufficient chemical bonding between fiber and matrix. But it is also likely that part of the failure arises from the inability to achieve intimate intermolecular contact between components. Strength of composites will thereby improve if one can modify the nature of the component surfaces so that their surface energies are more compatible with one another.

3.1 Cellulose Fiber Surfaces

The cellulose molecule is inherently extremely hydrophilic. The hydroxyl groups that cover the surface of a pulp fiber provide hydrogen bonds that are the ultimate forces that hold paper together and provide the basis of its mechanical strength properties (Caulfield, 1980). The good wetting of cellulose fibers by water is essential to the paper-making process, but also responsible for the poor strength properties of paper when wet. Water readily wets a cellulose fiber, whereas a non-polar material may interact very poorly with a cellulose fiber surface.

The necessary intimacy of contact essential for good adhesion in composite fabrication is achieved by good wetting of one component by the other. This compatibility between the surface energies of the two components and the resultant good adhesion are reflected in the mechanical strength of the composite (Westerlind and Berg, 1988). In order to improve the adhesion between cellulose fibers and polymer matrix, one may either modify the surface of the cellulose fiber or modify the polymer matrix in order to achieve the necessary compatibility of surface energies.

3.2 Surface Energies and Wettability

One way of characterizing surface energies is through wettability measurement, or measurement of contact angles formed at the interface between the solid, air and a liquid of known surface tension (Miller, 1985). In recent years techniques for measurement of wettabilities and contact angles of fibers have become well established (Berg, 1986; Young, 1986). The typical method uses an electronic microbalance to measure the forces. Techniques have been developed to deal with buoyancy effects and crimped fibers. In textile research the micro-balance method for contact angle measurement is now widely accepted for surface energy characterization. Additional difficulties arise when studying small natural fibers, but research has shown that meaningful measurements are possible even on wood pulp fibers (Young, 1976; Klungness, 1981).

Consider, for examplc, the simple depiction of two limiting cases. Case I is a fiber partially immersed in a fluid which tends to wet the fiber. The wetting angle is less than 90°. In Case II a fiber is partially immersed in a fluid that tends not to wet the fiber and the contact angle is greater than 90°. For Case I in which the liquid wets the fiber, the balance of the surface energies and surface tension is such that the net force on the fiber is one that draws the fiber further into the liquid. For Case II, the net force on the fiber is one that pushes the fiber from the liquid. The well established Wilhelmy equation relates the force, F, exerted on the fiber to the perimeter of the fiber, p, the surface, tension of the liquid, and the γ contact angle, θ, between the solid (fiber) and liquid (Wilhelmy, 1863).

$$F = p\gamma \cos\theta - Agh\rho$$

The second term accounts for the effect of buoyancy; where ρ is the liquid density, γ is the gravitational constant, A is the cross-sectional area of the fiber, and h is the depth of immersion. In most cases the second term can be safely ignored.

Using Wilhelmy's equation to measure surface tensions along with Young's equation allows the surface energy, or more generally the thermodynamic energy of interaction or the work of adhesion, to be evaluated. This work of adhesion may be viewed as the sum of terms corresponding to the contributions from the different types of interactions (Fowkes, 1972).

$$W = W^d + W^p + W^h + W^i$$

The superscript d refers to dispersive forces, p to dipole-dipole interactions, h to hydrogen bonding and i to induced dipole interactions. Since dipole and induction force contributions are usually small, this equation is more often written:

$$W = W^{lw} + W^{ab}$$

where the dispersion forces are combined with the dipole and induced dipole contributions and designated Lifshitz-van der Waals (lw) forces (Fowkes, 1987). All the other forces are described as acid/base interactions (ab). Lewis acid/base interactions cover any interaction that involves the sharing of an electron pair, most especially hydrogen bonding. In the case of cellulose, with a surface dominated by hydroxyl groups, the acid/base interactions can be described as predominantly hydrogen bonding.

Coupled surface energy and internal bond strength measurements indicate a strong correlation between the extent of hydroxyl-rich (acid/base) interface and good adhesive properties (Quillin, Caulfield, and Koutsky, 1992). To improve adhesion between cellulose and polypropylene (or other polyolefins), the general recommendation is to modify the hydrophobic polymer in a way that introduces chemical moieties on its surface that are capable of producing strong acid/base (hydrogen bonding) interactions with the groups on the surface of the cellulose.

3.3 Adhesion and the Interphase

The adhesion between the plastic matrix and polar lignocellulosic fibers is critical in determining the properties of the composite. Several different types of functionalized additives have been used to improve the dispersion and the interaction between cellulose-based fibers and polyolefins, polyvinyl chloride, polystyrene, etc. (Dalvag et al., 1985, Kokta, Raj, and Daneault, 1990).

The inherent polar and hydrophilic nature of the lignocellulosic fibers and the non-polar characteristics of the polyolefins result in difficulties in compounding and adhesion between the fibers and matrix. Proper selection of additives is necessary to improve the interaction and adhesion between the fiber and matrix phases. Maleic anydride (MA)-grafted polypropylene (MAPP) has been reported to function efficiently for lignocellulosic-PP systems. Earlier results suggest that the amount of MA grafted and the molecular weight are both important parameters that determine the efficiency of the additive (Felix et al., 1993; Sanadi et al., 1993). The maleic anhydride present in the MAPP not only provides polar interactions, but can covalently link to the hydroxyl groups on the lignocellulosic fiber (Figure 12.1). Han, Saka, and Shiraishi (1991) reported that the MAPP was localized on the cellulosic surface in a PP matrix: this was inferred from TEM-EDXA studies of osmium tetroxide-labeled MAPP. The formation of covalent linkages between the MA and the –OH of cellulose has been indicated through IR and ESCA analysis by Gatenholm et al. (1992).

The interactions between non-polar thermoplastics such as PP and any coupling agents such as MAPP are predominantly those of chain entanglement. Stresses applied to one chain can be transmitted to other entangled chains, and stress is distributed among many chains. These entanglements function like physical cross links that provide some mechanical integrity up to and above the glass transition temperature, T_g, but become ineffective at much higher temperatures (Neilsen, 1974). When polymer chains are very short, there is little chance of entanglement between chains and they can easily slide past one another (Neilsen, 1977). When the polymer chains are longer, entanglement between chains can occur and the viscosity of polymer becomes high. A minimum chain length or a critical molecular weight (M_e) is necessary to develop these entanglements and a typical polymer has a chain length

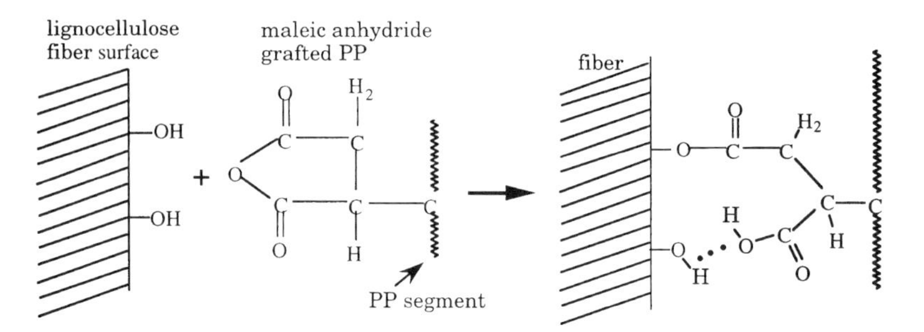

Figure 12.1 Reaction mechanism of maleated PP with the surface of the lignocellulosic fiber. Note potential of both covalent and H-bonding.

between entanglements equivalent to a M_e varying from 10,000 to about 40,000. The M_e varies depending on the structure of a polymer and, for example, linear polyethylene has a corresponding M_e for entanglements of about 4,000 (Neilsen, 1977), while for polystyrene it is about 38,000. Factors such as the presence of hydrogen bonding or the presence of side chains that affect the glass transition temperature of the polymer also will affect the M_e of the polymer melt. It is also important to note that the presence of the fiber surface is likely to restrict the mobility of the polymer molecules, and the minimum entanglement lengths will vary according the fiber surface characteristics.

Two major factors need to be considered while selecting additives to develop enough mechanical integrity of the interphase region so that there is sufficient stress transfer properties between the non-polar matrix and the polar fiber. Firstly, the functional additive present near the fiber surface should be strongly interacting with the fiber surface through covalent bonding and/or acid-base interactions (Figure 12.1). This means that sufficient functional groups should be present in the functional additive so that interactions can occur with the –OH groups on the fiber surface (Figure 12.2). Secondly, the polymer chains of the functional additive should be long enough to permit entanglements with the PP in the interphase (Figure 12.3).

In the case of MAPP interaction with the fiber surface there is a possibility of two or more MA groups, from the same MAPP molecule, interacting with different –OH groups on the fiber surface to form a tightly-bound MAPP molecule (Figures 12.2 and 12.3). Segmented loops are then formed between sites of MA and –OH interactions. It must be pointed out that although loop formation can and may occur, there will also be cases where a tail section of the MAPP 'sticks out' into the interface region. This section, we believe, has a greater possibility of interaction with the PP molecules because of greater entanglement possibilities as compared to any MAPP that has formed loops. It must be noted that a minimum segmented length is necessary for good interaction, whether through a tail or loop section of the MAPP. An ideal situation would be to create a molecule with a "head–tail" configuration, where one or more MA groups are grafted onto one end of the MAPP molecule, while the other end has no grafted MA molecules. This would maximize the length of entaglement with the PP molecules and also permit the MA groups to interact with the –OH groups on the fiber surface.

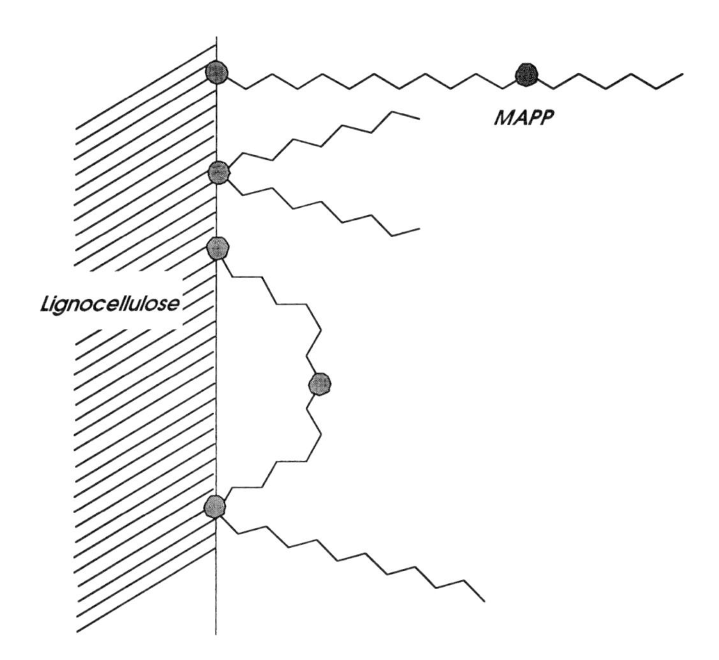

Figure 12.2 Different type of MAPP-fiber interactions. The dot represent maleic anhydride.

Using G-3002, a maleic anhydride-grafted PP, has resulted in efficient composites, and this is probably due to the higher molecular weight coupled with a high MA content (Sanadi et al., 1994a). The G-3002 from Eastman Chemical Co. is reported to have a number average molecular weight (M_n) of 20,000, a weight average molecular weight (M_w) of 40,000 and an acid number of 60. An acid number of 60 is about equivalent to a 6% by weight of maleic anhydride in the G-3002 (Eastman Chemical Company, 1992). However, there is some free anhydride present in the G-3002 and this complicates the understanding of the characteristics and function of the MAPP on the properties of the fiber-matrix interphase. The free MA may preferentially bond to available –OH sites on the fiber and reduce the interaction between the MAPP and the fiber. Furthermore, free MA bonded to the fiber surface can change the surface energetics of the fiber surface. Use of MAPP's with very higher molecular weights but low MA contents does not perform as well as the G-3002 (Sanadi et al., 1994b).

Theoretically, extremely long chains of MAPP with a lot of grafted MA would be an ideal additive in lignocellulosic-PP composites, creating both covalent bonding to the fiber surface and extensive molecular entanglement to improve properties of the interphase. However, extremely long chains may reduce the possibility of migration of the MAPP to the fiber surface because of the short processing times. If the MW of the MAPP is too high, the MAPP may entangle with the PP molecules so that the polar groups on the MAPP have difficulty "finding" the –OH groups on the fiber surface (Figure 12.3).

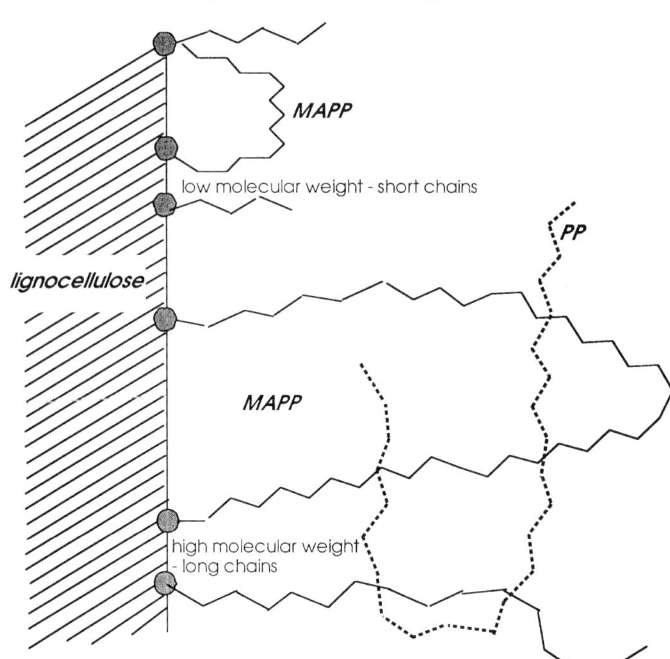

Figure 12.3 Schematic of possible PP molecules' entanglement with the longer chains of MAPP. Shorter chains of MAPP have less opportunity to entangle with the PP molecules.

Furthermore, the crystallization of the PP with MAPP depends on the amount of graft on the MAPP (Duvall et al., 1993, 1994). Lower anhydride (0.2% by weight) coupled with high molecular weight (M_n = 42,500 and M_w = 135,000) resulted in co-crystallization of a PP/MAPP blend and higher fracture strains as compared to a higher anhydride (2.7% by weight) coupled with low molecular weight (M_n = 16,100 and M_w = 52,100) that crystallized as a second phase (Duvall et al., 1993). It is again important to remember that the presence of the fiber surface and the potential of covalent attachment of the MAPP to the hydroxyl groups on the cellulose fiber surface will influence the morphological interaction between the MAPP and PP. Thermodynamic segregation of the MAPP to the fiber surface will result in the major part of the interaction between the MAPP and PP to occur in the interphase region.

Transcrystallinity around the fiber surface complicates the understanding since crystallites can act like cross links by tying many molecules together. The influence of the molecular weight on the interphase morphology and transcrystallinity are also important considerations. Crystallites have a much higher moduli as compared to the amorphous region and can increase the modulus contribution of the plastic matrix to composite modulus.

Previous studies are somewhat conflicting in the effects of surface characteristics on transcrystalline formation. Quillin, Caulfield, and Koutsky (1993) found that improvement in surface energetics between the cellulosic fibers and the PP resulted in the prevention of transcrystalline zones. Transcrystallinity was abundant when there was no surface modification of the polar fiber and the non-polar matrix. On the other hand, Gatenholm et. al. (1992) found that the higher the molecular weight of the MAPP, the greater were the number of nuclei formed at the fiber surface. It should be pointed out that both the above experiments were *model* single fiber (cellulose) tests with controlled cooling cycles. Vollenberg and Heikens' (1989a) studies on small particle size glass or alumina filled PP *composites* suggest that the use of sizing or coupling agent on the glass or alumina eliminated the formation of high modulus material near the fiber surface. This hypothesis was suggested for the annealed composite based on eliminating other possibilities. Solid state nuclear magnetic resonance experiments (Vollenberg and Heikens, 1989b) support this explanation.

4. PROPERTIES OF COMPOSITES

Cellulosic fillers/fibers have been incorporated in a wide variety of thermoplastics such as PP, PE, PS, PVC, and polyamides (Klason and Kubat, 1986 a, b). In general, dispersing agents and/or coupling agents, are necessary for property enhancement when fibers are incorporated in thermoplastics. Grafting chemical species onto the fiber surface has also been reported to improve the interaction between the fibers and matrix. Although grafting can improve the properties of the composite to a significant extent, this process increases the material cost the of system. The use of dispersing agents and/or coupling agents is a cheaper route to improve properties and makes more practical sense for high volume, low cost composite systems.

In general, cellulosic fillers or fibers have a higher Young's modulus as compared to commodity thermoplastics, thereby contributing to the higher stiffness of the composites. The increase in the Young's modulus with the addition of cellulosics depends on many factors such as the amount of fibers used, the orientation of the fibers, the interaction and adhesion between the matrix, the ratio of the fiber-to-matrix Young's modulus, etc. The Young's modulus of the composite can be crudely estimated through the simple rule of mixtures and other simple models if the Young's modulus of the filler–fiber is known (Hull, 1981). The use of dispersing or coupling agents can change the molecular morphology of the polymer chains both at the fiber-polymer interphase and also in the bulk matrix phase. Crystallites have much higher moduli as compared to the amorphous regions and can increase the modulus contribution of the polymer matrix to the composite modulus. A good understanding of the effect of dispersing agents and coupling agents on transcrystallinity at the fiber-matrix interphase and the corresponding effect on the composite Young's modulus is nonexistent. Therefore the influence and contribution of the molecular morphology on estimating the composite modulus through simple models are lacking.

In order to use any models to estimate composites properties, it is necessary to know the property of the fibers. In general, natural fibers such as kenaf and jute are in the form of strands that consist of discrete individual fibers, generally 2 mm to 6 mm long, which are themselves composites of predominantly cellulose, lignin, and hemicelluloses. Strand and individual fiber properties can vary widely depending on the source, age, separating techniques, moisture content, speed of testing, history of the fiber, etc. The properties of the individual fibers are very difficult to measure. Earlier work on a natural bast strand, sunn hemp (*Crotalaria juncea*) suggested that the strand properties ranged widely. The tensile strengths of the strands of sunn hemp varied from about 325–450 MPa, while the tensile modulus ranged from 27–48 MPa (Sanadi et al., 1985). In a natural fiber-thermoplastic composite, the lignocellulosic phase is present in a wide range of diameters and lengths, some in the form of short strands and others in forms that seem closer to the individual fiber. The high shearing energy of blending the strands and the polymer in a mixer results in fiber attrition but can also axially separate the strands into discrete individual fibers.

Cellulosic fillers/fibers can be classified under three categories, depending on their performance when incorporated in a plastic matrix. Wood flour and other low cost agricultural-based flour can be considered as particulate fillers that enhance the tensile and flexural moduli of the composite with little effect on the composite strength. Wood fibers and recycled newspaper fibers have higher aspect ratios and contribute to an increase in the moduli of the composite, and can also improve the strength of the composite when suitable additives are used to improve stress transfer between the matrix and the fibers. The improvement in modulus is not significantly different than the cellulosic particulate fillers. The most efficient cellulosic additives are some natural fibers such as kenaf, jute, flax, etc. The specific Young's modulus and specific flexural modulus, the ratio of the composite modulus to the composite specific gravity of composites with natural fibers such as kenaf are significantly higher than those possible with wood fibers. The specific moduli (the ratio of the composite modulus to the composite specific gravity) of high fiber volume fraction bast fibers-PP composites are high and in the same range as glass fibers-PP composites. The most efficient natural fibers are those that have a high cellulose content coupled with a low microfibril angle resulting in high mechanical properties.

Although several plastics have been used with cellulosic fibers, the major part of the work at the University of Wisconsin-Madison and the Forest Products Laboratory has been on polypropylene (PP). The following case study concentrates on this versatile plastic in combination with kenaf as a promising example of an agro-based fiber/plastic composite.

4.1 Experimental Methods

Kenaf strands harvested from mature plants were obtained from AgFibers Inc., Bakersfield, CA, and cut into lengths of about 1 cm. The fibers were not dried to remove any of the moisture present, and the moisture content of the fibers varied from 6% to about 9% by weight. In all our experiments the weight and volume percent reported is the amount of dry fiber present in the blend. The homopolymer was Fortilene-1602 (Solvay Polymers, 1991) with a melt flow index of 12 g/10 min. as measured by ASTM D-1238. A maleic anhydride grafted polypropylene (MAPP), Epolene G-3002, from Eastman Chemical Products, TN, was used as a coupling agent to improve the compatibility and adhesion between the fibers and matrix. The G-3002 had a number average molecular weight of 20,000 and a weight average molecular weight of 40,000 and an acid number of 60. An acid number of 60 is about equivalent to a 6% by weight of maleic anhydride in the G-3002 (Eastman Chemical Company, 1992).

The strands were not pulped prior to compounding as the former procedure can consume a significant amount of energy. The short kenaf fibers, MAPP and PP (the latter two in pellet form) were compounded in a high intensity kinetic mixer (Synergistics Industries Ltd., Canada) where the only source of heat is generated through the kinetic energy of rotating blades. The blending was accomplished at 4,600 rpm and then automatically discharged at 190°C. A total weight (fibers, PP and MAPP) of 150 g was used for each batch and about 1.5 kg of blended material was prepared for each set of experiments. Fiber weight varied from 20–60% and coupling agent weight varied from 0 to 3%. The total residence time of the blending operation depended on the proportions of fiber and PP present and averaged about 2 min.

The mixed blends were then granulated and dried at 105°C for 4 h. Test specimens were injection-molded at 190°C using a Cincinnati Milacron Molder and injection pressures varied from 2.75 MPa to 8.3 MPa depending on the constituents of the blend. Specimen dimensions were according to the respective ASTM standards. The specimens were stored under controlled conditions (20% relative humidity and 32°C) for three days before testing. Tensile tests were conducted according to ASTM 638-90, Izod impact strength tests according to ASTM D 256-90, and flexural testing using the ASTM 790-90 standard. The cross-head speed during the tension and flexural testing was 5 mm/min. Although all the experiments were designed around the weight percent of kenaf in the composites, fiber volumes fractions can be estimated from composite density measurements and the weights of dry kenaf fibers and matrix in the composite. The density of the kenaf present in the composite was estimated to be 1.4 g/cc. A flow chart of the experimental methods is shown in Figure 12.4.

4.2 Mechanical Properties

4.2.1 Effect of MAPP on Composite Properties

A small amount of the MAPP (0.5% by weight) improved the flexural and tensile strength, tensile energy absorption, failure strain, and un-notched Izod impact strength of the kenaf/PP composits. The anhydride groups present in the MAPP can covalently bond to the hydroxyl groups of the fiber surface. Any MA that has been converted to the acid form can interact with the fiber surface through acid-base interactions. The improved interaction and adhesion between the fibers and the matrix leads to better matrix to fiber stress transfer. There was little difference in the properties obtained between the 2 and 3% (by weight) MAPP systems. The drop in tensile modulus with the addition of the MAPP was probably due to polymer morphology.

Transcrystallization and changes in the apparent modulus of the bulk matrix can result in changes in the contribution of the matrix to the composite modulus and will be discussed later. There was little change in the notched impact strength with the addition of the MAPP, while the improvement in un-notched impact strength is

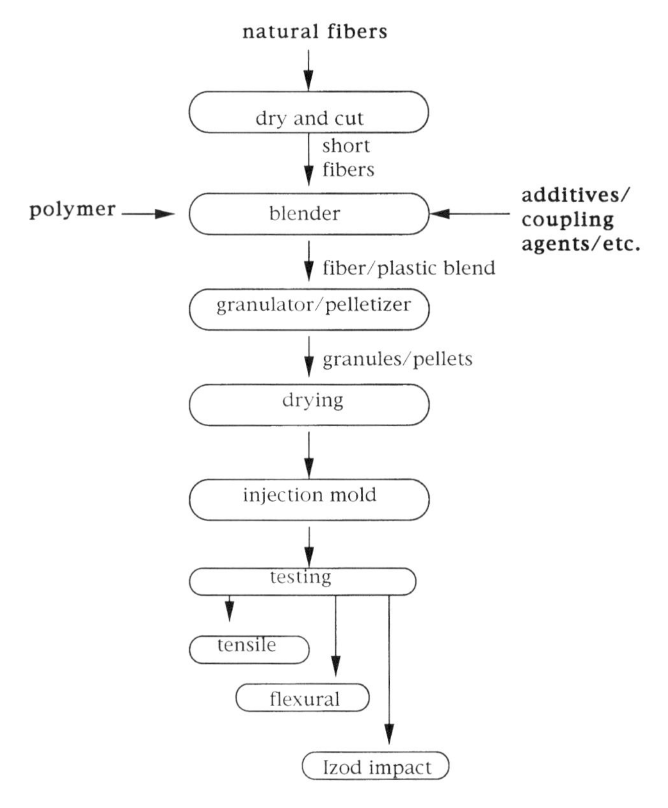

Figure 12.4 Flow chart of experimental processes for the blending and testing of composites.

significant. In the notched test, the predominant mechanism of energy absorption is through crack propagation as the notch is already present in the sample. Addition of the coupling agent had little effect in the amount of energy absorbed during crack propagation. On the other hand, in the un-notched test, energy absorption was through a combination of crack initiation and propagation. Cracks are initiated at places of high stress concentrations such as the fiber ends, defects, or at the interface region where the adhesion between the two phases is very poor. The use of the additives increases the energy needed to initiate cracks in the system, thereby resulting in improved un-notched impact strength values with the addition of the MAPP.

Use of the MAPP increased the failure strain and the tensile energy absorption. Thermodynamic segregation of the MAPP towards the interphase can result in covalent bonding to the –OH groups on the fiber surface. Entanglement between the PP and MAPP molecules results in improved interphase properties and the strain to failure of the composite. There is a plateau after which further addition of coupling agent results in no further increase in ultimate failure strain. The number average MW of the G-3002 is about 20,000 and the amount of entanglements between the PP molecules and MAPP is limited, and molecules flow past one another at a critical strain. Any further increase in the amount of G-3002 does not increase the failure strain past the critical amount. However, a minimum amount of entanglements is necessary through the addition of about 1.5% by weight for the critical strain to be reached.

4.2.2 Strength and Modulus

There is little difference in the tensile strength of uncoupled composites compared with the unfilled PP, irrespective of the amount of fiber present. This suggests that there is little stress transfer from the matrix to the fibers due to incompatibilities between the different surface properties of the polar fibers and non-polar PP. The tensile strengths of the coupled systems increase with the amount of fiber present and strengths of up to 74 MPa were achieved with the higher fiber loading of 60% by weight, or about 49% by volume. As in the case with tensile strength, the flexural strength of the uncoupled composites was approximately equal for all fiber loading levels, although there was a small improvement as compared to the unfilled PP. The high shear mixing using the thermokinetic mixer causes a great deal of fiber attrition. Preliminary measurements of the length of fibers present in the composite after injection molding showed that few fibers are longer than 0.4 mm. The strengths obtained in our composites were thus limited by the short fiber lengths. Higher strengths are likely if alternate processing techniques are developed that reduce the amount of fiber attrition while at the same time achieving good fiber dispersion.

The tensile modulus of the kenaf composites showed significant improvements with the addition of the fibers. The uncoupled composites exhibited some very interesting behavior, with moduli higher than the coupled systems at identical fiber loading. The possibility of a high stiffness transcrystalline zone forming around the fiber in the unmodified systems could lead to the high moduli observed. Although the possibility of different fiber orientations contributing to the higher modulus cannot be ruled out, preliminary studies on fiber orientation suggest that the difference in moduli cannot be explained exclusively by the difference in fiber orientations. Several studies suggest that the morphology of the polymer chains are affected by the presence of filler particles.

The specific tensile and flexural moduli of the 50% by weight kenaf coupled composites were about equivalent to or higher than typical reported values of 40% by weight coupled glass-PP injection molded composites. Table 12.1 shows typical commercial PP composites compared with kenaf-PP composites. Data on mineral filled systems from various sources, *Modern Plastics Encyclopedia* (1993) and *Machine Design: Materials Guide Issue* (1994), are included for comparison. The specific flexural moduli of the kenaf composites with fiber contents greater than 40% were extremely high and even stiffer than a 40% mica-PP composite.

4.2.3 Failure Strain and Tensile Energy of Absorption

The failure strain decreases with the addition of the fibers. Addition of a rigid filler/fiber restricts the mobility of the polymer molecules to flow freely past one another and thus causes premature failure. Addition of MAPP followed a similar trend to that of the uncoupled system, although the drop in failure strain with increasing fiber amounts was not as severe. Figure 12.5 (Sanadi et al., 1994c) shows typical stress strain curves of pure PP, uncoupled 50% by weight of kenaf-PP, and coupled systems with increasing amounts of kenaf in the composite. The decrease in the failure strain with increasing amounts of kenaf for the coupled systems is apparent. The non-linearity of the curves is mainly due to plastic deformation of the matrix. The distribution of the fiber lengths present in the composite can also influence the shape of the curve since the load taken up by the fibers decreases as the strain increases and detailed explanations are available elsewhere (Hull, 1981). The tensile energy absorption, the integrated area under the stress-strain curve up to failure, behaves in roughly the same manner as the tensile failure strain. The difference between the coupled and uncoupled composites increases with the amount of fibers present, although the drop in energy absorbed for the coupled composites levels off after the addition of about 35 volume % of fiber.

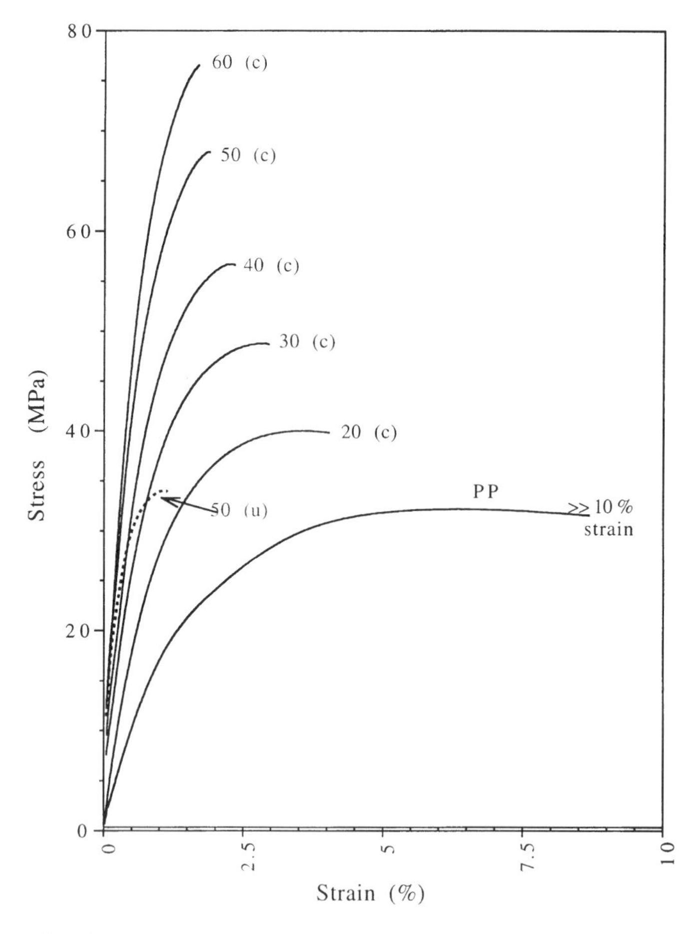

Figure 12.5 Tensile stress–strain curves of kenaf–PP. The numbers near the end of curves indicate kenaf weight % and (c) indicates coupled and (u) uncoupled composites. All coupled systems contained 2% by weight of MAPP. PP failure strain was >10%.

4.2.4 Impact Properties

The impact strength of the composite depends on the amount of fiber and the type of testing, i.e., whether the samples were notched or un-notched. In case of notched samples, the impact strength increases with the amount of fibers added until a plateau is reached at about 45% fiber weight, irrespective of whether MAPP was used or not. The fibers bridge cracks and increases the resistance of the propagation of the crack. Contribution from fiber pullout is limited since the aspect ratio of the fibers in the system are well below the estimated critical aspect ratio of about 0.4 mm (Sanadi et al., 1993). In the case of the unnotched impact values of the uncoupled composites, the presence of the fibers decreases the energy absorbed by the specimens. Addition of the fibers creates regions of stress concentrations that require less energy to initiate a crack. Improving the fiber-matrix adhesion through the use of MAPP increases the resistance to crack initiation at the fiber-matrix interface, and the fall in impact strength with the addition of fibers is not as dramatic. The impact strength can be increased by providing flexible interphase regions in the composite or by using impact modifiers and some work has been done in this area at the UW and FPL (Wieloch, Caulfield, and Sanadi, 1994). The use of an impact copolymer of PP (Amoco Impact Copolymer) improves the notched and un-notched impact resistance (Table 12.1), albeit with some reduction in modulus and strength of the composite.

4.3 Physical Properties

Water absorption and specific gravity of lignocellulosic fiber composites are important characteristics that determine end use applications of these materials. Water absorption could lead to a decrease in some of the properties and should be considered when selecting applications. It is difficult to entirely eliminate the absorption of moisture in the composites without using expensive surface barriers on the composite surface. Water absorption in lignocellulosic based composites can lead to a build-up of moisture *in* the fiber cell wall and also in the fiber–matrix interphase region. Moisture build-up in the cell wall could result in fiber swelling and affect the dimensional stability. If necessary, the moisture absorbed in the fiber cell wall can be reduced through the acetylation of some of the hydroxyl groups present (Rowell, Tillman, and Simonson, 1986) in the fiber, but with some increase in the cost (see Chapter 11). Good wetting of the fiber by the matrix and adequate fiber-matrix bonding can decrease the rate and amount of water absorbed in the interphasial region of the composite. A typical 50% by weight of kenaf-homopolymer PP blend absorbed about 1.05% by weight of water in a 24 h water soak test. This is considerably higher than any mineral filled systems. It is therefore very important to select applications where this high water absorption is not a critical factor such as in electrical housing components.

The specific gravity of lignocellulosic based composites is much lower than the mineral filled thermoplastic systems. The apparent density of the lignocellulosic fibers in PP is about 1.4 g/cc as compared to mineral fillers/fibers (about 2.5 g/cc). The specific gravity of a 50% (by weight) kenaf–PP composite is about 1.07, while

Table 12.1 Comparison of PP Composites

Filler/Reinforcement in PP	ASTM Standard	None	Kenaf	Kenaf-PP impact copolymer	Recycled newspaper fiber	Talc	$CaCO_3$	Glass	Mica
% filler by weight		0	50	50	40	40	40	40	40
% filler by volume (estimated)		0	39	39	30	18	18	19	18
Tensile Modulus, GPa	D638	1.7	8.3	7.5	4.4	4	3.5	9	7.6
Specific Tensile Modulus, GPa		1.9	7.8	7.0	4.5	3.1	2.8	7.3	6.0
Tensile Strength, MPa	D638	33	65	53	53	35	25	110	39
Specific Tensile Strength, MPa		37	61	50	54	28	20	89	31
Elongation at Break, %	D638	>10	2.2	2.5	3			2.5	2.3
Flexural Strength, MPa	D790	41	98		80	63	48	131	62
Specific Flexural Strength, MPa		46	92		82	50	38	107	49
Flexural Modulus, GPa	D790	1.4	7.3		3.9	4.3	3.1	6.2	6.9
Specific Flexural Modulus, GPa		1.6	6.8		4.0	3.4	2.5	5.0	5.5
Notched Izod Impact, J/m	D256A	24	32	74	21	32	32	107	27
Specific Gravity		0.9	1.07	1.07	0.98	1.27	1.25	1.23	1.26
Water Absorption %, 24 h	D570	0.02	1.05	1.3	0.95	0.02	0.02	0.06	0.03
Mold (linear) Shrinkage, cm/cm		0.028	0.003	0.004		0.01	0.01	0.004	

Data on mineral filled PP from various sources. (Modern Plastics Encyclopedia, 1993; and Material Design: 1994 Materials Selector Issue). Data for PP was using a homopolymer, unless otherwise mentioned.

that of a 40% (by weight) glass–PP composite is 1.23. The specific mechanical properties of kenaf–PP composites compare favorably to other filled commodity plastics. Since materials are bought in terms of weight and pieces or articles are, in general, sold by the number, more pieces can be made with lignocellulosic fibers as compared to the same weight of mineral fibers. This could result in significant material cost savings in the high volume and low cost commodity plastic market.

4.4 Effect of Fiber Type

Jute, kenaf, flax, sisal, sunn hemp, henequin, and coir are some of the many fibers being evaluated for their use in thermoplastics (Jacobson and Walz, 1994). Selection of the natural fibers to be used in plastics would depend on the availability of the fibers in the region and the properties of the composite needed for the selected application. For example, if high tensile and flexural strengths are needed, fibers such as flax or jute can be used. For applications where high impact toughness is necessary, fibers such as henequin can be utilized. Kenaf–PP composites have the advantage of being easier to compound and mold at high fiber loading and can result in very stiff composites at low material cost through the use of lower amounts of the plastics. The properties of coir–PP are slightly lower than even the newspaper fibers and the reinforcing efficiency of this fiber is low. The fiber may be useful as a filler in highly filled plastics. The low mechanical properties are probably due to the relatively low cellulose and high lignin content of the coir fibers. It is important to note that coir fibers are a byproduct of the coconut industry and not a plant/tree grown only for its fiber.

4.5 Recycling/Reprocessing

Agro-based fibers are less brittle and softer than glass fibers and are likely to result in composites that are easier to recycle than mineral based fibers. Although no post-consumer based recycling studies have been done on agro-based fibers, a short study on the effect of reprocessing has been conducted at the Forest Products Laboratory and the University of Wisconsin-Madison (Walz et al., 1994). Experimental details are as follows:

Short kenaf filaments were compounded with polypropylene (Fortilene-1602, Solvay Polymers) and MAPP using the thermokinetic mixer explained earlier in the text. The blend ratio was 50% kenaf to 49% PP to 1% MAPP, based on dry weight of material. The mixer was operated at 5200 rpm. A total of 2.25 Kg (15 batches of 150 g each) of material was blended for the experiment.

All the compounded material was then granulated, dried at 105°C for 4 hours and then molded at 190°C using the injection molder. Specimens were randomly selected to evaluate the tensile, flexural, and impact properties and five samples were used for each test: this first set of data was the control or virgin data denoted by "0" in Figures 12.6, 12.7, and 12.8. All the remaining non-tested specimens were once again granulated and injection molded. Once again, five specimens were randomly selected for mechanical property evaluation: this set was labeled as the 1st recycle data point. This procedure of injection molding and granulation was repeated for a total of nine recycle data points. Figures 12.6, 12.7, and 12.8 show that the repeated grinding and molding results in a deterioration of composite properties. The loss in properties is a combination of repeated fiber attrition and oxidative degradation of the polypropylene through chain scission.

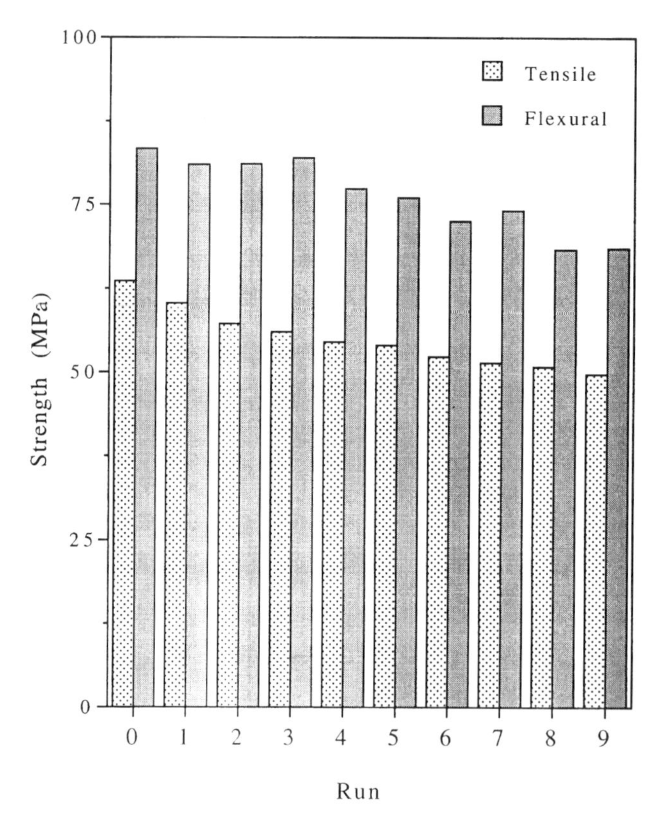

Figure 12.6 Effect of reprocessing 50% by weight kenaf–MAPP coupled–PP on composite tensile strength. The numbers in the abscissa indicate the number of times the composites were reprocessed.

5. CONCLUDING REMARKS ON ECONOMIC ASPECTS AND POTENTIAL MARKETS

The costs of natural fibers are, in general, less than those of the plastic matrix in bio-based composites, and high fiber loading can result in significant material cost savings. The cost of compounding is unlikely to be much more than for conventional mineral–inorganic based composite presently used by the plastics industry. Due to the lower specific gravity of the cellulosic based additives (approximately 1.4 as compared to about 2.5 for mineral based systems), there would be a definite weight advantage for these composites which may have implications in the automotive and other transportation applications. Furthermore, using the same

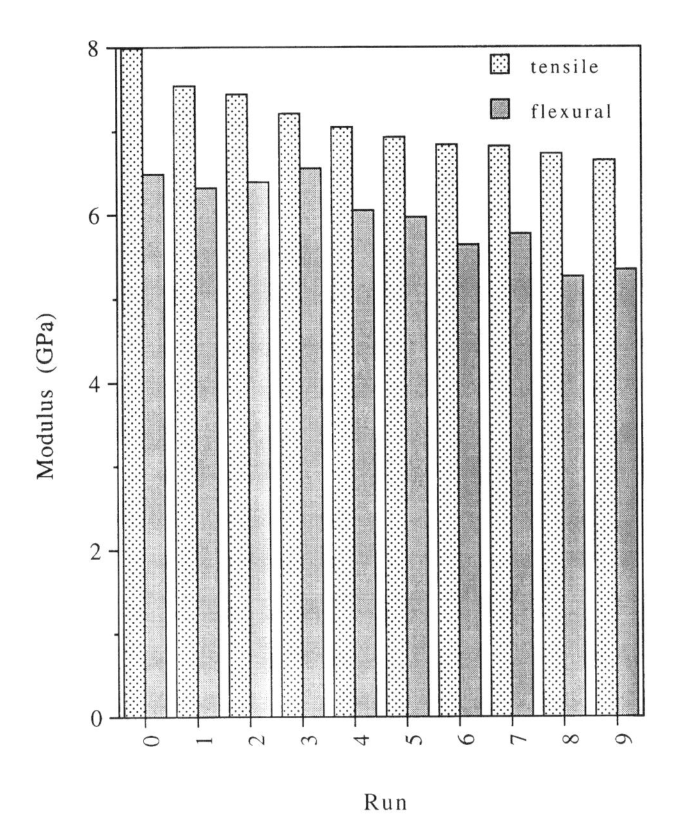

Figure 12.7 Effect of reprocessing 50% by weight kenaf–MAPP coupled–PP on composite tensile modulus. The numbers in the abscissa indicate the number of times the composites were reprocessed.

weight of plastic/natural fiber as, for example, plastic/glass fiber, about 20% more pieces are possible with the cellulosic based system. Cellulosic fibers are soft and non-abrasive and high filling levels are possible; 60% by weight of fiber has been successfully incorporated in PP-based composites. Reduced equipment abrasion and the subsequent reduction of re-tooling costs through the use of agricultural based fibers are definitely factors that will be considered by the plastics industry when evaluating these natural fibers. It is important to point out that we do not anticipate nor intend the total replacement of conventional based fillers/fibers with agricultural based fillers/fibers. We do, however, believe that these natural materials will develop their own niche in the plastics filler/fiber market in the future.

The volume of thermoplastics used in the housing, automotive, packaging and other low-cost, high volume applications is enormous. Recent interest in reducing the environmental impact of materials is leading to the development of newer materials or composites that can reduce the stress to the environment. In light of petroleum shortages and pressures for decreasing the dependence on petroleum products, there is an increasing interest in maximizing the use of renewable materials. The use of agricultural materials as a source of raw materials to the industry not

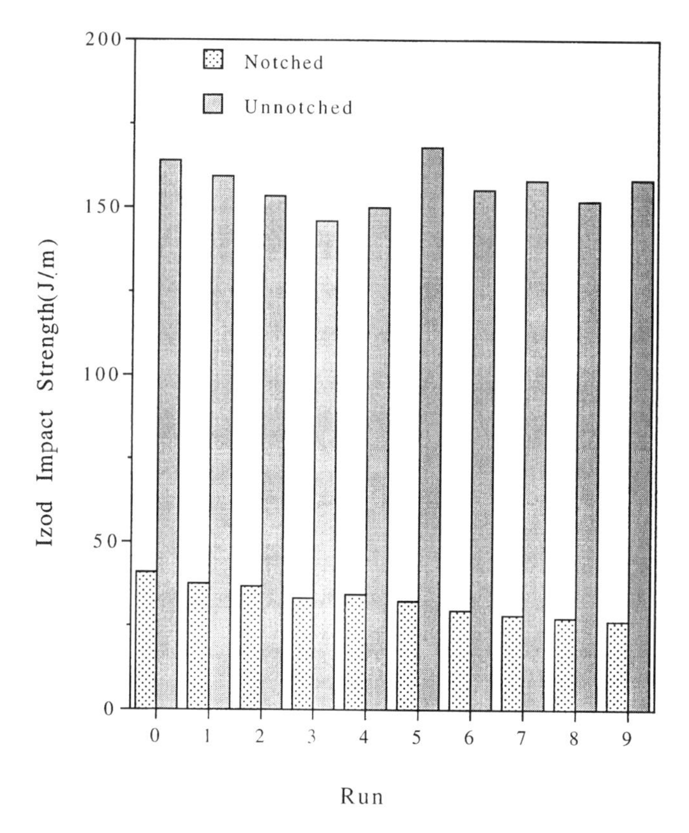

Figure 12.8 Effect of reprocessing 50% by weight kenaf–MAPP coupled–PP on composite Izod impact strength. The numbers in the abscissa indicate the number of times the composites were reprocessed.

only provides a renewable source, but could also generate a non-food source of economic development for farming and rural areas. Appropriate research and development funding in the area of agricultural based fillers/fibers filled plastics could lead to new value-added, non-food uses of agricultural materials. We believe that the amount of these fibers used in the automotive, furniture, housing, packaging, toy, and other industries could easily be in the range of hundreds of millions of pounds annually.

ACKNOWLEDGMENTS

The authors acknowledge partial support of this work from the USDA, Enhancing Value/Use of Agricultural Products Program (NRI Competitive Grants Program/ USDA award # 9303555). One of the authors (ARS) would like to acknowledge partial support by the Forest Products Laboratory, Madison, WI. The authors acknowledge the work of L. Wieloch and K. Walz, without whose help the data reported here could not have been compiled. The authors thank Gordon Fisher of

AgFibers, Inc. of Bakersfield, CA, J. M. Killough of Solvay Polymers, TX, and David Olsen of Eastman Chemical Products, TN, for their generous donation of materials. Testing was conducted by the staff of EML, Forest Products Laboratory, and help in this regard from M. Doran, D. Cuevas, and J. Murphy is much appreciated. The advice and assistance of J. Schneider and C. Clemons during processing of the composites are gratefully acknowledged.

REFERENCES

Bataille, P., Ricard, L., and Sappieha, S., Effect of cellulose in polypropylene composites, *Polym. Compos*, 10, 103, 1989.

Berg, J. C., The use and limitation of wetting measurements in the prediction of adhesive performance, *Composite Systems from Natural and Synthetic Polymers*, Salmen, L. et al., Eds., Elsevier, Amsterdam, 1986.

Bigg, D. M., Hiscock, D. F., Preston, J. R., and Bradbury, E. J., High performance thermoplastic matrix composites, *J. Thermoplastic Compos. Materials*, 1, 146, 1988.

Caulfield, D. F., Interactions at the cellulose-water interface, *Paper Science and Technology: The Cutting Edge*, Institute of Paper Chemistry, Appleton, WI, 70, 1980.

Czarnecki, L. and White, J. L., Shear flow rheological properties, fiber damage, and mastication characteristics of aramid-, glass-, and cellulose-fiber-reinforced polystyrene melts, *J. Appl. Polym. Sci.*, 25, 1217, 1980.

Dalvag, H., Klason, C., and Stromvall, H. E., The efficiency of cellulosic fillers in common thermoplastics, Part II, Filling with processing aids and coupling agents, *Intern. J. Polymeric Mater.*, 11, 9, 1985.

Duvall, J., Selliti, C., Myers, C., Hiltner, A., and Baer, E., Effects of compatibilizer structure on polypropylene/polyamide blends, *Tomorrow's Materials Today, Alloys, Blends and Modified Polymers*, SPE Technical Conference, Oct. 5–6, 1993.

Duvall, J., Selliti, C., Myers, C., Hiltner, A., and Baer, E., Interfacial effects produced by crystallization of polypropylene with polypropylene-g-maleic anhydride compatibilizers, *J. Appl. Polym. Sci.*, 52, 207, 1994.

Eastman Chemical Company, *Publication AP-31*, Aug., 1992.

Felix, J. M., Gatenholm, P., and Schreiber, H. P., Controlled interactions in cellulose-polymer composites. I. Effect on mechanical properties, *Polym. Compos.*, 1993, 14, 449, 1992.

Fowkes, F. M., *J. Adhesion*, 4, 155, 1972.

Fowkes, F. M., *J. Adhesion Sci. and Tech.*, 1, 7, 1987.

Gatenholm, P., Felix, J., Klason, C., and Kubat, J., Cellulose-polymer composites with improved properties, *Contemporary Topics in Polymer Science, Vol. 7,* Salamone, J. C., Riffle, J., Eds., Plenum Press, New York, 1992.

Han, G-S., Saka, S., and Shiraisi, N., Composites of wood and polypropylene. morphological study of composite by TEM-EDXA, *Mokuzai Gakkaishi*, 3, 241, 1991.

Hull, D., *An Introduction to Composite Materials,* Cambridge Univ.; Cambridge, 1981.

Jacobson, R. and Walz, K., *Internal Report*, Forest Products Laboratory Madison, WI, 1994.

Karmarker, A., *Internal Report*, Forest Products Laboratory Madison, WI, 1994.

Katz, H. S., and Milewski, J. V., *Handbook of Fillers for Plastics*, Van Nostrand Reinhold, New York, 1987.

Klason, C. and Kubat, J., Cellulose in polymer composites, *Composite Systems from Natural and Synthetic Polymers*. Salmen, L.; de Ruvo, A.; Seferis, J. C.; Stark, E. B. Eds., Elsevier Science, Amsterdam, 1986a.

Klason, C. and Kubat, J., Cellulosic fillers for thermoplastics, *Polymer Composites*, Sadlacek, B. Ed., Walter de Gruyter & Co., Berlin, 153, 1986b.

Klungness, J., *TAPPI*, 64(12), 65, 1981.

Kokta, B. V., Raj, R. G., and Daneault, C., Use of wood flour as filler in polypropylene; studies on mechanical properties, *Polym.-Plast. Tecnol. Eng.*, 28, 247, 1989.

Machine Design: Materials Selector Issue, Penton Pub., Cleveland, OH, 1994.

Milewski, J. V., Whisker and short fiber technology, *Polym. Composites*, 13, 223, 1992.

Miller, B., Experimental aspects of fiber wetting and liquid movement between fibers, *Absorbency*, Chatterjee, P., Ed., Elsevier, New York, 1985.

Modern Plastics Encyclopedia, McGraw-Hill, New York, 1993, 269.

Myers, G. E., Clemons, C. M., Balatinecz, J. J., and Woodhams, R. T., Effects of composition and polypropylene melt flow on polypropylene – Waste newspaper composites, *Proceed. Annual Technical Conference, Society of Plastics Industry*, 602, 1992.

Neilsen, L. E., *Mechanical Properties of Polymers and Composites*, Marcel Dekker, Inc., New York, 1974.

Neilsen, L. E., *Polymer Rheology*, Marcel Dekker, Inc., New York, 1977.

Quillin, D.T., Caulfield, D. F., and Koutsky, J. A., *Materials Research Society Symposium Proceedings*, Vol. 266, 113, 1992.

Quillin, D. T., Caulfield, D. F., and Koutsky, J. A., Crystallinity in polypropylene system. I. Nucleation and crystalline morphology, *J. Appl. Polym. Sci.*, 50, 1187, 1993.

Raj, R. G. and Kokta, B. V., Compounding of cellulose fibers with polypropylene; Effect of fiber treatment on dispersion in the polymer matrix, *J. Appl. Polym. Sci.*, 38, 1987, 1989.

Rowell, R. M., Tillman, A. M., and Simonson, R., A simplified procedure for the acetylation of hardwood and softwood flakes for flakeboard production, *J. Wood Chem. Tech.*, 6, 427, 1986.

Sanadi, A. R., Prasad, S. V., and Rohatgi, P. K., Sunn hemp fiber-reinforced polyester composites: Analysis of tensile and impact properties. *J. Materials Science*, 21, 4299, 1985.

Sanadi, A. R., Rowell, R. M., and Young, R. A., Interphase modification in lignocellulosic fiber-thermoplastic composites, *Engineering for Sustainable Development: AICHE Summer National Meeting*, paper 24f, 1993.

Sanadi, A. R., Young, R. A., Clemons, C., and Rowell, R. M., Recycled newspaper fibers as reinforcing fillers in thermoplastics: Analysis of tensile and impact properties in polypropylene, *J. Rein. Plast. Compos.*, 13, 54, 1994a.

Sanadi, A. R., Caulfield, D. F., and Rowell, R. M., Reinforcing polypropylene with natural fibers, *Plast. Eng.*, April, 27, 1994b.

Sanadi, A. R., Caulfield, D. F., Jacobson, R. E., and Rowell, R. M., Renewable agricultural fibers as reinforcing fillers in plastics: Mechanical properties of kenaf fiber-polypropylene composites, Submitted *Indust. Eng. Chem. Research*, 1994c.

Solvay Polymers, *Fortilene Polypropylene-* Product Line, Solvay Polymers, Texas, 1991.

Thermoplastic Molding Compounds, *Material Design: 1994 Material Selector Issue*, Penton Publishing, OH, 184, 1994.

Vollenberg, P. H. T. and Heikens, D., Particle size dependence of the Young's modulus of filled polymers, 1, Preliminary experiments, *Polym.*, 30, 1656, 1989a.

Vollenberg, P. H. T. and Heikens, D., Particle size dependence of the Young's modulus of filled polymers, 2, Annealing and solid-state nuclear magnetic resonance experiments, *Polym.*, 30, 1663, 1989b.

Walz, K., Jacobson, R., and Sanadi, A. R., Effect of reprocessing/recycling on the mechanical properties of kenaf-PP composites, *Internal Report*, University of Wisconsin–Madison and Forest Products Laboratory, 1994.

Weiloch, L., Caulfield, D. F., and Sanadi, A. R., Toughness improvement in natural fiber reinforced polypropylene, *Internal Report,* University of Wisconsin-Madison and Forest Products Laboratory, 1994.

Westerlind, B. S. and Berg, J. C., *J. Appl. Polymer Sci.*, 36, 523,1988.

Wilhelmy, J., *Ann. Physik*, 119, 177, 1863.

Woodhams, R. T., Thomas, G., and Rodgers, D. K., Wood fibers as reinforcing fillers for polyolefins, *Polym. Eng. Sci.*, 24, 1166, 1984.

Woodhams, R. T., Law, S., and Balatinecz, J. J., Properties and possible applications of wood fiber – Polypropylene composites, *Proc. Symposium on Wood Adhesives*, Madison, WI, May 16–18, 1990.

Yam, K. L., Gogoi, B. K., Lai, C. C., and Selke, S. E., Composites from compounding wood fibers with recycled high density polyethylene, *Polym. Eng. Sci.*, 30, 693, 1990.

Young, R. A., Wettability of wood pulp fibers. Applicability of methodology, *Wood and Fiber,* 8, 120, 1976.

Young, R. A., Structure, swelling and bonding of cellulose fibers, *Cellulose: Structure, Modification and Hydrolysis,* Young, R. A., and Rowell, R. M., Eds., Wiley-Interscience, New York, 1986, Chap. 6, 91.

CHAPTER 13

Filters, Sorbents, and Geotextiles

Brent English

CONTENTS

1. AGRO-BASED FILTERS

Filtering is the process of separating solid particles from liquids and gases by passing the fluid through a porous, fibrous or granular substance. Fluids are filtered: for clarification, with the solids being discarded; to remove solids from a fluid, with the fluid being discarded; or to separate the fluid from the solids, both being retained. Mathematical models have been developed that can approximate a given filter media's effectiveness with a given filtrate, but actual experiments or pilot trials are needed to determine definitive results. Some of the important factors affecting filtration include:

1. Effective filter area
2. Pressure drop across the filter
3. Resistance of filter medium to fluid flow
4. Swelling effect of solvent on filter medium
5. Compressibility of filter medium under fluid pressure
6. Sizes of suspended particles
7. Tendency of particles to flocculate
8. Viscosity of slurry
9. Temperature of slurry
10. Rate of formation of filter cake
11. Resistance of filter cake to fluid flow
12. Consistency of the slurry (Eaton-Dikeman, 1960)

Many materials have been used for filter media, including a range of inorganic and organic materials. Agro-based materials used for filters include woven fabrics of cotton, flax and hemp; nonwoven fibers in mats or columns made of straw, bagasse, kenaf and coir; paper made from wood or cotton; and charcoals made from coir or other materials. Depending on the filter function, these materials are often used in combination and in conjunction with filter aids.

Filters made from intertwined fibrous agro-based materials form intricate network capillary passages that are long when compared to their cross section. In addition, they are nonuniform and tortuous as a result of the formation and variations in individual fibers. Usually the material to be filtered is a fluid in which particles of assorted shapes and sizes are suspended. The filtration process removes these solids by several different methods. Often the solids removed are much smaller than the size of the capillaries would indicate. In addition to the above factors, Eaton–Dikeman (1960) identified the ability of a filter to retain solids to be further influenced by:

1. The physical properties of the liquid, such as temperature and viscosity.
2. The chemical properties of the liquid. For example, agro-based materials will swell to different degrees depending upon the liquid with which it contacts. Thus the pore size distribution of the filter is dependent on the liquid to be filtered.
3. The duration of the filtering operation.
4. The size, shape, and chemical nature of the suspended solids.

1.1 Filter Mechanisms

Three distinct modes of filtration have been identified. Direct sieving occurs when particles are retained by the entrances or constrictions in the filter pores by an actual physical blocking mechanism. Cake filtration also occurs by a physical blocking mechanism, but it is affected by particles previously retained in the filter by other means. In other words, the particles that have been filtered out of the liquid now serve to filter additional particles. Cake filtration can occur both in and above the filter. Standard blocking is the mechanism whereby particles smaller than the filter pore size are attached to the fibers along or within the pores or to other particles previously retained. Standard blocking is also thought to occur when small particles are trapped in stagnation points in the liquid flow behind individual fibers in the filter. In any case, standard blocking is facilitated by molecular forces. Theoretically, all of the above modes could occur simultaneously in a filter. From a practical standpoint, however, standard blocking and direct sieving tend to occur first, with cake filtration starting as the others are beginning to end.

Fluid mechanics teaches that there are two main types of flow: turbulent and streamline. Filters generally operate under streamline flow. In a filter, the rate of flow can be related to the pressure drop and to the properties of the liquid and the solid particles. The capacity of a filter to permit a liquid to pass through it is generally expressed in terms of the permeability or permeability coefficient.

Streamline flow in a channel of constant circular cross section is governed by Poiseuille's law. Modified by Kozeny (1927) and Carmen (1937), Poiseuille's law can be applied to liquid flow through a porous media:

$$q = \left[e^3/kS_v^2(1-e)^2\right]\left[A\Delta P\, g_c/\mu L\right]$$

q = Volumetric rate of flow
A = Cross sectional area of porous bed
L = Thickness of bed
e = Void fraction of porosity, dimensionless
S_v = Specific surface of solid particles on a volume basis
k = A dimensionless constant
μ = Viscosity of the fluid
ΔP = Pressure drop causing the flow
g_c = The gravitational constant

This formula states that the rate of flow is directly proportional to the pressure drop across the bed and to the area of the bed perpendicular to the direction of flow, and inversely proportional to the viscosity of the liquid and the thickness of the bed in the direction of flow. The constant k is determined empirically, and most materials range from 3– 6. The Kozeny–Carmen relation can be used to accurately predict the flow of liquids in simple systems; however, it has limitations when randomly oriented, highly variable, agro-based materials are considered. The permeability coefficient, K, of a particular filter system is represented by the Kozeny–Carmen relation and can be expressed as:

$$K = \left[e^3 / kS_v^2 (1 - e)^2\right]$$

Two types of empirical models have been developed to predict the variation of pressure drop in the course of deep filtration of liquids. The first model assumes that the internal surfaces are uniformly coated by small particles. The second assumes that the filter is gradually clogged by large particles. Both models can be fitted to experimental data; they cannot be used, however, to predict the pressure drop in new situations. An empirical equation for constant rate filtration can be written in the following form (Hudson, 1948):

$$\Delta P / \Delta P_0 = 1 / (1 - jR)^m$$

P = Piezometric pressure
R = Retention (volume of deposited particles/unit filter volume)
j, m = Empirically determined constants

The reported values for j and m range from about 30 into the hundreds, with most values falling between 30 and 80. Herzig et al. (1970) used the following formula for determining pressure drop when uniform coating of the internal pore surface is assumed:

$$\Delta P / \Delta P_0 = \left[1 / (BR/E)^{-2}\right]$$

where B = Inverse compaction factor of the retained particles.

1.2 Geotextile Filter Applications

DeBerardino (1993) reported on filtration applications for geotextiles. Geotextile filters are used primarily in civil engineering applications for retaining soil and allowing water to pass. Four filtration criteria were established:

1. Soil retention to control piping
2. Sufficient water passage capability to handle excess hydrostatic pore pressure

3. Ability of the filtration system to resist clogging over long periods
4. Survivability and durability of the geotextile (design life of the system is especially important for agro-based applications).

All voids in soils are connected to neighboring voids, making flow possible through the densest of soils (Lambe and Whitman, 1969). The following equation was developed by H. Darcy in the 1850s for fluid flow through soils:

$$Q = KiA$$

Q = The rate of flow
K = Coefficient of permeability
i = The gradient
A = The total cross sectional area of the sample.

Giroud (1988) determined that a criterion to determine acceptable water passage through a soil geotextile/filter system could be expressed as:

$$K_g > i_s K_s$$

K_g = Coefficient permeability of the geotextile
i_s = Hydraulic gradient of the soil
K_s = Coefficient permeability of the soil.

1.3 Removal of Heavy Metal Ions

Randall and Hautalla (1975) reported on the use of agricultural byproducts to remove heavy metal ions from waste solutions. They reported that agricultural residues that were high in tannins, such as peanut skins, walnut expeller meal, redwood bark, and western hemlock bark had a strong affinity for lead, copper, mercury, and cadmium. Residues that were low in tannin but high in cellulose, like peanut hulls, wood, and straw had little or no affinity.

In their research, Randall and Hautalla reported that a redwood bark column had been effectively used to remove lead from 38,000 liters of waste water from a lead battery plant. The water was clean enough to be discharged into a municipal sewer. Residues that contain soluble organic matter may need to have these soluble materials washed away if used for filters to prevent discolored water and biochemical oxygen demand (BOD) discharge.

1.4 Biofilters for Volatile Organic Compounds

European countries have used biofilters extensively for emission control, while in the United States, their use has generally been limited to odor control in applications such as waste water treatment and animal rendering plants. Langseth and Pflum (1994) reported that biofilters could be used to remove volatile organic compounds (VOCs) from a wood composite panel mill with an estimated 95 percent efficiency. Two filters are being used on a pilot scale, handling about 16 percent of the press output gas and 25 percent of the flake dryer exhaust. Each filter is approximately 6.1 m wide × 30.5 m long × 1.2 m deep.

The VOCs from this mill consist primarily of alcohols, aldehydes, organic acids, and small amounts of low molecular weight volatile organics such as benzene and toluene. The VOCs exist in very dilute amounts, generally around 500 ppm in air. The VOCs enter the biofilter from underneath through a vented concrete slab floor. The bottom 15–20 cm of the biofilter consists of new pine and poplar bark; the remainder consists of partially composted bark and end trimmings amended with 2–4 percent agricultural lime. In the biofilter, the VOCs are attacked by bacteria that occur naturally in decaying wood. The bacteria convert pure hydrocarbon compounds into CO_2 and H_2O. Retention times are quick, with a 90 percent reduction in light molecular weight VOCs in around 10 seconds. Heavier molecular weight VOCs are generally broken down in 30 to 60 seconds. The life of the biofilter media is thought to be 1–2 years, after which pressure drop builds to unacceptable levels.

2. AGRO-BASED SORBENTS

The use of agro-based resources for absorption certainly predates written record. Folk medicine remedies include the use of various plants and their fibers for the stopping of wounds, the production of poultices, and for other purposes as well. When agriculture began and animals were domesticated, we can be certain that animals were bedded, then, as they are now, in agricultural residues. Today, simple economics and environmental concerns are expanding the use of agro-based materials for absorbents.

Short of direct experimentation, no direct way to accurately predict the sorptive capacity of an agro-based material has been determined. Several important factors that influence sorbtive capacity, however, are density, porosity, selectivity, and retention.

2.1 Density

There are two types of density relative to absorption: true density and bulk density. True density is a measure of solids only, regardless of any internal voids or intersticial areas, and, once determined, can be considered constant for a given material. To determine true density, both mass and volume must be known. Volume can be determined by submersing a known mass of the material into a container of known volume of a wetting liquid. The liquid must be allowed to fully penetrate the material and displace any entrapped gasses. It may be necessary to augment the penetration with vacuum or agitation. True density can then be determined by the formula:

$$D_s = D_L \frac{M_s}{M_1 + M_s - M_2}$$

D_s = Density of the solid
D_L = Density of the wetting liquid
M_s = Mass of the solid
M_1 = Mass of liquid required to fill the measuring container without fiber
M_2 = Mass of liquid required to fill the measuring container with M_s in the container.

Bulk density is a measure of density including solids, pores, and interstices, and may vary depending on compaction. Bulk density is a simple measure of mass/unit volume. For fibrous materials, bulk density will vary widely. Uncompacted fibrous materials may have bulk densities as low as, say, 0.01 gm/cm^3. Compacted bulk densities can approach true density.

2.2 Porosity

Porosity, or void volume, is a measure of how much volume is available in a system for absorption. Like bulk density, porosity may vary depending on compaction. Porosity can be expressed as:

$$P_R = \frac{V_t - V_s}{V_t}$$

P_R = Porosity (%)
V_t = Total volume of the system
V_s = Volume of the solid.

At first glance, porosity may appear to be an indicator of absorptive capacity. In fact, it is a measure of a system's capacity only under ideal conditions. In practice, a sorbent's capacity is often less than its porosity. Fibrous materials often have porosities of 90– 95%, while granular material's porosity is often less than 40%.

Porosity is also often expressed as void ratio, p. Void ratio is the ratio of the void volume to the solid volume and is expressed as:

$$p = \frac{V_t - V_s}{V_t}$$

2.3 Selectivity

Selectivity is the ability of a sorptive material to preferentially absorb one material over another. For instance, most agro-based materials will, to varying degrees, selectively absorb oil over water. This makes these materials attractive sorbents in oil spills caused by tanker and off shore oil rig leaks.

The degree of selectivity is influenced by the sorbent's pore size, wettability, and capillary pressure. Past history of the sorbent is also important in selectivity. For instance, in the case of oil spills, the sorbent's ability to preferentially absorb oil over sea water is affected by whether the sorbent was exposed to the oil first or the water.

2.4 Retention

Retention is the ability of a saturated sorptive material to retain fluid when conditions are conducive to drainage. Retention is important because, in practice, sorbents are often used in one location and transported to another for disposal or fluid removal. A sorbent with a high degree of retention will be able to transport more fluid. Retention levels are based largely on conditions related to capillary action.

Generally, at equilibrium, a saturated sorbent will hold more fluid after drainage than an unsaturated sorbent will take in. This is because the capillary nature of sorbent systems, especially those made of irregular agro-based materials, is not regular. During drainage, fluid will stop when a neck in the capillary system is small enough so that the pressure difference is enough to keep the liquid column in place. During absorption, the liquid will only enter the sorbent until it reaches a wide area where the pressure drop is insufficient to take it any further into the sorbent system. Because agro-fibers vary widely, the equilibrium point will vary throughout the sorbent system.

2.5 Agro-Based Sorbent Applications

Coghlan (1992) reported that researchers at Virginia Polytechnic Institute and State University in Blacksburg, Virginia were able to absorb more oil with kenaf, cotton, and milkweed floss sorbent systems than with commercial polypropylene fiber systems. The researchers reported that milkweed floss in particular was a good sorbent, absorbing about 40 times its weight in oil, compared to ten times with polypropylene fibers. The great ability of milkweed floss to absorb oil was attributed to a waxy coating on the milkweed fiber surface. It was reported that the milkweed floss retained 75 percent of its ability to absorb oil after three cycles of soaking the fibers in oil and then mechanically removing the oil by squeezing. Agro-based absorbents are seeing commercial use as oil absorbent socks and booms for oil spill clean up.

Using modified wheat straw to remove emulsified oil from water was reported on by Fanta et al. (1986). The wheat straw was treated by first heating the wheat straw in a sodium hydroxide solution and then subjecting it to an ion exchange reaction with hexadecyltrimethylammonium bromide (CTAB). The researchers showed that sufficient NaOH is needed to disrupt the straw particles to produce a high surface area sorbent, but that too much NaOH removed hemicellulose. Minimizing hemicellulose removal is important because the uronic acid substituents of hemicellulose are responsible for much of the ion exchange capacity of the straw. The researchers thought that the simplicity of their straw-CTAB preparation might make the process commercially attractive.

3. AGRO-BASED GEOTEXTILES

Geotextiles are any textile like material, either woven, non-woven, or extruded, used in civil engineering applications to increase soil structural performance. The main structural performance functions geotextiles provide are aggregate separation, soil reinforcement and stabilization, filtration, drainage, and moisture or liquid barriers (Dewey, 1993). The market for geotextiles is growing, with worldwide sales of over 700 million square meters annually (Sen Gupta, 1991). Polypropylene is the material of choice for about 80% of all geotextiles. Polyester accounts for about 15%, and polyethylene and nylon about three percent. The remaining two percent consists of agro-based materials. Even then, only a few are 100% natural fiber-based. Many of these contain some portion of synthetic material to hold the geotextile together. This is normally a polypropylene net or polyester scrim sheet that sandwiches the agro-based component. Most geotextiles applications are permanent, such as landfill liners and roadway construction, and the use of agro-based, biodegradable materials would have adverse results. Other applications, though, are temporary or short term, and the use of agro-based, biodegradable materials is worthy of consideration.

Commercially, there is a good variety of geotextiles available that contain a majority percentage of agro-based materials. The agro-based materials are used because of their low cost, biodegradability, moisture-holding ability, and environmentally friendly image. Most are used in erosion control where they serve to stabilize the soil surface while natural vegetation is established. Some of them contain seeds to accelerate and control the re-growth. Other applications include mulches, and the related products of filters and sorbents.

Data conflicts on the amount of erosion control geotextiles sold, with estimates for North America ranging from 22–73 million square meters. According to Homan (1994), agro-based materials make up about 60 percent of the erosion control market. Regardless of the number, erosion control geotextiles markets are growing at 10–15% a year overall.

3.1 Erosion Control

Sediment accounts for roughly two thirds of pollution in United States waterways. Most commercial agro-based geotextiles target this problem (Figure 13.1) As mentioned elsewhere, erosion control geotextile markets are expanding rapidly. Spurring this growth is increasing regulation concerning run-off from all manners of construction sites. Typical applications include roadbank stabilization, reinforcement of waterways, construction site slope stabilization, and silt fences (Figure 13.2).

Figure 13.1 Bio-based geotextiles are typically installed to control erosion.

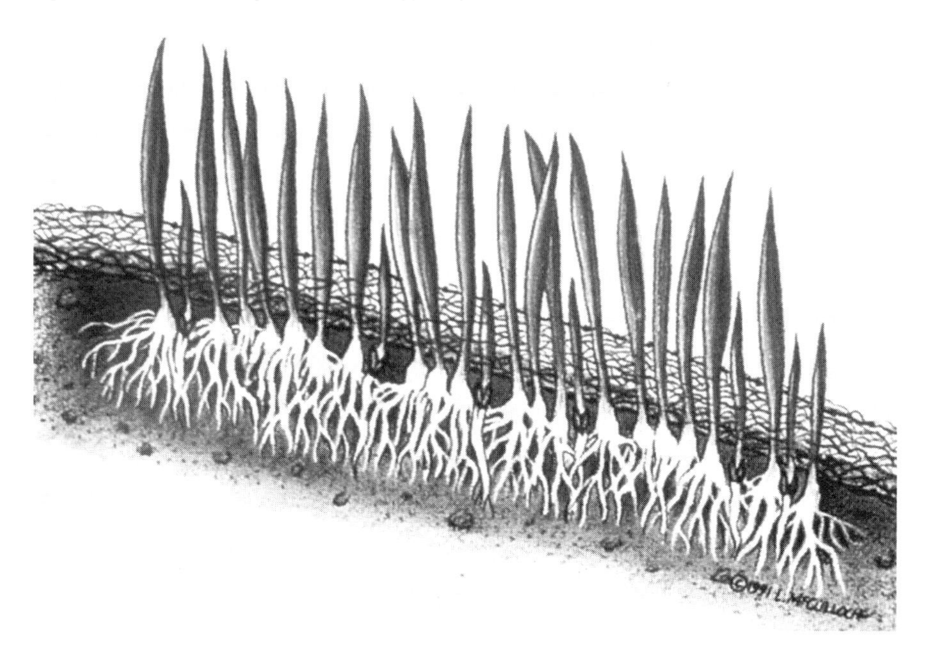

Figure 13.2 As geotextiles degrade, they provide mulch and conserve moisture for plant growth. Eventually, plant growth takes over the soil protection function.

3.1.1 *Mechanism of Erosion*

Erosion is caused by glaciers, marine activity, rivers, wind, and water. Erosion control geotextiles are being used to provide protection against all but glacial. Of these, control of erosion caused by water flow relating to rain is the most common. Rain-caused soil erosion is a function of soil detachment by raindrop impact and transport capacity of thin sheet flow (Meyer et al., 1975).

When raindrops impact the soil surface, they dislodge and lift soil particles. If the soil is level, the net dislodgment is zero. On a slope however, the dislodged soil particles tend to go down the slope, resulting in soil movement. If the rainfall intensity is sufficiently high, the ability of the soil to absorb the rainfall will be compromised, and thin sheet overland flow will occur. Overland flow carries away soil particles dislodged by raindrop impact and also particles dislodged from the soil surface by imparted shear stress. This condition is known as "sheet erosion."

Gully erosion occurs when overland flow becomes concentrated in grooves or channels in the soil surface. The concentrated flow becomes turbulent, and soil particles are dislodged from the channel sides and bottom, leading to larger and larger channels.

3.1.2 *Role of Erosion Control Systems*

The role of erosion control systems is to prevent sheet and gully erosion by any or all of the following strategies: reducing raindrop impact energy, reducing overland flow velocities, protection of the bare soil to prevent the surface layer from being washed away, containment and reinforcement of the top soil layer, or retainment of moisture to create a highly saturated and heavy surface zone (Rustom and Wegget, 1993). Erosion control systems exist on a continuum from natural vegetative cover, to hybrid natural/synthetic systems, to entirely synthetic systems. An example of a purely synthetic system would be rubble placed along a river bank or a concrete lined drainage system. A hybrid system might consist of a naturally reinforcing vegetative root system augmented by a nondegradable synthetic geogrid placed in the soil. Selection of an erosion control systems is based on a variety of factors, soil type, degree of slope, flow conditions, budget, and others. The role of agro-based biodegradable geotextiles is to provide erosion control while vegetation is established to perform the erosion control function.

3.1.3 *Design of Erosion Control Systems*

Ingold and Thomson (1990) suggest using the Universal Soil Loss Equation (USLE) for selecting the type of erosion control strategy to use. The USLE has been expressed in several different forms. One common one is:

$$E = R \times K \times L \times S \times P_C \times C$$

E = Mean annual soil loss (mass/area) acceptable
R = Rainfall erosivity index

K = Soil erodibility index
L = Slope length factor
S = Slope steepness factor
P_C = Conservation practice factor
C = Crop factor.

The crop factor, C, represents the ratio of soil loss or yield under a given crop to that for bare soil. Often quoted C or yield factors for various crop and forest cover range from 0.001–0.02. Various agro-based geotextiles and mulches are often given the value 0.01, although the actual number must be determined experimentally. It should be noted that agro-based systems are designed to degrade, and thus become less effective erosion control systems, as vegetation is established to take over the erosion control function.

Two things affect the yield factor of any erosion control system. They are the run off factor of the system and the erodibility factor of the underlying soil. It is possible for a particular system, such as a tightly woven fabric, to have a high run off factor, but because much of the water is carried on top of the textile, little soil is eroded. Conversely, very porous systems may allow all of the water to pass through, and if ground contact is not good, sheet and gully erosion can take place under the geotextile.

Water diversion channels are often designed around a maximum condition known as a 100-year storm event. In severe conditions bed scour and undermining of slopes must be prevented by proper channel design. Austen and Theisen (1994) reported that the U.S. Soil Conservation Society's volumetric approach can be used to estimate watershed flow. From this estimate, water velocity can be calculated using the Continuity Equation expressed as:

$$Q = AV_{ave}$$

Q = Design discharge in the channel, (m^3/sec)
A = Flow area in the channel, (m^2)
V_{ave} = Average velocity in the cross section, (m/sec).

After an initial cross section area is selected, actual flow conditions and depth of flow can be calculated using Mannings Equation expressed as:

$$V_{ave} = 1/n\, R^{2/3} S_f^{1/2}$$

V_{ave} = Average velocity in the crosssection (m/sec.)
n = Mannings roughness coefficient
R = Hydraulic radius, equal to the cross-sectional area, A, divided by the wetted perimeter, P
S_f = Friction slope of the channel approximated by the average bed slope (for uniform flow conditions).

The velocity of flow is not the only engineering factor used to determine the erosion control system. Hydraulic tractive forces, or shear stress imparted by the moving water must be taken into account. Shear stress is calculated using the Tractive Force Equation:

$$Y_{ave} = WRS_f$$

Y_{ave} = Average shear stress in crosssection (kg/m^2)
W = Unit weight of water (9.8 KN/m^3)

Most erosion control system manufacturers can provide useful data for using the above equations to determine erosion control system application. In addition, some manufacturers provide computer software to help design systems using their products. However, most testing is not yet standardized and use and installation of systems are very site specific. In addition, installation contractors may not always use manufacturer recommended installation techniques. As such, safety factors are normally incorporated into any design function. Complete discussion of all of the factors in the above equations is beyond the scope of this document, and they are presented solely to give a picture of the relationship between the variables affecting erosion and related control systems.

3.2 Agro-Based Erosion Control Systems

Commercially, there are three main agro-based erosion control systems. They are geotextiles, hydromulches and silt fences. Geotextiles include various mats, blankets or nets, of both woven and non-woven materials (Figure 13.3). Geotextiles are supplied as roll goods and are installed in the field with various anchoring schemes (Figure 13.4). Hydromulches are applied *in situ*. They consist of various agro-based materials in a water slurry that are spray applied. Hydromulches almost always include seeds, while not all geotextiles do. Silt fences are temporary "curtains" placed around a construction site to trap or filter sediment from run-off. Table 13.1 shows general agro-based geotextile application parameters. A laundry list of various agro-based erosion control systems might look like this: .

3.2.1 Coir (Coconut Husk) Netting

Coir has the highest tensile strength of any natural fiber and retains much of its tensile strength when wet (Figure 13.5). Coir is very long lasting as well, with in-field service life of 4 to 10 years. These properties make it uniquely suited for more rigorous erosion control applications that still require biodegradability. Coir netting has an open area of 40 to 70 percent, and it is often used in conjunction with other erosion control systems. For instance, hydroseeding can be performed before or after installation, or individual plantings can be placed through the net. Because of its high tensile and wet strength, coir netting can be used in very high flow velocity

Figure 13.3 A kenaf-based geotextile. The soy oil was added to improve processing of the kenaf fiber.

Figure 13.4 Installation of bio-based geotextiles on a steep slope. Metal staples or wood stakes are used to anchor.

Table 13.1 General Bio-Based Geotextile Application Parameters

Geotextile	Durability (seasons)	Maximum Slope	Maximum Flow Resistance	Seeds Incorporated
Coir Netting	4–10	>1:1	very high	no
Jute Netting	1–2	>1:1	very high	no
Straw Mats	1	3:1	moderate	optional
Wood Wool Mats	2–3	1:1	moderate to high	optional
Hybrid Synthetic/Bio-Based Systems	indefinite	>1:1	very high	optional
Silt Fences	1–2	3:1	moderate to high	no

Note: The above table is intended to offer some very general guidelines; variance is likely due to site conditions.

Figure 13.5 Installation of a coir based mat along a drainage ditch.

conditions. Some use of coir in nonwoven geotextiles is found as well, with polyproplene netting used to hold the mat together. Coir netting can be applied on slopes greater than 1:1

3.2.2 Jute Netting

Jute netting is similar to coir netting in appearance and application. The main differences are that its service life is less, generally 1– 2 years, it cannot be used in the high flow conditions that coir can, and it is usually not recommended for slopes greater than 1:1.

3.2.3 Straw Mats

Straw mats are a very popular means to provide temporary erosion control to moderate slopes. The straw source may be from wheat, oat, rice or other grains or grasses. Straw is cheap, easy to work with and readily available in most localities. Straw lasts about one growing season, so the site is usually seeded before the mat is installed. Straw works by reducing rainfall impact and retaining moisture to provide an environment for seed germination and plant growth. Some straw mats contain extraneous weed seeds, which may be unwanted in some situations.

All straw mats need some secondary means to hold the straw into the blanket. Various schemes exist. Most have a lightweight polypropylene netting on one or either side of the mat. The netting is generally held in place by tackifiers or lightweight cotton chain stitching. The polypropylene netting is often billed as "photo-biodegradable." Some rare instances have been reported of birds or small animals being entangled in the netting, but this does not seem to be a major problem. A larger problem is mower entanglement. This can happen on landscaped sites where the polypropylene netting does not sufficiently degrade before the vegetation is established and mowing begins.

Sometimes straw is used in conjunction with more durable biodegradable or nondegradable grids or nets to form a hybrid system. In these systems, the straw provides the rainfall impact resistance and seed growth environment, while the more durable portion stays in the soil to provide additional reinforcement to the vegetative root system. Lightweight straw mats are effective on 4:1 and 3:1 slopes under low to moderate flow velocities. Many hybrid systems can be used on 2:1 and 1:1 slopes under medium to moderately high flow velocities.

3.2.4 Wood Wool (Excelsior) Mats

Wood wool, or excelsior, consists of thin strands or shavings of virgin wood, usually aspen. For erosion purposes, the wood wool is usually 0.5–2 mm in cross sectional dimension and 10–20 cm long. Used for many years as a packing material, wood wool is also used in building materials and filters. Wood wool is more expensive than straw; straw is a crop residue and wood wool is a product manufactured for a specific end use.

Wood wool mat construction is similar to straw mats, with most using stitching or hot melt glue to attach the wood wool to a polypropylene net. Performance differs in longevity and erosion resistance. Wood wool has an effective life of 2–3 growing seasons and can be used effectively on steeper slopes, sometimes up to 1:1, depending on mat construction and flow conditions. Manufacturers offer wood wool in hybrid systems more frequently than straw, probably because of its better performance in more demanding applications.

3.2.5 Hydromulch

Hydromulch and hydroseeding are methods of using water to form a biodegradable mat *in situ* (Figure 13.6). Straw, virgin, and recycled wood fiber, and recycled

Figure 13.6 Hydromulching is the *in situ* production of a geotextile. Hydromulching offers the ultimate in ground conformance. Seeds, fertilizers, herbicides, and tackifiers are frequently incorporated into the hydromulch.

waste paper are all used in hydromulching applications. To apply hydromulch, agro-based materials, seeds, and water are blended together and sprayed onto the site using a water cannon. Water cannons can spray the hydromulch up to 70 meters. Sometimes the seeds are pre-applied; other times all of the materials are applied simultaneously. Typical application rates are 200 kg of seed and 1,200–2,000 kg of mulch/hectare. Fertilizers, pH adjusters, moisture retention agents, and herbicides can be applied simultaneously with the seed and mulch. In some short term construction applications, such as when top soil is stored in pile for later reapplication, seeds are not incorporated, and hydromulch is applied solely to prevent erosion losses.

Tackifiers are often added or post-applied to the hydromulch to increase erosion resistance and windblown movement of the dried mulch (Wolf et al., 1984). Typical tackifiers include guar gum, seaweed extractives, asphalt emulsions (always post-applied), and synthetic polymers. For large installations, hydromulches are lower in cost than geotextiles, and they offer the ultimate in ground conformance. Hydromulching is especially suitable for steep and irregular slopes where erosion control mat installation would be dangerous or nearly impossible (Salkever, 1994). They are often applied to slopes much greater than 1:1 and can tolerate low to moderate velocity water flow. Application of seeds, fertilizers, mulch, and other ingredients can be mixed by the tankfull, and thus, very site specific. Hydromulches are generally considered to be good for one season.

3.2.6 Silt Fences

Silt fences are designed to catch sediment around construction sites. Essentially, they are temporary porous "dams" designed to slow down and catch run off so that the sediment settles out and the water either passes through the fence or percolates into the underlying soil. Almost all commercial silt fences are made from woven polypropylene fabric. From an agro-based perspective, effective silt fences could probably be made from woven jute fabric.

3.3 Geotextile Seed Incorporation Methods

In normal non-agricultural seeding practices, like those found around construction sites, after the soil is prepared, seed is broadcast-spread and raked into the soil. This technique is moderately effective where no erosion takes place and growing conditions are favorable. In practice though, seeds are often dislodged and redistributed by soil or wind erosion. In addition, animal predation can seriously reduce the seed count. The effects of these conditions can often be remedied by the timely application of a mulch in the form of a biodegradable geotextile.

In more severe applications though, where the degree of slope is great enough to make even seed distribution difficult, or where heavy rains or high water flow velocity could redistribute the seeds under the geotextile, seed incorporated geotextiles have application (Figure 13.7). In this respect, the geotextile has three functions. The first function is as a vehicle for the even distribution of seeds. The second function is to protect the seed placement from wind, rain, and predation after distribution. The third function is to degrade and provide mulch while the vegetation is being established.

Several methods are used to incorporate seeds into biodegradable geotextiles. Commercial seed incorporated kenaf or wood geotextiles are available that use non-woven polyester scrim sheets on both sides of the mat to hold the seeds in place. These lightweight mats are sold primarily to the home owner for lawn installation and repair. Other seed incorporated sheet mulches are made using cereal straw, kraft paper, and polypropylene netting. The seeds are distributed and held in place between two lightweight sheets of kraft paper, and are stitched in place to the straw and polypropylene. Experimentally, seeds have been incorporated into geotextiles using starch-based glues to hold the seeds in place on a kraft paper backing (English, 1994). The kraft paper is then needlepunched or stitched to the geotextile. Fertilizers and herbicides can also be incorporated using any of the described methods. Commercially, mats using grass seeds or blends of wildflowers are available. Most manufacturers will also make custom blends.

3.4 Geotextiles for Mulching Applications

Mulching, the spreading of materials around the base of a plant to mitigate adverse temperatures or moisture loss, control weeds or enhance soil fertility and tilth, has been a common horticultural practice for centuries (Waggoner et al., 1960). This practice is usually accomplished by spreading loose agricultural residue

Figure 13.7 Application of wildflower seeds to a geotextile backing sheet in the laboratory. Seeds are often incorporated into commercial geotextiles.

between rows of crops. Like other agricultural crops, the survival of tree seedlings is a major concern to tree growers everywhere. Over the last 30 years forest managers in the United States have evaluated the use of mulches for their biological effectiveness, longevity and cost in the establishment of tree seedlings (McDonald and Helgerson, 1990) (Figure 13.8).

Commonly, seedlings are overplanted and thinned as they mature. In low survival rate areas, seedlings may have to be replanted. These practices are costly and time consuming. Environmental factors that affect seedling survival include moisture, temperature, light, chemical presence or absence, and mechanical damage. Mulches can be used to control most of these factors. Mulches work mainly by suppressing weed growth. This enables the seedling to make full use of light, moisture, and nutrients. The mulch also acts as a soil insulator and as a vapor block. As a soil insulator, the mulch helps keep the soil warm in the early and late part of the growing season. As a vapor barrier, the mulch acts to suppress evaporation.

Historically, the effectiveness of mulches on seedling survival has varied widely. Soil conditions, light, and the longevity of the mulch contribute to this; in especially adverse conditions, survival has increased from near 0% to more than 90%. In more typical situations, survival increases from 40–60%.

Mulch materials can be categorized two ways: loose mulches and sheet mulches. Successful application of loose mulches, like bark, sawdust, or straw, is largely dependent on application thickness. Because they are bulky, loose mulches are most successful when seedling access is good, such as in an orchard or nursery. As such, their application in remote sites is limited.

Figure 13.8 Geotextiles can be used around the base of seedlings to increase survival and enhance growth. The top two are lignocellulosic-based.

Sheet mulches, in the form of small geotextiles, on the other hand, can be rolled up or folded to allow packing into remote sites. Sheet mulches consist of woven or non-woven materials, or plastic film. Film mulches and most woven plastic mulches are less bulky than non-woven mulches made from agro-based materials, but they have several disadvantages. Plastic mulches need to be very well anchored to keep them from being dislodged by wind or animals. If this happens, the mulch litters the forest or folds over and smothers the seedling. Plastic mulches suppress weed growth reasonably well, but if not perforated, rainfall can be diverted from the seedling. Degradability of the plastics is also limited.

Key mulch characteristics for seedling survival have been identified. The mulch should be dark to create temperatures hot enough to kill germinants and sprouts that emerge under the mulch, and also possess good insulative characteristics. The mulch should be porous for water infiltration, yet still retard water loss from underneath it. The mulch should be strong and durable enough to last until the seedling is well established, usually about three years. Good ground conformance would keep the mat from being dislodged. A biodegradable mulch will limit forest litter and save removal costs and may increase mulch effectiveness. The mulch should be low in cost and lightweight for ease of transportation and installation. Special properties related to agro-based geotextiles and geotextile mulches in particular are water permeability, absorption and run off.

Of these three properties, permeability is the most important, as this relates to the amount of rainfall immediately available to the seedling. High run-off levels would indicate little water available to the seedling, although if the seedling is on a slope, water might run under the mulch from up the slope. A degree of absorption is probably important; however, an overly absorbent material might trap all of the water made available during a light rain.

Research was conducted at the USDA Forest Products Laboratory to determine the permeability of a variety of commercial and prototype geotextile mulches (English, 1994). Most of the commercial mulches let less than one fourth of the water pass. Of these, a few let no water pass through at all. Best permeability values were obtained by a perforated polypropylene sheet mulch with 81.2% of applied water passing through, and a grass/straw mat held together with a tackifier (making it similar to hydromulch) with 90.0%. Relatively high values for various other agro-based commercial and prototype mulches were found in the 40 to 60% range. Field experience, however, is what really counts, and while this research is still ongoing and for the most part, unpublished, several general trends and conclusions can be identified.

First and foremost, the successful application of mulches to enhance seedling survival is site specific. If the site is especially remote, the bulkiness of the non-woven agro-based geotextiles was a hindrance, and the plastic mulches were preferred for installation, although they are more difficult to anchor. If reasonably accessible, agro-based mulches offered good ground conformance and biodegradability. Few of the mats that are durable enough to last until the seedling is established have had any adverse effect on seedling survival, although if improperly installed, some plastic mulches will abrade the seedling stem and kill it. Most of the agro-based geotextiles were considered durable enough to last until the seedling is established, although a few did not last one season. The economics of mulch installation are somewhat gray, and again, site specific. Generally, if survival rates are reasonably good and the site is accessible for maintenance, there is no reason to apply mulches. However, if the site has an inherently low survival rate, and is not readily accessible, the use of mulches may have merit.

3.5 Test Methods for Geotextiles

ASTM has established 15 practices and tests to evaluate the properties of geotextiles (1994). Industry often uses the results of the following tests for determining applications (IFAI, 1991).

ASTM D 3786-87 *Hydraulic Bursting Strength of Fabrics.* This test uses a Hydraulic Diaphragm Bursting Tester, commonly called a Mullen Burst Test, to determine resistance to bursting. The test specimen is grabbed by a ring which is located over a diaphragm. The diaphragm is then inflated so that it pushes upon the geotextile until it bursts.

ASTM D 3787 *Puncture Strength of Geotextiles.* Like ASTM D 3786-87, this test supports the test specimen in a ring. The specimen is then punctured using a 8 mm rod.

ASTM 4355-84 *Deterioration of Geotextiles From Exposure to Ultraviolet Light and Water.* The level of deterioration is determined by testing the tensile strength of the composite after exposure.

ASTM D 4491 *Water Permeability of Geotextiles by Permittivity.* This test determines water permeability of the geotextiles in an uncompressed state.

ASTM D 4533-85 *Trapezoid Tearing Strength of Geotextiles.* In this test the trapezoid tear method is used to determine tear strength.

ASTM D 4595-86 *Tensile Properties of Geotextiles by the Wide-Width Strip Method.* Used primarily for reinforcement applications, this tensile test method utilizes grips that extend the full width of the test specimen.

ASTM D 4632-86 *Breaking Load and Elongation of Geotextiles (Grab Method).* This test method uses two 2.5 cm by 5 cm test grips to grab and pull a 10 cm by 20 cm test specimen to determine elongation and strength properties.

ASTM D 4751 *Apparent Opening Size of Geotextiles.* Various sized glass beads are sieved by the geotextile to determine the apparent size opening.

For the short term erosion control applications of agro-based geotextiles, tensile and elongation properties are probably the most important. When placed on a slope, gravity tends to pull the geotextile down a slope. When the weight of the geotextile is increased by water absorption and trapped solids from run-off, this condition is increased. If poorly or improperly anchored, the geotextile can tear and slide down the slope, resulting in failure of the intended erosion control function. Importantly, standard test procedures for erosion control have yet to be agreed upon by industry (Allen and Lancaster, 1994). Although much of the testing is done by independent laboratories, different manufacturers may have different test procedures and failure criteria.

For filtration and absorption functions, water permeability and apparent opening size may be the most important properties. Deberardino (1994) effectively argues that for most purposes, the Mullen Burst test described above is an unneeded specification because geotextiles are very rarely subjected to these exact forces. For the instances where it may happen, the value can be directly correlated from tensile data.

Regardless of the testing conducted, the geotextile industry has begun to standardize the reporting of the results of these and other tests as a Minimum Average Roll Value (MARV) (Wayne, 1994). MARVs are roughly estimated as the mean value minus two standard deviations. MARVs; then give a 95% confidence level that can be used when engineers specify geotextiles for critical applications.

REFERENCES

Allen, S. and Lancaster, T., An important first step, *Geotechnical Fabrics Report*, 12(1), 38, 1994.

ASTM, Soil and Rock (II): D4943-latest; Geosynthetics in 1986 Annual, *Book of ASTM Standards Volume 4.09*, American Society of Testing and Materials, Philadelphia, PA, 1994.

Austin, D. N. and Theisen, M. S., BMW extends vegetation performance limits, *Geotechnical Fabrics Report,* 12(3), 8, 1994.

Carman, P. C., *Trans. Inst. Chem. Engrs.*, (London) 15, 150, 1937.

Coghlan, A., Waxy weeds sup on oil slicks, *New Scientist*, 134(1821), 20, 1992.

Deberardino, S., Filtration design: A look at the state-of-the-art practice, *Geotechnical Fabrics Report*, 11(8), 4, 1993.

Deberardino, S., The role of mullen burst in geotextile specifications, *Geotechnical Fabrics Report,* 12(3), 41, 1994.

Dewey, C. S., Use of geotextiles on federal lands highway projects, *Engineering Field Notes*, 26, 17, 1993.

Eaton–Dikeman Company, *Handbook of Filtration*, The Eaton–Dikeman Company, Mt. Holly Springs, PA, 1960.

English, B., Production of biobased, biodegradable geotextiles: a USDA Forest Service research update, in *Proceedings, Pacific Rim Bio-Based Composites Symposium*, Vancouver, BC, Canada, Nov. 6–9, 1994.

Fanta, G. F., Abbott, T. P., Burr, R. C., and Doane, W. M., Ion exchange reactions of quaternary ammonium halides with wheat straw. Preparation of oil-absorbents, *Carbohydrate Polymers,* 7, 97, 1986.

Giroud, J. P., Review of geotextiles filter criteria, in *First Indian Geotextiles Conference on Reinforced Soil and Geotextiles*, India, 1, 1988.

Herzig, J. P., Leclerc, D. M., and LeGoff, P., "Flow through porous media," 129, *American Chemical Society, Publ.,* Washington, D.C., 1970.

Homan, M., 1993 Erosion control market, *Geotechnical Fabrics Report,* 12(1), 34, 1994.

Hudson, H. E., Jr., A theory of the functioning of filters, *J. Am. Water Works Assoc.,* 868, 1948.

IFAI, *Geosynthetics,* Informational brochure published by the Industrial Fabric Association International, Geotextile Division, 345 Cedar Street, St. Paul, MN, 1991.

Ingold, T. S. and Thomson, J. C., A design approach for preformed erosion control systems, *Geotextiles, Geomembranes and Related Products,* Balkema, Rotterdam, ISBN 90-6191-1192, 375, 1990.

Kozeny, J., Akad. Wiss, Wien, *Math. Naturw.* Klasse 136 (Abt. 11a): 271, 1927.

Lambe, T. W., and Whitman, R. V., *Soil Mechanics*, John Wiley and Sons, New York, 1969, p. 251.

Langseth, S., and Pflum, D., Weyerhauser tests large pilot biofilters for VOCs removal, *Panel World,* March, 1994.

McDonald, P. M. and Helgerson, O. T., Mulches aid in regenerating California and Oregon forests: Past, present and future, *USDA General Technical Report PSW-123*, USDA, Washington, D.C., 1990.

Meyer, L. D., Foster, G. R., and Romkens, M. K. M., Source of soil eroded by water from upland slopes, in *Present and Prospective Technology for Predicting Sediment Yields and Sources*, Proceedings of the Sediment-Yield Workshop, USDA Sedimentation Laboratory, Oxford, MI, Nov. 28, 1972. Agricultural Research Service, ARS-S-40, 1975.

Randall, J. M. and Hautala, E. I., Removal of heavy metal ions from waste solutions by contact with agricultural by-products, in *Proceedings: Industrial Waste Conference*, Purdue University, Lafayette, IN, 30, 412, 1975.

Rustom, R. N. and Wegget, J. R., A laboratory investigation of the role of geosynthetics in interill soil erosion and sediment control, *Geotechnical Fabrics Report*, 11(3), 16, 1993.

Salkever, A., Hydroseeding makes its mark on erosion control, *Erosion Control*, 1(2), 22, 1994.

Sen Gupta, A. K., Geotextiles: Opportunities for natural-fibre products, *International Trade Forum*, Jan.–Mar., 10, 1991.

Waggoner, P. E., Miller, P. M., and DeRoo, H. C., *Plastic Mulching–Principles and Benefits,* Bull. No. 634, New Haven: Connecticut Agricultural Experiment Station, 1960.

Wayne, M. H., Defining MARV, *Geotechnical Fabrics Report*, 12(3), 32, 1994.

Wolf, D. D., Blaser, R. E., Morse, R. D., and Neal, J. L., Hyro-application of seed and wood-fiber slurries to bind straw mulch, *Reclamation and Revegetation Research*, 3(2), 101, 1984.

Index

A

B

C

D

G

H

I

J

N

O

P

Q

R

S